Managing Global Genetic Resources: Agricultural Crop Issues and Policies

Managing Global Genetic Resources: Agricultural Crop Issues and Policies

Contributors

Luciano Lourenço Nassl, Mário Sérgio Sigrist et al.

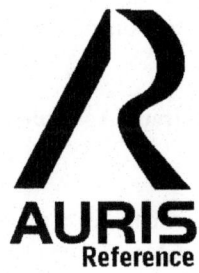

AURIS
Reference

www.aurisreference.com

Managing Global Genetic Resources: Agricultural Crop Issues and Policies

Contributors: Luciano Lourenço Nassl, Mário Sérgio Sigrist et al.

Published by Auris Reference Limited

www.aurisreference.com

United Kingdom

Copyright 2016

Printed in 2017 for Sale in the Indian Subcontinent

Managing Global Genetic Resources: Agricultural Crop Issues and Policies

ISBN: 978-1-78154-960-5

British Library Cataloguing in Publication Data
A CIP record for this book is available from the British Library

Printed in the United Kingdom

Exclusively distributed by CBS Publishers & Distributors Pvt. Ltd.

Sales & Distribution Rights only for India, Pakistan, Bangladesh, Sri Lanka, Nepal and Bhutan. This book is not to be sold outside these territories.

Contents

List of Abbreviations .... *vii*

List of Contributors.................................. ....*ix*

Preface... ....*xv*

Chapter 1   **Genetic Resources: The Basis for Sustainable and Competitive Plant Breeding** 1

Chapter 2   **Agriculture Germplasm Resources: A Tool of Conserving Diversity ... 21**

Chapter 3   **Climate-Smart Agriculture Global Research Agenda: Scientific Basis for Action** 47

Chapter 4   **Ex Situ Conservation of Biodiversity with Particular Emphasis to Ethiopia** 131

Chapter 5   **Citizens' Preferences for the Conservation of Agricultural Genetic Resources**.. 155

Chapter 6   **Agricultural Biotechnology Development and Policy in China**......... 175

Chapter 7   **Cryopreservation of Spices Genetic Resources**................................. 201

Chapter 8   **Diversity and Genetic Erosion of Ancient Crops and Wild Relatives of Agricultural Cultivars for Food: Implications for Nature Conservation in Georgia (Caucasus)** 233

Citations 283

Index.. 285

List of Abbreviations

APSIM	Agricultural Production Systems Simulator
AIC	Akaike information criterion
ASC	Alternative-specific constants
BIC	Bayesian information criterion
CT	Canopy temperature
CE	choice experiment
CSA	Climate-smart agriculture
CGIAR	Consultative Group on International Agricultural Research
CWR	Crop wild relatives
FGB	field gene banks
CGN	Genetic Resources of the Netherlands
GMO	Genetically Modified Organisms
IAC	Instituto Agronômico de Campinas
IFPRI	International Food Policy Research Institute's
IMPACT	International Model for Policy Analysis of Agricultural Commodities and Trade
KSEP	Key Science Engineering Program
LN	liquid nitrogen
LCFS	Low carbon fuel standard
MOA	Ministry of Agriculture
MOST	Ministry of Science and Technology
MS	Murashige and Skoog
NAGS	National Active Germplasm Sites
NBPGR	National Bureau of Plant Genetic Resources
NGOs	nongovernmental organizations
NDVI	Normalized Difference Vegetation Index
PGR	plant genetic resources
PGRFA	Plant Genetic Resources for Food and Agriculture
PCR	polymerase chain reaction
PEs	Primary somatic embryos
SFTPRC	Special Foundation of Transgenic Plants Research and Commercialization
SDPC	State Development and Planning Commission
SSTC	State Science and Technology Commission
SGSV	Svalbard Global Seed Vault
SGRP	System-Wide Genetic Resources Programme
TCD	Trinity College Dublin
VPD	vapour pressure deficit
WTP	willingness to pay

List of Contributors

Luciano Lourenço Nass
Empresa Brasileira de Pesquisa Agropecuária (Embrapa), Secretaria de Relações Internacionais, 70.770-901, Brasília, DF, Brazil

Mário Sérgio Sigrist
DuPont Pioneer - Pesquisa Soja, Rodovia DF 250, km 20, 73.301-970, Brasília, DF, Brazil

Cláudia Silva da Costa Ribeiro
Embrapa Hortaliças, Rodovia Brasília/Anápolis BR 060, km 09, 70.359-970, Gama, DF, Brazil

Francisco José Becker ReifschneiderI
Embrapa Hortaliças, Rodovia Brasília/Anápolis BR 060, km 09, 70.359-970, Gama, DF, Brazil

Rukhsar Ahmad Dar
KVK, Neyoma, Ladakh, India

Mushtaq Ahmad
Division of Genetics and Pant Breeding, Sher-e-Kashmir University of Agricultural Sciences and Technology, Kashmir Shalimar campus, Srinagar -191 121, India

Sanjay Kumar
KVK, Neyoma, Ladakh, India

Monica Reshi
KVK, Neyoma, Ladakh, India

Kerri L Steenwerth
Crops Pathology and Genetics Research Unit, c/o Department of Viticulture and Enology, United States Department of Agriculture (ARS/USDA), Agricultural Research Service

Amanda K Hodson
Department of Land, Air and Water Resources, University of California at Davis

Arnold J Bloom
Department of Plant Sciences, University of California at Davis

Michael R Carter
Department of Agricultural and Resource Economics, University of California at Davis

Andrea Cattaneo
Climate Smart Agriculture Project, Food and Agriculture Organization of the U.N., Viale delle Terme di Caracalla

Colin J Chartres
eWater, University of Canberra Innovation Centre, University Drive South

Jerry L Hatfield
National Laboratory for Agriculture and the Environment, ARS/USDA

Kevin Henry
Where the Rain Falls, CARE France
School of Global Envirfonmental Sustainability, Colorado State University

Jan W Hopmans
Department of Land, Air and Water Resources, University of California at Davis

William R Horwath
Department of Land, Air and Water Resources, University of California at Davis

Bryan M Jenkins
Department of Biological and Agricultural Engineering, University of California at Davis

Ermias Kebreab
Department of Animal Science, University of California at Davis

Rik Leemans
Environmental Sciences, Wageningen University

Leslie Lipper
Agricultural and Development Economic Analysis Division, Food and Agriculture Organization of the U.N., Viale delle Terme di Caracalla

Mark N Lubell
Department of Environmental Science and Policy, Center for Environmental Policy and Behavior, University of California at Davis

Siwa Msangi
Environment and Production Technology Division, International Food Policy Research Institute (IFPRI)

Ravi Prabhu
World Agroforestry Center (ICRAF)

Matthew P Reynolds
Plant, International Maize and Wheat Improvement Center, Consultative Group on International Agricultural Research (CGIAR) Apdo, Postal

Samuel Sandoval Solis
Department of Land, Air and Water Resources, University of California at Davis

William M Sischo
Food- and Water-borne Disease Research Program, College of Veterinary Medicine, Washington State University

Michael Springborn
Department of Environmental Science and Policy, University of California at Davis

Pablo Tittonell
Plant Sciences, Wageningen University

Stephen M Wheeler
Department of Landscape Architecture, University of California at Davis

Sonja J Vermeulen
Department of Plant and Environmental Sciences, Climate Change, Agriculture and Food Security, Consultative Group on International Agricultural Research (CGIAR), University of Copenhagen

Eva K Wollenberg
Gund Institute for Ecological Economics and Rubenstein School of Environment and Natural Resources, University of Vermont

Louise E Jackson
Department of Land, Air and Water Resources, University of California at Davis

Mohammed Kasso
Department of Zoological Sciences, Addis Ababa University, Addis Ababa, Ethiopia

Mundanthra Balakrishnan
Department of Zoological Sciences, Addis Ababa University, Addis Ababa, Ethiopia

Eija Pouta
MTT Agrifood Research Finland, Helsinki, Finland

Annika Tienhaara
MTT Agrifood Research Finland, Helsinki, Finland

Heini Ahtiainen
MTT Agrifood Research Finland, Helsinki, Finland

Jikun Huang
Chinese Academy of Sciences

Qinfang Wang
Chinese Academy of Sciences

K. Nirmal Babu
Indian Institute of Spices Research, Kerala

D. Minoo
Providence Women's College, Kerala India

G. Yamuna
Indian Institute of Spices Research, Kerala

K. Praveen
Indian Institute of Spices Research, Kerala

P.N. Ravindran
Indian Institute of Spices Research, Kerala

K.V. Peter
Indian Institute of Spices Research, Kerala

Maia Akhalkatsi
Ilia State University Georgia

Jana Ekhvaia
Ilia State University Georgia

Zezva Asanidze
Ilia State University Georgia

Ola T. Westengen
Nordic Genetic Resource Center (NordGen), Alnarp, Sweden
Centre for Development and the Environment, University of Oslo, Oslo, Norway

Simon Jeppson
Nordic Genetic Resource Center (NordGen), Alnarp, Sweden

Luigi Guarino
The Global Crop Diversity Trust, Bonn, Germany

Preface

The text Managing Global Genetic Resources: Agricultural Crop Issues and Policies examines the structure that underlies efforts to preserve genetic material, including the worldwide network of genetic collections, the role of biotechnology, and a host of issues that surround management and use. First chapter emphasizes the importance of plant genetic resources and discusses about aspects of costs involved in conservation and suggest recommendations for strengthening the area in Brazil. Second chapter focuses on agriculture germplasm resources. Third chapter deals with climate-smart agriculture global research agenda. Fourth chapter discusses ex situ conservation of biodiversity with particular emphasis to Ethiopia. The aim of fifth chapter is to emphasize the importance of in situ conservation of native cattle breeds and plant varieties in developing conservation policies. The objectives of sixth chapter are to review the status of China's agricultural biotechnology research and commercialization, and to gain a better understanding of China's policies governing both agricultural biotechnology research and its applications (or commercialization). Cryopreservation of spices genetic resources has been presented in seventh chapter. Diversity and genetic erosion of ancient crops and wild relatives of agricultural cultivars for food have been proposed in last chapter.

Chapter 1

GENETIC RESOURCES: THE BASIS FOR SUSTAINABLE AND COMPETITIVE PLANT BREEDING

Luciano Lourenço Nass[1], Mário Sérgio Sigrist[2]; Cláudia Silva da Costa Ribeiro[3]; Francisco José Becker Reifschneider[1,3]

[1]Empresa Brasileira de Pesquisa Agropecuária (Embrapa), Secretaria de Relações Internacionais, 70.770-901, Brasília, DF, Brazil

[2]DuPont Pioneer - Pesquisa Soja, Rodovia DF 250, km 20, 73.301-970, Brasília, DF, Brazil

[3]Embrapa Hortaliças, Rodovia Brasília/Anápolis BR 060, km 09, 70.359-970, Gama, DF, Brazil

ABSTRACT

Plant genetic resources are the fuel for breeding, which in the search for higher yield and adapted genotypes, manipulates genes in order to meet the needs of farmers, and especially, of the current market. However, the use of accessions available in germplasm banks is low. Topics discussed in this paper emphasize the importance of plant genetic resources, and warn about problems related to genetic vulnerability; also, they discuss about aspects of costs involved in conservation and suggest recommendations for strengthening the area in Brazil.

INTRODUCTION

The Importance of Agrobiodiversity, an Unexplored Treasure

Agriculture and biodiversity are intimately linked to each other. Biodiversity in agriculture is essential for human development. Agricultural biodiversity, or agrobiodiversity, is a generic term which basically includes all components of biological diversity (plants, animals, microorganisms) that are important for food and agriculture itself, as well as all components of ecological biodiversity that make up agroecosystems. The importance of agricultural biodiversity is evident, since it supplies food, wood, fiber, oil, medicine, and fuel. Furthermore, agrobiodiversity contributes to ecosystem services, being the conservation of water and soil the most known, as well as pollination.

The beginning of agriculture 10,000 years ago is undoubtedly one of the major events in the history of human life on Earth. Approximately 300,000 species of plants have been described, of which 3,000 have been used by humans for food. Currently, around 300 species are used, and out of these, only 15 are responsible for 90% of all human food. The 15 most used species are rice, wheat, corn, sorghum, barley, sugarcane, beet, potato, sweet potato, cassava, common bean, soybean, peanut, coconut, and banana (Goodman 1990). Together, rice, potato, corn and wheat, represent 60% of this total and are the staples of human consumption.

Brazil, as a megadiverse country, has extensive economic opportunity for the development of new food, fiber, drugs products; these unique opportunities also come accompanied by greater responsibility (Silva et al. 2011) to national and international society. The importance and the economic potential of this heritage for current and future Brazilian generations are monumental. For a significant portion of the population, agricultural biodiversity is the primary income source. In 2010, Brazil's agricultural gross domestic product (agricultural GDP) reached US$ 821 billion, a total GDP of US$3.7 trillion (http://www.cepea.esalq.usp.br/pib/); and millions of Brazilians rely on jobs related to agriculture. Historically, Brazil's agriculture has been one of the primary engines of economic growth. In 2010, the three main export items were soybean, iron ore and oil. Besides, coffee and meats were also among the most important agricultural export products.

Agriculture, when properly carried out, produces a range of environmental services and contributes significantly to the conservation and use of biodiversity. At the same time, agriculture can generate negative impacts to this same biodiversity, by the expansion of the agricultural frontier, the misuse of available technologies, or simply by greed, which is not controlled by the political system in which we live. This loss of diversity is a concern, since it puts at risk the agriculture and ecosystem services provided by it.

Brazil is the leading country among the megadiverse countries, holding in its territory an estimate of 13% of all species on the planet. Over 20% of the world flora, in a total of approximately 55,000 described species, are found in Brazil. Despite having the largest biodiversity on the planet, with a huge range of native species,Brazilian agriculture is highly dependent on exotic species[1], both from the Americas and from other continents.

Brazil's agriculture is based, among others, on: sugar cane, from New Guinea; coffee, from Ethiopia; rice, from the Asian continent; soybeans and oranges, from China; corn, from Mexico; wheat, from Minor Asia. Several native species are important for human consumption with regional and local importance, such as cassava, pineapple, peanuts, cocoa, cashew, cupuaçu, passion fruit, guaraná, among others. Native forage species also contribute to the support of a good part of the livestock sector in Brazil (MMA 2011). Livestock depends on cattle from India and forage grasses brought originally from Africa. Fish farming depends on tilapia, from Eastern Africa, and carp, brought from China; moreover, apiculture and pollination of major crops are based on Africanized bees. However, this is not a Brazilian weakness, since the dependence of exotic genetic resources is a global phenomenon.

Brazilian agriculture would never have reached today's stage without a systematic and growing import of these genetic resources for food, oil, fiber and energy. There are many good examples of this rich history (Reifschneider et al. 2010): it is noteworthy the introduction of cattle in Brazil, and the germplasm introduction and adaptation gardens of the 16th century and beyond, exemplified by the Quinta do Tanque, demonstrating that the concern with agrobiodiversity in Brazil is historic!

The «Quinta do Tanque» The Quinta do Tanque, located in Salvador, Bahia, is considered one of the most important civilian monuments of Brazil. Its history begins in 1555, when the Jesuits built a cottage for the college of Bahia. A Quinta, measuring about two acres, had large gardens and a dam, or a tank. There, the Jesuits cultivated fruits and vegetables in an orchard irrigated by springs and precious water reservoirs, which is the origin of the name Quinta do Tanque. At the Quinta, plants from Europe, Asia and America were also cultivated. Later, the Quinta also served as experimental garden, in which species from all over the world were tested and selected. The cultivation of cinnamon in Brazil started at the Quinta, and cocoa started to be grown in Bahia (which was previously grown in Maranhão) by intervention of the Jesuits (Reifschneider et al. 2010).

The genetic vulnerability of agriculture nowadays and the successful use of native and exotic germplasm

The use of 15 species that represents 90% of all human food makes clear the narrow basis on which we depend. A limited number of species that we use for our livelihood presents a huge concern about the genetic vulnerability of agriculture; a concern that once belonged to researchers, but which today permeates society. Despite the conservation of a vast amount of genetic variability in germplasm collections or banks, the economically important crops continue becoming more uniform. Thus, despite all the efforts made in the establishment of germplasm banks in the international arena, this was not enough to make the world agriculture less vulnerable to diseases and pests. The existence of this vulnerability is due to the use of uniform genotypes in extensive farming areas, and there are many examples well-studied examples of problems arising from such genetic uniformity. The disaster involving the use of potato clones susceptible to the fungus *Phytophthora infestans* in Ireland, the ruin of the grapevines caused by an insect parasite that feed on root of grapevines (*Phylloxera*) observed in France, and the use of corn hybrids with only one source of male-sterility susceptible to *Helminthosporium maydis* race T fungus are classic examples of genetic vulnerability.

Nevertheless, there are many examples of successful use of native and exotic germplasm to support the development of national agriculture. The research in plant breeding and genetic resources is one of the most relevant innovation activities for the country, having produced results that have contributed significantly to the main qualitative and quantitative gains achieved by Brazilian agriculture - and Brazilian development - over the past decades. Plant breeding in Brazil is among the best in the world, with significant contributions, highlighting the well-trained human resources and the development of a large diversity of plants adapted to tropical conditions (Queiroz and Lopes 2007). Thus, it is important to mention the tropicalization of soybean, which allowed its cultivation in low latitude regions; the development of numerous varieties of cassava and beans with distinctive traits, including resistance to several pathogens; and the new cultivars of *Capsicum* peppers and peanuts (*Arachis*) for forage.

All this work aiming at the sustainable use of germplasm is based on the existence of collections or germplasm banks, used by the present generation and maintained for the future generations, and properly characterized so they can be effectively used in plant breeding. The availability of germplasm is of fundamental importance for the improvement of any species. Thus, every breeding program depends, ultimately, on genetic resources from germplasm banks or from those that are in use by farmers. Nass (2001) discussed in details the use of plant genetic resources for breeding.

The Importance of Plant Genetic Resources

Plant genetic resources (PGR) are the basis of food security and global energy. It is essential that these resources are properly preserved and characterized for the current and future demand, since they serve as raw material for plant breeding. Given their importance, it is expected that issues related to PGR are discussed frequently in various media in order to encourage debate in society. The degree of importance that society gives to certain subjects can be estimated, for example, through field surveys using structured questionnaires. However, a research of this kind has a high cost to obtain acceptable margins of error. Alternatively, the degree of exposure of a particular subject in the media can be used as a proxy for its significance to society. This latter approach was used here to illustrate the attention given to PGR by scientific journals. We evaluated papers published by five scientific journals of wide circulation nationally and internationally: PAB (Pesquisa Agropecuária Brasileira), CBAB (Crop Breeding and Applied Biotechnology), Horticultura Brasileira, Crop Science and HortScience. All articles published from January 2008 to the most recent volume were evaluated, regardless of the section in which the articles were allocated, making up a total of 4,777 papers analyzed. For each volume, the total number of published papers and the number of papers related to PGR were counted and classified by scanning the titles of articles, seeking for keywords such as *germplasm, accession* and *pre-breeding*. Abstracts were read in case of doubt. Articles related to PGR were then divided into three categories in order to identify the main objective of the study: *conservation methods, evaluation/characterization* and *use in breeding*. Manuscripts involving tissue culture, collection techniques, or improved cryopreservation protocols used in germplasm banks were classified as*conservation methods*. In turn, articles included in the category *evaluation/characterization* presented as objectives the characterization and evaluation of collections or sets of materials using molecular markers or agromorphological descriptors, and also the evaluation of genotypes regarding biotic and abiotic stresses. Finally, examples of manuscripts in the category use in breeding involved the use of native or exotic germplasm to develop populations and lines from backcrossing programs.

In general, there was a variation in the quantity of articles associated to PGR in relation to the journals. However, when the same journal was analyzed over the years, the number of articles related to genetic resources remained almost stable (Figure 1). These results suggest that, given the time horizon considered, there were no major changes related to the importance given to the subject PGR. Nevertheless, the results reflect the differences of the proposed objectives for each journal, which may be more focused on genetics and conservation or on the management and cultivation of crops. In absolute terms, the international journals evaluated showed a greater amount of published articles on genetic resources. In part, this happens due to the fact that these journals have specific sections for PGR in each volume, which indicates that the importance of this subject is consolidated in such journals. However, the evaluation of the importance given to the topic brings biased results in favor of international journals, as they also showed higher total number of papers. Thus, it is necessary to evaluate the importance given in relative terms (Figure 2). In this case, CBAB had the highest proportion of articles related to the topic, around 30%, followed by international journals, with approximately 20% each.

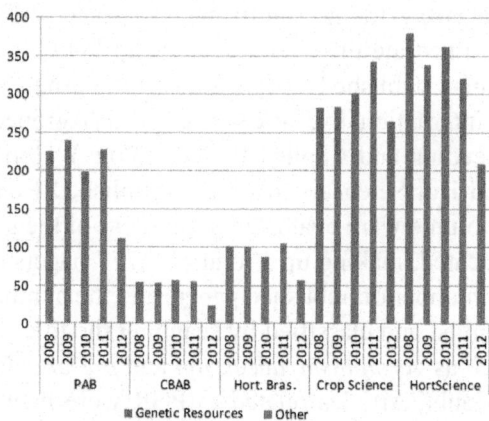

Figure 1: Amount of articles regarding plant genetic resources in relation to the total number of papers published according to the journal and year.

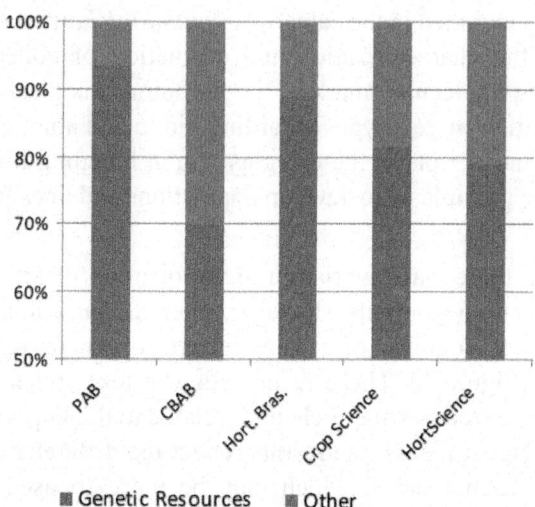

Figure 2: Percentages of articles related to plant genetic resources by journal, grouping all the years evaluated.

Considering only those papers which discussed genetic resources, it was verified that the majority aimed at the evaluation or characterization of the genetic diversity of germplasm, followed by the use of the PGR in breeding, and finally, the development of methods and techniques of conservation (Figure 3). In general, this ranking of objectives remained consistent, both over the five years of evaluation as well as in relative terms within each journal (Figure 4).

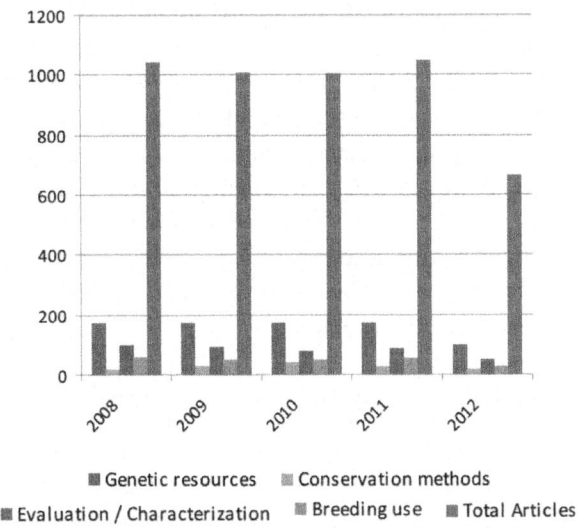

Figure 3: Annual quantity of articles related to plant genetic resources according to the proposed objectives, grouping all journals.

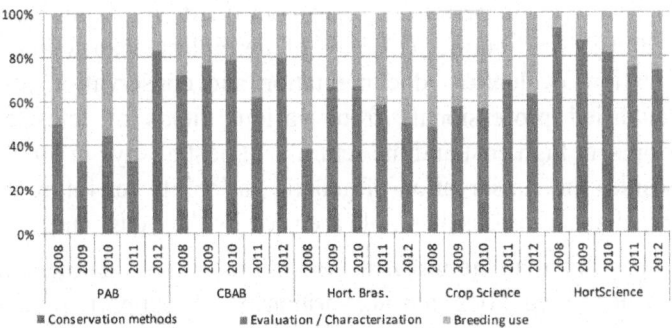

Figure 4: Percentage of articles' objective related to plant genetic resources according to the year and the journal.

Based on the results, it can be considered that, although the journals evaluated dedicate a space for studies related to PGR, the number of manuscripts focused on this topic could definitely increase. In a way, this should be the likely trend for the coming years, mainly due to the increasing importance of PGR in relation to intellectual property and as an alternative to the search for genes resistant to biotic and abiotic factors to face the expected consequences of t climate change. Considering the papers which were focused on PGR, it is clear the need for a greater number of studies aimed at improvement and innovation in conservation methods and techniques, which is important for reducing operating costs in germplasm collections and banks. The major weight of studies aimed at the characterization of germplasm was expected due to the amount of unknown accessions and reduced costs of using molecular markers. However, it is necessary to question the real need to characterize all genotypes stored in banks, although the characterization is one of the conditions to promote the use of PGR. Based on the amount of accessions currently stored and on the costs involved in phenotypic and molecular characterization, it is necessary to set priorities for a more effective allocation of the limited financial resources of the collections (Koo and Wright 2008).

Brazilian Germplasm Collections - a Snapshot

Until the 1940s, the centers of origin of cultivated plants were considered unlimited sources of genetic variability. The expansion of the agricultural frontier without the concern for preserving the environment and natural resources endangered wild relatives of many domesticated species. The possibility of exhaustion of genetic variability of many species and related wild species led the scientific community, in the late 1960s, to promote the conservation of the hereditary material of many plant species, especially those of agricultural importance, under controlled environmental conditions (Ribeiro 2000). The routine activities of germplasm banks like collection, characterization, evaluation, documentation and conservation of accessions require qualified professionals from various fields of knowledge; these activities present high cost and the return is almost always in long term (Nass 2007). Besides the conservation of genetic variability for future use, another objective is that accessions available are used by their customers.

The first Brazilian germplasm collections were organized in research institutes and universities aiming primarily to support several existing breeding programs. Among them, the contribution of the Instituto Agronômico de Campinas (IAC) in the development of agriculture in São Paulo and in Brazil is indisputable and is due to the research on genetic resources and the creation of active germplasm banks (AGBs) of various plant species, such as coffee, sugarcane, rubber, cassava, rice, soybean, common bean, various fruits, vegetables, and ornamental plants. IAC›s coffee AGB was established in 1932 at Fazenda Santa Elisa, in Campinas, SP, aiming the research in different areas, including genetic improvement, besides conservation of germplasm.

IAC's coffee AGB (exotic species - example of successful use of the collection in breeding program, with the release of several cultivars important for the development of Brazilian coffee) IAC›s program for genetic improvement started about 80 years ago, and it has released several cultivars, which have been recommended for planting in different regions of the country. It is estimated that 90% of the Arabica type coffee in Brazil come from cultivars developed at the Institute. The increase of genetic diversity of the collection is a concern of researchers from IAC, and it has had partnership with various institutions in order to establish exchange of coffee germplasm. The short stature cultivars «Catuaí Amarelo» and «Catuaí Vermelho» have modified systems of coffee production in the country and allowed the introduction of cultivation in large areas of Cerrado in São Paulo, Minas Gerais and Goiás.

Embrapa Vegetables' Capsicum AGB (some autochthonous species - an example of successful use of CapsicumAGB in the breeding program of Embrapa Vegetables)

The Capsicum breeding program led by Embrapa Vegetables started about 30 years ago; it has a collection of more than 4,000 germplasm accessions and large genetic variability. The program involves researchers from different fields and various units of Embrapa, research and educational institutions, cooperatives, producers and processing companies. In the last 20 years lines with resistance to diseases have been made available; also, several cultivars of different types of peppers important for Brazilian agriculture have been released. One of the released cultivars is today responsible for over 50% of all the pepper sauce produced in the country.

Although the private sector has participated significantly in the development and availability of cultivars and hybrids of corn in the Brazilian market, in the public sector, IAC was one of the pioneer institutions to develop double hybrids in the late 1930›s to the 60›s (Queiroz and Lopes 2007). The corn breeding program of the Department of Genetics of the Escola Superior de Agricultura «Luiz de Queiroz» (ESALQ/USP) also made a great contribution, allowing the creation of one of the first corn germplasm collections in the country, in 1952. Research at ESALQ involved both the collection of corn landraces and Brazilian indigenous varieties and the development of cultivars. In 1975, the database was transferred to Embrapa Maize and Sorghum, in Sete Lagoas, Minas Gerais (Andrade et al. 1994). The Department of Genetics of ESALQ also pioneered the establishment of germplasm collections of vegetables, which served as the basis for genetic improvement of various vegetables, with emphasis on onion and brassica programs (Queiroz and Lopes 2007). The Universidade Federal de Viçosa (UFV) was one of the first national institutions to work with soybean breeding. The program started in 1963 and over 20 cultivars have been developed. IAC and Embrapa Soybean began studies with adaptation of soybean cultivars in the 1970s (Borém 2001, Queiroz and Lopes 2007). UFV also keeps some germplasm collections, for example, beans, soy and various vegetables.

The Empresa Pernambucana de Pesquisa Agropecuária (IPA) stood out in the national scenario for developing breeding programs oriented to the Northeast, such as cowpea (Vigna unguiculata), onion and tomato for the industry. Several bean and tomato cultivars were developed and occupied significant acreages due to different traits such as fruit setting at high temperatures, fruit firmness and resistance to several diseases (Queiroz and Lopes 2007), attributes essential for a successful crop in the Northeast. IPA maintains germplasm collections of cowpea (Assunção et al. 2005), sweet potato (Ritschel and Huaman 2002), several tropical fruits regionally important like Brazilian cherry (Eugenia uniflora), Brazilian guava (Psidium araça), yellow mombin (Spondias mombin), star fruit (Averrhoa carambola), jackfruit (Artocarpus heterophyllus), soursop (Annona muricata), sugar apple (Annona squamosa), pomegranate (Punica granatum), «umbu» (Spondias tuberosa), guava (Pisidium guajava), sapodilla (Manilkara sapota), among others (Ferreira et al. 2005).

Since 1972, conservation of genetic resources and genetic improvement of cotton, rice, coffee, winter cereals, beans, temperate fruits, cassava, corn and sorghum have been the main lines of research at the Instituto Agronômico do Paraná (IAPAR).

According Valls (2007), the creation of the National Research Center for Genetic Resources (Cenargen, now Embrapa Genetic Resources and Biotechnology), in 1974, by the Empresa Brasileira de Pesquisa Agropecuária (Embrapa), enabled the interaction of this center, with other State and Federal institutions, in order to consolidate the philosophical basis on enrichment and long-term conservation of genetic variability. Such interaction has generated mutual benefits, being it the possibility of establishing a vast network of active germplasm banks (Table 1), or by creating safe conditions for storage of accessions in the long term, under appropriate infrastructure built in Brasília-DF. Currently, the long term collection of Embrapa (Colbase) maintains approximately 120,000 accessions, representing 222 genera and approximately 700 species. It is estimated that, in Brazil, the total number of accessions available in the active germplasm banks is around 300,000.

Using the Treasure

The development of more competitive, better adapted cultivars is a continuous process which requires adaptation to new and often unexpected challenges faced by modern agriculture. Despite the great genetic variability in germplasm collections, underutilization of these resources in genetic breeding programs have been registered in Brazil and other countries, and may have several reasons: practical problems in processing and shipping of samples; legal aspects and quarantine-related issues; lack of information and description of accessions in the banks; restricted adaptability of accessions; insufficient number of breeders to meet the large amount of crops/plant species in the country (even having an available germplasm collection), low seed quality and/or insufficient number of seeds kept in collections; difficulty of crossing exotic germplasm and improved genotypes (Nass 2001, 2011, Valls 2007).

Many plant breeders do not make direct use of the genetic resources since they consider their working collections significant, or because most accessions, despite showing some characteristic of agronomic interest, bring together many undesirable traits that are difficult to manage (linkage drag). However, pre-breeding programs have been shown to be efficient in the use of genetic resources available in AGBs for enlarging the genetic breeding programs of different species (Nass and Paterniani 2000, Nass 2001, 2011, Nass et al. 2007, Nass and Sigrist 2012). The second Brazilian report on the use of genetic resources in Brazil emphasizes several examples of the use of Brazil's collection in pre-breeding programs (Mariante et al. 2009).

Economics of conserving plant genetic resources

The demand for proper valuation of PGR has increased due to new applications offered by biotechnology. Biodiversity held in germplasm banks is the essential raw material for agribusiness and for society in general. When assigning value, one should keep in mind the expected benefit for society. Besides the use related to the production of food, fiber, medicines and bioenergy, the total value of PGR should also consider the cultural, aesthetic and even psychological aspects of the awareness of conserving something that can be extinguished (Brown 1991). But the determination of this value is not trivial, as is determining the cost/benefit of preservation.

Although there is little information about these issues, some studies have sought to reconcile the knowledge of biodiversity with economic principles to enhance and optimize the use and conservation of genetic resources (Evenson and Gollin 2003). For example, given that the budget of germplasm bank is often limited, it is necessary to decide what to conserve. If all accessions in a bank present equal value, accessions with higher maintenance costs would be undesirable. In case all of them have the same cost of conservation, materials which are most likely to be used in the present would be favored. Additionally, similar accessions have less value when compared with rare accessions (Brown 1991, Smale and Koo 2003).

Benefits: Valuation of Plant Genetic Resources

The total value resulting from the conservation of a particular accession can be categorized into non-use value and use value. Non-use, or existence value, reflects the satisfaction of individuals or societies in simply knowing that something exists and is maintained regardless of use. In turn, use value can be divided into direct and indirect. The indirect use value reflects the contribution of PGR for habitats or ecosystems, while the direct use value is the value generated by the development of food, fiber, biofuels and drugs. Both values have dimensions in the present and in the future. A third value is known by option value, which considers the value in having flexibility of use in relation to a future unexpected demand. Based on these definitions, it is noteworthy the complexity in defining, with confidence, the total value of a single accession or whole germplasm banks. Although the theory for estimating these categories have already been proposed (Brock and Xepapadeas 2003), ultimately, many model parameters may be subjective, especially in relation to non-use, indirect use and option values. This presents a difficulty, since one of the main functions of germplasm banks is to meet future challenges. Therefore, these important components of the total value of a germplasm collection are usually underestimated.

Based on various assumptions, it can be estimated the current value of future benefits of germplasm in breeding by combining the probability of finding a useful accession with its expected benefit (e.g. yield increase). However, the time necessary to explore and incorporate useful genes into agronomically elite materials affects the magnitude of the expected benefits due to the time value of money. Of all the values associated with PGR, the most tangible in practical terms is its direct use, usually measured based on financial gain generated by increased production, improved quality, resistance to biotic and abiotic stresses or other characteristics incorporated into elite materials. These benefits are originated from the use of wild relatives and landraces to introduce desirable traits by breeders, using accessions and information generated by curators of germplasm banks. Estimating the benefits of plant breeding by separating the contribution of breeders to the direct use value of germplasm is a complex task, so most studies aim at estimating the value of genetic improvement as a whole, and not only the use value of genetic resources *per se* (Rubenstein et al. 2005). In general, the direct increase in productivity achieved with the new variety, for example, can be estimated by multiplying the productivity excess, in relation to a control variety, at market price and quantity sold in a given period. This value is then compared with the total cost of the breeding program for obtaining the variety, resulting in a rate of return. Although these studies present an overview of genetic gain for production, they do not always define correctly the use value and benefits derived from genetic resources. Considering a characteristic controlled by a single gene, the estimate is easily obtained, but there is greater complexity as an increased amount of genes is introduced. Additionally, many of the estimates are based on experimental data, under controlled environmental and management conditions different from those faced by farmers (Alston et al. 1995). Finally, the supply chain and demand in agribusiness is variable, so that a single estimate of gain for the entire agribusiness system is unlikely.

In any case, the results considering the contribution of genetic resources and genetic improvement as a whole emphasize the economic importance of the activity. In most cases, the rate of return obtained from plant breeding has been extremely positive, adding value to the entire production chain. Byerlee and Traxler (1995) estimated a rate of return of 52% in an international cooperation for wheat breeding in developing countries. Pardey et al. (1996) evaluated data from two centers of CGIAR, reporting benefit and cost ratios of 48:1 for rice and 190:1 for wheat. Brennan et al. (1997) suggested that 64% of the gains from rice breeding in Australia were due to the acquisition of international germplasm, generating benefits of US$ 848 million. Evenson and Gollin (1997) estimated that without the efforts of an international network for evaluation of rice germplasm, 20 new commercial varieties would not have been developed. In this case, the present value of production loss for a period of 20 years would be US$ 1.9 billion.

Aiming at identifying the intrinsic use value of genetic resources, several empirical approaches have been proposed (see Evenson et al. 1998). However, such methods are poorly used in practice, based on real data. Evenson and Gollin (1997) examined the genealogies of rice varieties produced by IRRI between 1965 and 1990, concluding that the addition of 1,000 accessions to the bank was associated with the obtainment of 5.8 varieties. Assuming a period of 10 years after acquisition and a discount rate of 10%, gains were estimated at US$ 325 million. Using similar methodology, Lamarié and Evenson (1998) estimated that rice production in India was 5.6% higher due to the incorporation of landraces in the collection throughout the 1980s.

As seen, the theoretical basis for estimating the total value of genetic resources is still being developed. Since PGRs are considered public goods that are not directly marketed, the biggest limitation is to define a total value for the accessions or genes preserved in a germplasm collection, that considers both use and non-use values.

Costs: The Management of Collections

Since quantifying the benefits (products) generated from PGR has shown to be a challenging task, a more effective approach to increase the value of these resources is to correctly manage the costs involved in *ex-situ*conservation. Cost analysis is derived from microeconomic theory of production, in which a genebank can be treated similarly to any other company (Pardey et al. 2001). Germplasm banks are organized to produce outputs, represented by viable accessions and their information. Based on this simple statement, it is concluded that genebank managers often have to decide about the financial resources needed to produce a certain amount of outputs. As part of the decision process, managers should seek to minimize operational costs using the same infrastructure and maximize production using the same budget. In both cases, the final result will be gain in efficiency.

The total costs of operating a germplasm bank are generally classified into human capital, variables (production labor and operating expenses) and fixed (Koo et al. 2003). The costs involved in the operations of a germplasm bank depend on a number of factors, such as the biological characteristics of the species preserved, preservation methods used, social and economic conditions of the country and local edaphoclimatic factors. Consequently, cost surveys should be performed for each genebank, since conditions and conservation objectives are rarely similar (Koo et al. 2004). For example, the cost of long term conservation of species with vegetative propagation or recalcitrant seeds, which demands special techniques of tissue culture or *in vivo* conservation, is usually greater than the orthodox seed storage (Smale and Horna 2010).

Despite the difficulties, all germplasm banks share a set of basic activities, which can be summarized to better estimate the components of total operating cost on an annualized basis (Table 2). Variable costs are easier to estimate, since the manager is aware of costs of supplies purchase and employees payment. In turn, the fixed costs can be estimated based on the purchase price of the good, which is depreciated by a determined interest rate (Koo and Smale 2003).

Based on cost information in Table 2, the average cost can be calculated by dividing the total cost by the total number of accessions. For comparison purposes, the mean cost for each activity can also be calculated, or the mean costs of different years of the same activity could be compared to measure the efficiency. For example, the average annual cost of storage is calculated as the total cost of storage in a given year divided by the total number of accessions stored in the collection. The average annual cost can be divided into average variable cost and average fixed cost depending on the components of the total cost considered (Smale and Koo 2003). The mean fixed cost tends to decrease when increasing the number of accessions stored, unless it is necessary to purchase new equipment or expand facilities. Moreover, the average variable cost tends to decrease to a certain quantity of stored material due to efficiency gain (economies of scale). However this cost increases again due to excessive use of variable resources under the same fixed production factors. At this point, managers must make a decision of expanding the facilities or reducing the number of stored accessions. Further clarification on the costs associated with *ex-situ* conservation is available in a study by Horna and Smale (2010), where costs of conservation of PGR in CGIAR centers were estimated.

Cost analysis has proven to be an essential tool for better management of activities and financial resources spent on the conservation of genetic resources. An inefficient financial management of collections incurs not only in excessive cost, but also in the incapability of the bank in disseminating feasible accessions containing all the information necessary to meet the demand. Aiming at better management of collections worldwide the System-Wide Genetic Resources Programme (SGRP) of the Consultative Group on International Agricultural Research (CGIAR) has recently developed a database and tools (including a decision-support tool) to assist PGR managers, the Crop GeneBank Knowledge Database (http://cropgenebank.sgrp.cgiar.org/).

A Vision on the Future - Recommendations for Strengthening the Area

Agrobiodiversity has been, since its discovery, one of the pillars of national development. In recent years, the development of national agriculture, both for corporate and family farming, had a great improvement through the adoption of appropriate public policies, higher and more easily accessible credit, continuous and exemplary entrepreneurship of the Brazilian producer, increasing external and internal demand for products traditionally produced in the country, availability of appropriate technologies to tropical agriculture and many other factors. However, the basis of this agriculture has used just a little of our agrobiodiversity and it depends, substantively, on a limited number of species.

An optimistic view of the future:

- Predicts increasing use of agrobiodiversity in favor of Brazilian society and the world's population;
- Predicts the launch of a national program of recovery of germplasm collections of Federal and State public institutions, renewing the seeds and vegetative materials, strengthening the characterization of accessions, and modernizing and integrating the databases so they can be easily accessible to public and private users;
- Predicts the release of a call for proposals, by Federal and State agencies that support research, in a coordinated way, in order to strengthen the

capacity of young professionals in national and foreign institutions that have expertise in the subject;

- Predicts the simplification of legislation, which due to its complexity has tremendously hindered the use of plant genetic resources for the benefit of Brazilian society;

- Continues the strengthening program of Active Germplasm Banks of Embrapa (the largest AGB collection of Brazil), which started with the Agroverde program, in 2010, which supported the AGBs of cassava, cowpea, grape, rice and soybeans;

- Estimates the inclusion of a growing number of plants and their derivatives in the Brazilian diet, which are now forgotten, unknown, unfamiliar, or restricted to specific regions, and makes use of genetic variability to increase the quality of food by the use of traditional techniques and new biotechnology;

- Projects an increase in national and global agriculture production and productivity, based on the rational use of agrobiodiversity and technology, meeting the demand for food, oil, fiber, energy and nutrients of the world's population, and particularly for the less favored ones; and finally,

- Predicts the establishment of ConservaBrasil©, an audacious and futuristic vision for conservation and use of agrobiodiversity. ConservaBrasil©'s mission is to preserve the long-term national biodiversity, in particular the collections of species of agricultural interest, both native and exotic, for future use for the benefit of Brazilian society; in its vision, ConservaBrasil©will have under its tutelage, in 2020, the largest collection of Brazilian germplasm for future use by national agriculture. This initiative will be responsible for keeping thousands of plant, animal and microbial accessions used in Brazil. The germplasm collection maintained by ConservaBrasil©will be one of the largest and best preserved in the world. The collection will have captured significant fraction of the genetic diversity of species relevant for Brazilian agriculture. The results of research on long-term conservation of germplasm will enable the Brazilian society to ensure that the varieties, lines, clones, hybrids, semen, ova and microorganisms of interest are being preserved for future use with the highest technical accuracy. Ongoing operations in 2020 will ensure, in the future, the supply of accessions fundamental for maintaining the competitiveness of domestic agriculture. In this process, Brazilian society, with support from Embrapa, will be participant in the efforts of conservation and use of germplasm that will ensure the future of Brazilian agriculture.

REFERENCES

1. Alston JM, Norton GW and Pardey PG (1995) Science under scarcity: principles and practice for agricultural research evaluation and priority setting. Cornell University Press, Ithaca, 618p.

2. Andrade RV, Azevedo JT, Borba CS and Andreoli C (1994) Banco ativo de germoplasma. In Embrapa (ed.)Relatório técnico anual do Centro Nacional de Pesquisa de Milho e Sorgo 1992-1993. Embrapa, Sete Lagoas, p. 276-277

3. Assunção IP, M-Filho LR, Resende LV, Barros MCS, Lima GSA, Coelho RSB and Lima JAA (2005) Genes diferentes podem conferir resistência ao Cowpea severe mosaic vírus em caupi Fitopatologia Brasileira 30:274-278.

4. Borém A (2001) Melhoramento de plantas. 3rd ed., UFV, Viçosa, 500p.

5. Brennan J, Singh IP and Lewin LG (1997) Identifying international rice research spillovers in New South Wales.Agricultural Economics 17:35-44

6. Brock WA and Xepapadeas A (2003) Valuing biodiversity from an economic perspective: a unified economic, ecological and genetic distance approach. American Economic Review 93:1597-1614.

7. Brown GM (1991) Valuation of genetic resources. In Orians GH, Brown Jr. GM, Kunin WE and Swierbinski JE (eds.)The preservation and valuation of biological resources. University of Washington Press, Seattle, p. 203-245.

8. Byerlee D and Traxler G (1995) National and international wheat improvement research in the post green revolution period: evolution and impacts. American Journal of Agricultural Economics 77:268-278.

9. Evenson R and Gollin V (1997) Genetic resources, international organizations and improvement in rice varieties.Economic Development and Cultural Change 45:471-500.

10. Evenson R and Lemarié S (1998) Optimal collection and search for crop genetic resources. In Smale M (ed.)Farmers, gene banks and crop breeding: economic analyses of diversity in wheat, maize and rice. Kluwer, London, p. 79-82.

11. Evenson R, Gollin D and Santaniello V (1998) Agricultural values of plant genetic resources. CABI, Wallingford, 285p.

12. Ferreira MAJF, Wetzel MMVS and Valois ACC (2005) El estado del arte de los recursos genéticos en las Américas: conservación, caracterización y utilización. Embrapa Recursos Genéticos e Biotecnología e Programa

Cooperativo de Investigación y Transferencia de Tecnología para los Trópicos Suramericanos (PROCITROPICOS), Brasília, 100p

13. Gollin D and Evenson R (2003) Valuing animal genetic resources: lessons from plant genetic resources. Ecological Economics 45:353-363.

14. Goodman MM (1990) Genetic and germplasm stocks worth conserving. Journal of Heredity 81:11-16.

15. Horna D and Smale M (2010) Evaluating cost-effectiveness of collection management: a methodological framework. Available at <http://cropgenebank.sgrp.cgiar.org/images/file/management/DST/framework_dst.pdf> Assessed on Oct 14, 2012.

16. Koo B and Smale M (2003) Economic costs of genebank operation. In Engels JMM and Visser L (eds.) A guide to effective management of germplasm collections. IPGRI, Rome, p. 93-106 (IPGRI handbooks for genebanks, 6).

17. Koo B and Wright BD (2008) The optimal timing of evaluation genebank accessions and the effects of biotechnology. American Journal of Agricultural Economics 82:797-811.

18. Koo B, Pardey PG, Wright BD, Bramel P, Debouck D, van Dusen ME, Jackson MT, Rao NK, Skovmand B, Taba S and Valkoun J (2004) Saving seeds: the economics of conserving crop genetic resources ex-situ in the future harvest centres of the CGIAR. CABI Publishing, Oxfordshire, p. 7-20.

19. Koo B, Pardey PG and Wright BD (2003) The economic costs of conserving genetic resources at the CGIAR centres.Agricultural Economics 29:287-297.

20. Mariante AS, Sampaio MJA and Inglis MCV (2009) The state of Brazil's plant genetic resources. Embrapa Technological Information, Brasília, 163p.

21. MMA (2011) Quarto relatório nacional para a convenção sobre diversidade biológica: Brasil/Ministério do Meio Ambiente. MMA, Brasília, 248p.

22. Nass LL (2001) Utilização de recursos genéticos vegetais no melhoramento. In Nass LL, Valois ACC, Melo IS and Valadares-Inglis MC (eds.) Recursos genéticos e melhoramento: plantas. Fundação MT, Rondonópolis, p. 30-55.

23. Nass LL (2007) Recursos genéticos vegetais. Embrapa Recursos Genéticos e Biotecnologia, Brasília, 858p.

24. Nass LL (2011) Pré-melhoramento vegetal. In Lopes MA, Fávero AP, Ferreira MAJF, Faleiro FG, Folle SM and Guimarães EP (eds.) Pré-

melhoramento de plantas; estado da arte e experiências de sucesso. Embrapa Informação Tecnológica, Brasília, p. 23-38.

25. Nass LL and Paterniani E (2000) Breeding: a link between genetic resources and maize breeding. Scientia Agricola 57:581-587

26. Nass LL and Sigrist MS (2012) Wild species: potential use in pre-breeding. In Borén A, Lopes MTG, Clement CR and Noda H (eds.) Domestication and breeding: amazon species. Suprema, Viçosa, p. 101-115.

27. Nass LL, Nishikawa MAN, Fávero AP and Lopes MA (2007) Pré-melhoramento de germoplasma vegetal. In: Nass LL (ed.) Recursos genéticos vegetais. Embrapa Recursos Genéticos e Biotecnologia, Brasília, p. 683-744

28. Pardey PG, Alston JM, Christian JE and Fan S (1996) Hidden harvest: U.S. benefits from International Research Aid. International Food Policy Research Institute, Washington, 17p.

29. Pardey PG, Bonwoo Koo, Wright BD, Van Dusen ME, Skovmand B and Taba S (2001) Costing the conservation of genetic resources: CIMMYT›s ex situ maize and wheat collection. Crop Science 41:1286-1299

30. Queiroz MA and Lopes MA (2007) Importância dos recursos genéticos vegetais para o agronegócio. In Nass LL (cd.) Recursos genéticos vegetais. Embrapa Recursos Genéticos e Biotecnologia, Brasília, p. 61-119.

31. Reifschneider FJB, Henz GP, Ragassi CF, Anjos UG and Ferraz RM (2010) Novos ângulos da história da agricultura no Brasil. Embrapa Informação Tecnológica, Brasília, 112p.

32. Ribeiro CSC (2000) Criando novas variedades. In Reifschneider FJB (org.) Capsicum: pimentas e pimentões no Brasil. Embrapa Transferência de Tecnologia/Embrapa Hortaliças, Brasília, p. 68-80

33. Ritschel PS and Huamán Z (2002) Variabilidade morfológica da coleção de germoplasma de batata-doce da Embrapa-Centro Nacional de Pesquisa de Hortaliças. Pesquisa Agropecuária Brasileira 37:485-492

34. Rubenstein KD, Heisey P, Shoemaker R, Sullivan J and Frisvold G (2005) Crop genetic resources: an economic appraisal. USDA, Washington, 41p. (Economic Information Bulletin, 2).

35. Silva JAA, Nobre AD, Manzatto CV, Joly CA, Rodrigues RR, Skorupa LA, Nobre C, Ahrens S, May PH, Sá TDA, Cunha MC and Rech Filho EI (2011) O código florestal e a ciência: contribuições para o diálogo. Sociedade Brasileira para o Progresso da Ciência, São Paulo, 124p

36. Smale M and Koo B (2003) Biotechnology and genetic resource policies: what is a genebank worth?International Food Policy Research Institute (IFPRI), Brief 7:1-5.

37. Sousa GS (1879) Tratado descriptivo do Brazil em 1587. 2nded., João Ignácio da Silva, Rio de Janeiro, 382p.

38. Valls JFM (2007) Caracterização de recursos genéticos vegetais. In Nass LL (ed.) Recursos genéticos vegetais. Embrapa Recursos Genéticos e Biotecnologia, Brasília, p. 281-342.

Chapter 2

AGRICULTURE GERMPLASM RESOURCES: A TOOL OF CONSERVING DIVERSITY

Rukhsar Ahmad Dar[1],Mushtaq Ahmad[2],Sanjay Kumar[1],Monica Reshi[1]

[1]KVK, Neyoma, Ladakh, India
[2]Division of Genetics and Pant Breeding, Sher-e-Kashmir University of Agricultural Sciences and Technology, Kashmir Shalimar campus, Srinagar -191 121, India

ABSTRACT

Three major physical resources in the world comprise land, water and the biological diversity. Agricultural biodiversity is an important component of biodiversity, which has a more direct link to the well being and livelihood of mankind than other forms of bio¬diversity. In fact, it is one of our most fundamental and essential resources, one that has enabled farming systems to evolve since the birth of agriculture about 10,000 years ago. Food plant and animal species have been collected, used, domesticated and improved through traditional sys¬tems of selection over many generations. The resulting diversity of genetic resources developed by early farmers now forms the basis on which modern high yielding and disease resistant varieties have been produced to feed the growing human population, expected to reach 9.1 billion by 2050. According to the Convention on Biological Diversity (CBD), "agricultural biodiversity in¬cludes all components of biological diversity of relevance to food and agriculture, and all components of biological biodiversity that constitute agro-ecosystems: the variety and variability of animals, plants and micro-organisms, at the genetic, species and ecosystem levels, which are neces¬sary to sustain key functions of the agricultural ecosystem, its structure and processes". The effective conservation and use of agricultural biodiversity is very important in ensuring sustainable increases in the productivity and produc¬tion of healthy food by and for mankind as well as contrib¬uting to increased resilience of agricultural ecosystems.

INTRODUCTION

There are many threats or drivers of changes on biodiversity that have been recognized and intensified in recent years (Millennium Ecosystem Assessment, 2005). With regard to agriculture the most important ones include changes in land use, replacement of traditional varieties by modern cultivars, agricultural intensification, increased population, poverty, land degradation and environmental change (including climate change) (FAO, 2010). It is predicted that climate change will have a significant impact on agriculture with temperatures rising on average by 2 to 4°C over the next 50 years, causing significant changes in regional and seasonal patterns of precipitation (Burke et al., 2014). Climate change will also impact agricultural biodiversity in a major way. Model projections carried out by Lane and Jarvis (2007) based on global distribution of suitable cultivated areas of 43 crops, highlight that more than 50% may decrease in extent. Evidence based on bioclimatic modelling suggests that climate change could cause a marked contraction in the distribution ranges of CWR. In the case of wild populations of peanut (Arachis spp.), potato (Solanum spp.) and cowpea (Vigna spp.), studies suggest that 16 to 22% of these species may go extinct by 2055, with most species possibly losing 50% of their range size (Jarvis et al., 2008). These threats or drivers of change are leading to large scale degradation and loss of agricultural biodiversity and consequently its genetic variability (Millennium Ecosystem Assessment, 2005; van de Wouw et al., 2009). Information regarding the threat and rate of genetic erosion among various components of agricultural biodiversity is important, yet very little work has been carried out to quantify the magnitude of any trends. The availability of large gene pools, including CWR, is becoming even more important as farmers will need to adapt to changing conditions that result from these pressures. It is likely that many of the genetic traits which will be necessary to adapt our crops to changing climate will be found in CWR. Hence, it is widely urged that such strategies be adopted which may be used to get maximum crop stand and economic returns from adverse environments. Major strategies which may be used to overcome the adverse effects of such stressful environments may include screening and selection of well adopted existing germplasm of potential crops (Ahmad et al., 2014).

There are two main strategies for conserving agricultural biodiversity, namely ex situ and in situ conservation, both of which are equally important and should be regarded as complementary (Thormann et al., 2006; Engelmann and Engels, 2002). Ex situconservation is the conservation of components of biodiversity outside their natural habitats. It is generally used to safeguard populations that are at present or potentially under threat and need to be collected and conserved in genebanks in the form of seeds, live plants, tissues, cells and/or DNA materials. Article 2 of the CBD defines in situ conservation as "the conservation of ecosystems and natural habitats and the maintenance and recovery of viable populations of species in their natural surroundings and, in the case of domesticated or cultivated species, in the surroundings where they have developed their distinctive properties" (UNCED, 1992). It thus refers to the maintenance of a species in its natural habitat. This can be either on farm, requiring the maintenance of the agro-ecosystem along with the cultivation and selection processes on local varieties and landraces, or in the wild, which involves the maintenance of the ecological functions that allow species to evolve under natural conditions.

STATUS AND TRENDS OF AGRICULTURAL BIODIVER-SITY

Little is known about the global status of agricultural biodiversity. Although the CBD recognize genetic diversity as one of the fundamental levels of biodiversity, actions to protect genetic diversity are lacking (Laikre et al., 2010). Policy makers and scientists require a better understanding of how the intraspecific diversity is changing over time and space in order to make informed decisions for their conservation. However, there is no routine global scale monitoring of genetic diversity over time (Frankham, 2010; Laikre et al., 2010), except for a few target species at national level (Laikre et al., 2008). A major challenge remains to develop simple inexpensive means to monitor genetic diversity at a global scale (Frankham, 2010). Several efforts under the 2010 Biodiversity Indicators Partnership (http://www.twentyten.net) have been made to identify indicators useful to detect changes in species and ecosystem diversity, but there are only two initiatives that are explicitly working on developing indicators that deals with genetic variation for agricultural biodiversity (Laikre, 2010; Walpole et al., 2009).

The only authoritative account of agricultural biodiversity status at the global level is represented by the First and Second reports on the State of the World's Plant Genetic Resources for Food and Agriculture published by the Food and Agriculture Organization of the United Nations (FAO, 1998, 2010). The Second Report mention that there are about 7.4 million accessions conserved in over 1750 gene banks around the world in either seed banks, field collections, and in vitro and cryopreservation conditions (FAO, 2010). This represents an increase of more than 1.4 million accessions added to ex situ collection since publication of the first report on theState of the World's Plant Genetic Resources for Food and Agriculture. Although reportedly over-represented, a large part of the genetic diversity of major food crops is stored in ex situ collections. The exact proportion is still uncertain, but estimates suggest that more than 70% of the genetic diversity of some 200 to 300 crops is already conserved in gene banks (SBSTTA, 2010). In addition there are over 2,500 botanic gardens maintaining samples of some 80,000 plant species (FAO, 2010). However, regeneration of gene bank accessions remains a major problem, threatening collections (FAO, 1998). In the past decade there have been significant advances made in regenerating collections at risk, in part due to efforts made by the Global Crop Diversity Trust (CGDT) in supporting regeneration programmes of globally important priority gene bank collections for 22 priority crops for which crop specific regeneration guidelines have recently been produced (Dulloo et al., 2013). Another major achievement has been the creation of the Svalbard Global Seed Vault (SGSV) in 2008, established to serve as the ultimate safety net for seeds samples from the world's most important collections (GCDT, 2010).

Great efforts for the conservation of many CWRs and wild species have been made by the Millennium Seedbank (MSB) at Wakehurst Place, Royal Botanic Gardens, Kew, UK which aims to house up to 10% of the world's seed-bearing flora, principally from arid zones by 2010. Genetic erosion has also been prevented by the significant amount of crop genetic diversity in the form of traditional varieties and neglected and underutilized species (NUS) that continues to be maintained on-farm. Yet, in spite of these advances, important reservoirs of adaptive variation such as CWR, landraces and NUS, which are increasingly recognized by the global scientific community as key resources for the maintenance of agro biodiversity, remain under-represented (FAO, 2010). CWR in particular, which have avoided the genetic bottleneck of domestication, contain greater genetic variation than their cultivated relatives and represent an important reservoir of genetic resources for breeders (Maxted and Kell, 2009). Yet to retain the genetic characteristics that make them so valuable for crop improvement, it is now widely recognized that populations of CWR are best conserved in situ, in their wild habitats, where they can continue to adapt and evolve along with their natural surroundings, thus ensuring new variation is generated in the gene pool and the continued supply of the novel genetic material critical for future crop improvement. The underpinning of the conservation strategy of most countries is a protected areas system and this is reflected in the CBD, where the main thrust of biodiversity conservation is in situ, through the development of such protected systems. Populations of many CWR occur in existing protected areas, but this alone does not in many cases represent effective in situconservation without some degree of management or intervention targeted at the populations of the particular target species, particularly if the species is threatened. Despite protected areas being in existence for many years we still have not been able to undertake significant actions to conserve the CWR they contain, except a few cases.

Despite this, the in situ conservation of CWR has gained increasing attention in many countries, as demonstrated by their inclusion in the many national reports drafted for the Second report on the State of the World's Plant Genetic Resources for Food and Agriculture (FAO, 2010). Unfortunately, little quantitative data were provided by countries on the changing status of CWR, but several reports indicated that specific measures had been taken to promote their conservation. The Second report also mentions that surveys and inventories of CWR were carried out in at least 28 countries and many new priority sites for conserving CWR in situhave been identified over the last decade. There is also evidence that public awareness of the importance of CWR, and neglected and under-utilized species such as traditional vegetables and fruits, is growing both in developing and developed countries (FAO, 2010). This has been furthered by a number of global initiatives aimed at conserving CWR, such as the proposed establishment of a global network for the in situ conservation of CWR (Maxted and Kell, 2009), and more concretely by the creation of web-based international platforms for the exchange of CWR information and data. These include the European platform "An Integrated EuropeanIn Situ Management Work Plan: Implementing Genetic Reserves and On Farm Concepts" (AEGRO) and the CWR Global Portal, developed as part of the UNEP/GEF Crop Wild Relative Project, that provides access to CWR information and data at the global level (Thormann et al., 2012). The significant increase in number of scientific articles published on CWR and on specific actions targeting their conservation is also a testimony to the renewed interest in CWR, however, to the best of our knowledge few of the recommendations have been implemented, largely due to a lack of funds and capacity.

Over the last decade, the number and coverage of protected areas has increased by approximately 30% (United Nations, 2010), yet limited efforts have been made to target CWR, whose conservation remains unplanned and largely an indirect effect of protecting flagship species or threatened habitats. For example, despite the increase in isolated activities targeting CWR conservation, the formal recognition and/or the adoption of appropriate management regimes to protect CWR are largely lacking. Furthermore, considering that national parks and other conservation areas cover only 12 to 13% of the earth's surface, it is clear that these areas alone will not be able to ensure the continued existence of CWR species, the majority of which occur in marginal lands outside protected areas, where no form of legal protection is offered. If protected areas are to ensure the long-term survival of CWR they will need to become more flexible in size and scale and a connected network of habitats will need to be established to allow species to migrate and adjust their ranges in response to global change and anthropogenic disturbances, along with the development of effective management strategies targeting their conservation (that is, off-reserve management). The success of this strategy will depend largely on promoting more biodiversity-sensitive management of ecosystems outside protected areas, and successfully engaging private landowners and local communities living around protected areas in the conservation process. Finally, more effective policies, legislation and regulations that take into account the impacts of global changes on future species distribution and that govern the in situ conservation of CWR, both inside and outside of protected areas, are needed, along with closer collaboration and coordination between the agriculture and environment sectors.

Formidability of Genetic Resources

Wild plants have often played an important role in many diets due to their higher nutritional value than cultivated species. These are, at the same time, hardy and resilient. Crop varieties are improved by the suitable recombination of genes from the wild, made more productive, resistance to biotic stresses, tolerant to abiotic stresses, and better nutritional and keeping quality. Such characteristics, needed to improve crop varieties, may be found in a range of cultivated as well as wild plants. This broad variability provides essential link in the food chain, which, in turn, provides the basic for world food and nutritional security. Plant genetic resources essentially constitute the prime components of the food chain ever since the dawn of agriculture. In the history of some 12,000 years, nearly 30,000 edible plant species have been utilized as a source of food. However, merely a hundred odd plant species out of these have been propagated to provide about 90% of the world food and, further, only three species among these, namely, rice, wheat and maize produced the two-third. An assessment of the contribution of different plant sources towards the dietary energy supply at the global level shows predominance of only two crops, that is, rice (26%) and wheat (23%) (FAO, 1996). The search for new diversity is, therefore, important.

In the developing and the economically weaker parts of the world, the discovery of wild species for food may have coincided with the hunger season, such as, those preceding the crop harvest particularly when drought or flood situations occurred. Mother Nature provided food for people at such junctures when they badly needed it and the resultant discovery of plant species or their diversity became the automatic human choices for further propagation. Even today, though agriculture has advanced so much, humans still gather many wild and semi-wild plants or plant parts like fruit, leaf root, seed, nut, wood etc. for use. About 80,000 species of plants have been used to meet the routine needs by the human beings. Of these, 30,000 species so far have been identified as edible and about 7,000 species have been cultivated and/or collected for food at one time or the other. Presently only 20 to 30 crops, such as cereals (wheat, rice, maize, millets, sorghum), root/tuber crops (potato, sweet potato, cassava), legumes (pea, beans, peanut, soybean) and sugarcane, sugar beet, coconut and banana are mainly used to feed the world (NAS, 1975).

Synoptic View of Plant Diversity

There are 425,000 species in living plants in the plant kingdom from unicellular algae to the highly evolved flowering plants. The flowering plants are a diverse group which are seed producing plants that have evolved in synchronization with the evolution of insects which help the plants in cross-pollination assuring heterozygous population. Hence, this group of flowering plants (about 250,000 species) have developed great plasticity for adaptation for different climatic regimes. They consist of a variety of life forms from the minute Wolffia (1 mm long) to the largest Eucalyptus regnant growing to height of 100 m. This spectrum of flowering plants includes humble herbaceous species, beautiful orchids, parasitic Rafflesia arnoldii having largest flowers (1 m across), plants of medicinal value, trees of timber importance, food plants, fodder species, gums and resin producing plants. The 250,000 flowering plant species are packed in about 17,000 genera and 300 to 400 families. Some of the economically important families which hold life supporting food sustenance species are Gramineae, Legumonosae, Criciferae, Cucurbitaceae, Rosaceae, Brassicaceae and Rutaceae. The drug yielding families cover a spectrum of alkaloids producing crops such as, Apocynaceae, Papavaraceae, Asteraceae, Cannabinaceae, Piperaceae, Zingiberaceae and Rubiaceae. Gums and resins occur in several families as Eurphorbiceae, Dipterocarpaceae, Mimosaceae and Sapotaceae. The families vary in size from monotypic ones to large families having 25,000 to 35,000 species. The family Orchidaceae has about 25,000 to 35,000 species, Compositae has about 20,000 species, Legumionceae has about 14,00 species and Gramineae has about 8,000 species, while there are about 35 families which has only one species.

Plant Diversity in India

Indian subcontinent has a rich and varied heritage of biodiversity, encompassing a wide spectrum of habitats from tropical rainforests to alpine vegetation and from temperate forests to coastal wetlands. It is one of the eight centres of origin (Vavilov, 1951) and is one of the 12 mega gene centres of the world; possess 11.9% of world's flora. About 33% of the country's recorded flora are endemic to the region and are concentrated mainly in the North-East, Western Ghats, North-West Himalays and Andaman and Nicobar islands, nurish one third of the human population on this earth (Damania, 2002; Mayres et al.,

2000) have brought out an updated list of 25 global hotspots of diversity out of which 8 hotspots are in figured in India. The Indian sub-continent is a centre of domestication and diversification of several economically useful wild plant species comprising about 3,000 plants of edible value, 4,000 species having known reputed medicinal value, 700 plants of traditional and social significance, 500 fibre yielding species, 400 fodder species, 40 species having insectivorous uses, 300 gum and dye yielding plants, 100 aromatic and essential oil yielding species (MoEF, 1994; Chowdary and Murti, 2000). Indian diversity comprises of 49,219 higher plant species, out of which, 5,725 are endemic and belonging to 141 genera under 47 families. Of these endemic species, 3,500 are found in Himalayas and adjoining regions and 1,600 in the Western Ghats alone.

India is a homeland of 167 cultivated species and 329 wild relatives of crop plants (Arora, 1991). It has around 30,000 to 50,000 landraces of rice, wheat, Pigeonpea, mango, turmeric, ginger, sugarcane, etc. and ranks 7th in terms of contribution to world agriculture. Further, around 1,000 wild edible plant species are exploited by native tribes. These include 145 species of roots and tubers, 521 of leafy vegetables/greens, 101 of buds and flowers, 647 of fruits and 118 of seeds and nuts (Arora and Pandey, 1996). In addition, nearly 9,500 plant species of ethano-botanical uses are reported from the country of which around 7,500 are of ethano-medical importance and 3,900 are multipurpose edible species.

The endemic plant wealth of the country has also been supplemented with the species/forms that had been introduced from abroad. These species got naturalized over time and have undergone the process of domestication on being isolated climatically and spatially. Prominent among these aer apple, pear, peach, apricot, grape, almond, date palm, maize, potato, sweet potato, tomato, bean, onion, garlic, chilli, lentil, clove, coriander, cumin, fennel, coffee, cocoa, cashew nut, litchi, cinchona, strawberry, blueberry, tea, rubber and pineapple.

BIODIVERSITY IN JAMMU AND KASHMIR

The State of Jammu and Kashmir has been regarded as heaven on earth, and is also called the bio-mass state of India. This area, located in the far north of the Indian republic, is a mountainous zone in the north-west Himalayas that shares international boundaries with Pakistan in the west, Chinese autonomous region of Xinjiang in the north and Tibet in the north-east. The North-western-Himalayan region being the rich repository of biological heritage, particularly in respect of agri-horticultural crops and was recognized that collection and maintenance of germplasm is essential to provide genetic diversity within a crop and to reduce chances of genetic vulnerability. Exploration and collection of native biodiversity, particularly in agri-horticultural crops of the region, including the wild relatives, rare/endangered plants together with the documentation of related ethano-botanic information for exercising to concomitant with regeneration and preliminary evaluation of the collected genetic resources to ensure their long term conservation as well as use in crop breeding programmes recognizing the fact that improvement and sustenance of cultivated crop species requires variability.

A total of 1911 germplasm accessions comprising local cultivars that were in cultivation before introduction of improved cultivars , old varieties, land races, wild crop relatives and under-utilized crops of agri-horticultural significance were collected in respect of various field, vegetable , and horticultural crops as well as medicinal and aromatic plants. The collected biodiversity included 742 accessions in cereals, 38 in pseudo cereals, 28 in millets,71 in oilseeds, 358 in pulses, 377 in vegetable crops, 21 in spices and condiments, 13 in fodder crops, 204 in medicinal and aromatic plants, 55 in fruits crops and 4 in others. The collection of agro-biodiversity in different crops has not only helped in ensuring their conservation on a long term basis but their use may also increase productivity, food security and economic returns. The valuable biological resources will make the farming systems more stable and sustainable. By establishing suitable linkages with user scientists in the university and sister institutions in the region a total of 382 accessions in cereals, 135 pulses, 78 vegetables, 26 horticultural crops, 110 in medicinal and aromatic plants were made available for use in respective crop improvement programmes. Their eventual use in the development of varieties with high yield potential and improved quality characteristics may diversify production and income opportunities for the end user.

Conservation of Germplasm

Global concern about loss of valuable genetic resources prompted international action. Programs for conservation of plant genetic resources for food and agriculture were thus initiated and gene banks established in many countries. The main objective was to collect and maintain the genetic diversity in order to ensure its continued availability to meet the needs of different users. The concept of germplasm conservation demands that collection methods initially capture maximum variation and subsequently, conservation and regeneration techniques minimize losses through time. To this effect, plant genetic resources (PGR) conservation activities comprise collecting, conservation and management, identification of potentially valuable material by characterization, and evaluation for subsequent use.

There are two approaches for conservation of plant genetic resources, namely in situ and ex situ. In situ conservation involves maintaining genetic resources in the natural habitats where they occur, whether as wild and uncultivated plant communities or crop cultivars in farmers' fields as components of the traditional agricultural systems. Ex situ conservation on the other hand, involves conservation outside the native habitat and is generally used to safeguard populations in danger of destruction, replacement or deterioration. Approaches to ex situ conservation include methods like seed storage, field gene banks and botanical gardens. DNA and pollen storage also contribute indirectly to ex situ conservation of PGR.

Ex situ Conservation Approach

Ex situ conservation refers to the conservation of germplasm away from its natural habitat. This complementary approach for conservation had begun on a wide scale about three decades ago and is now practised, to some extent, in almost all countries as a means to conserve crop species diversity for posterity. This strategy is particularly important for crop gene pools, and can be achieved by propagating/ maintaining the plants in genetic resource centre, botanical gardens, tissue culture repositories or in seed gene banks. The Second Report mention that there are about 7.4 million accessions conserved in over 1750 genebanks around the world in either seed banks, field collections, and in vitro and cryopreservation conditions, (FAO, 2010).

Various approaches are employed for the ex situ conservation depending upon the mode of reproduction and nature of plants to be conserved. Ex situ conservation approach generally comprises the following methods: seed storage, field gene banks, in vitro storage, pollen storage, DNA storage and botanical gardens.

Seed Storage

In the past, many collections were maintained without the help of storage facilities which would affect the viability of seeds. Due to this, the conserved accessions had to be regenerated very frequently leading to loss of genetic diversity in gene banks (Frankel and Hawkes, 1975). In maintaining genetic purity of the conserved accessions, problems arise due to differential survival in storage, selection during regeneration, out-crossing with other entries and genetic drift (Allard, 1970). Good storage conditions coupled with proper grow-outs are expected to reduce the effects of such problems (Rao, 1980).

Storing orthodox seeds at low moisture content and at subzero temperature is the most convenient and widely used method of genetic conservation.

Orthodox seeds are the seeds which can withstand dehydration without damage. This type of seeds can be stored in the dry state on long term basis (indefinite period) which can be prolonged by decreasing their moisture content and storage temperature (at sub zero temperature).

The number of seed storage facilities has increased dramatically over the last two decades. Today, according to the WIEW – World Information and Early Warning System on Plant Genetic Resources for Food and Agriculture – databases of the FAO, there are 1320 national, regional and international germplasm collections in the seed form, 397 of which are maintained under long- or medium-term storage conditions. Over 6.1 million accessions have been conserved as seeds.

As opposed to common orthodox seeds, there are a number of species whose seeds are unable to withstand desiccation, that is, cannot be dried to low levels for optimum storage. Such seeds are referred to as 'recalcitrant' seeds (Roberts and King, 1986). Mainly these seeds originate from the plants grown in tropical and sub-tropical regions. These seeds can be stored for short duration (up to several months) by imbibed storage (at higher levels of seed moisture/hydrated state) and relatively warm conditions (well above zero temperatures because they are often chilling sensitive. e.g., rubber, cocoa, coconut. Seeds such as oil palm and coffee, showing intermediate storage behaviour (Ellis et al., 1990, 1991) were grouped as recalcitrant until recently. Careful adjustment of the desiccation level and storage conditions allowed their storage for increased period (1 to 2 years).

Very low temperature storage using liquid nitrogen, called cryo-preservation, appears to be promising, with a more extended life span, described as long-term storage (-20°C). Another area in which considerable work is required is on storage of ultra dry seeds (dried to seed moisture content of 2 to 5%) at room temperature conditions and in hermetically sealed containers (Zhou et al., 1995). However, more research will be necessary before ultra dry seed technology can be adopted (Zheng et al., 1998). Prior to embarking on any seed conservation programme, a decision is to be made on how long it will be necessary to maintain the germination capacity of the seed lots, because longer storage requires more exacting storage conditions. This shall be determined by the objective of the conservation which could be research, introduction, breeding, etc.

Field Gene Bank Conservation

Many important varieties of field, horticultural and forestry species are either difficult or impossible to conserve as seeds (that is, no seeds are formed or if formed, the seeds are recalcitrant) or reproduce vegetatively. Hence they are conserved in field gene banks (FGB). FGBs provide easy and ready access to conserved material for research as well as for use. It is one of the options of a complementary strategy for the conservation of germplasm of many plant species. The conservation of germplasm in field gene bank involves the collecting of materials and planting in the orchard or field in another location. Field gene bank has traditionally been used for perennial plants, including:

•Species producing recalcitrant seeds;

•Species producing little or no seeds or sterile seeds;

•Species that are preferably stored as clonal material;

•Species that have a long life cycle to generate breeding and/or planting material.

Field gene banks are commonly used for such species as cocoa, rubber, coconut, coffee, sugarcane, banana, cassava, sweet potato, yam, tropical and temperate fruits, vegetatively propagated crops (e.g. wild onion and garlic) and forage grasses (e.g. sterile hybrids or shy seed producers). This is the traditional method of conservation to keep the germplasm in plantations as mature trees. It provides mature material for vegetative propagation, hybridization and characterization. The site for a field gene bank should have a suitable climate and soil for the species and should have an adequate water supply. The site should be chosen in a location with little or no threat of pests, diseases, bush fire and vandalism.To avoid loss of vigour as well as to prevent the incidences of attack by pests the plants have to be replanted routinely, and this adds to the cost further.

Botanical Gardens

There are about 2500 botanic gardens and arboreta worldwide. It is estimated that these gardens maintaining samples of some 80,000 plant of threatened species in botanical gardens and arboreta. The objectives of most of the gardens include:

•Maintaining essential ecological processes and life support systems,

•Preserving genetic diversity, and

•Ensuring sustainable utilization of species and ecosystem.

However, the botanical gardens may play a limited role in the context of conservation and propagation and probably a greater role in public awareness and education. Botanical gardens may mainly be used to display a great number of different and exotic species. As the number that can be maintained in this manner is limited, it cannot reflect or conserve genetic diversity. There is a possibility that a few well-managed gardens can emphasise on conservation of certain groups of species as living collections (that is, field gene banks). Often botanical gardens focus their conservation efforts on wild, ornamental, rare and endangered species. Indeed botanical garden conservation could be considered as field gene bank and/or seed gene bank, depending on the conservation method being used. The living plant collections in botanic gardens and arboreta may be considered as field collections, but the original purpose of the gardens and arboreta is not for germplasm conservation. Most of the germplasm conserved in botanical gardens do not belong to the PGRFA.

In vitro Storage

Research on finding solutions to better conserve these difficult-to-store seeds has focused on the use of biotechnology (Engelmann and Engels, 2002). In vitro slow-growth conservation methods, involving culturing different parts of the plant (meristem, tissues, cells) into pathogen-free sterile culture in a synthetic medium with growth retardants have been cited as good ways of complementing and providing backup to field collections. It has long been known that in vitro slow growth method suffers high risks of somaclonal variation (Withers, 1993) and also from the need to develop individual maintenance protocols for the majority of species (Thormann et al., 2006).

Slow growth

Slow growth procedures allow clonal plant material to be held for 1 to 15 years under tissue culture conditions with periodic sub-culturing, depending on species. There are several methods by which slow growth can be maintained. In most cases, a low temperature often in combination with low light intensity or even darkness is used to limit growth. Temperatures in the range of 0 to 5°C are employed with cold tolerant species, but for tropical species which are generally sensitive to cold, temperatures between 15° and 20°C are used. It is also possible to limit growth by modifying the culture medium, mainly by reducing the sugar and/or mineral elements concentration and reduction of oxygen level available to cultures by covering explants with a layer of liquid medium or mineral oil (Withers and Engelmann, 1993). Although slow growth procedures have been developed for a wide range of species, they are routinely used for conservation of genetic resources of only a few species including Musaspp., potato, sweet potato, cassava, yam, Allium spp. and temperate tree species. About 37,600 accessions are reportedly conserved by in vitro techniques in gene banks, worldwide (FAO, 1996).

Cryopreservation

Cryopreservation, the process in which living tissues are conserved at very low temperatures (-196°C) in liquid nitrogen (LN) or in vapour phase (-150°C) to arrest mitotic and metabolic activities, provides a more promising option (Thormann et al., 2006). Significant progress has been made in cryopreservation research over the past twenty years and much of that research has been focusing on understanding the desiccation sensitivity of recalcitrant seeds and on the underlying mechanism of desiccation tolerance (Engelmann and Panis, 2009). The techniques for cryopreservation currently in use are quite varied and include the older classical techniques based on freeze-induced dehydration of cells as well as newer techniques based on vitrification (Engelmann, 2000). In classical techniques, tissues are cooled slowly at a controlled rate (usually 0.1-4°C/min) down to about -40°C, followed by rapid immersion of samples in liquid nitrogen. Slow freezing is carried out using a programmable freezing apparatus. Cryoprotectants are added to the freezing mixtures to maintain membrane integrity and increase osmotic potential of the external medium. Classical cryopreservation procedures have been successfully applied to undifferentiated culture systems such as cell suspensions and calluses (Kartha and Engelmann, 1994). However, in case of differentiated structures, they have been employed for storage of apices or embryonic axes of only cold-tolerant species (Reed and Chang, 1997), and their utilization for tropical species has been limited (Escobar et al., 1997). Vitrification-based procedures involve removal of most or all free able water by physical or osmotic dehydration of explants, followed by ultra-rapid freezing which results in vitrification of intracellular solutes, that is, formation of an amorphous glassy structure without occurrence of ice crystals which are detrimental to cellular structural integrity. These techniques are more appropriate for complex organs like embryos and shoot apices; they are also less complex and do not require a programmable freezer, hence are suited for use in any laboratory with basic facilities for tissue culture.

DNA Storage

With the rapid development in the field of molecular genetics and genomics, DNA material is becoming more and more in demand for molecular studies and is one of the most requested materials from gene banks (Andersson, 2006). The establishing of a DNA storage facility as a complementary "back-up" to traditional ex situ collections has been suggested (Dulloo et al., 2013), but little effort has been made to collect and conserve DNA as a genetic resource. Some efforts have been made to establish DNA banks for endangered animals (Ryder et al., 2000) and a few plant DNA banks including Missouri Botanic

Garden, Royal Botanic Gardens - Kew, Australian Plant DNA Bank and Trinity College Dublin (TCD) (Rice et al., 2006; Hodkinson et al., 2007). The Global Biodiversity Information Facility (GBIF) in Germany has establish a DNA Bank Network in 2007, last accessed 22 September 2010, and offers a worldwide central web portal, providing DNA samples of complementary collections (microorganisms, protists, plants, algae, fungi and animals). The GBIF Germany DNA network would provide a good mechanism to link both to the scientific community conserving genotypes in genebanks and to breeders and molecular biologists that use the resources for genetic improvement.

Pollen Storage

Pollen storage was mainly developed as a tool for controlled pollination of asynchronous flowering genotypes, especially in fruit tree. Even if it may not be considered to be a viable method for meaningful genetic conservation of genotypes, cryopreservation is likely to be more successful than other storage techniques routinely employed for pollen. Pollen can be easily collected and cryopreserved in large quantities in a relatively small space. In addition, exchange of germplasm through pollen poses fewer quarantine problems compared with seed or other propagules.

The pollen longevity of different species varies between minutes and years depending on the taxonomic status of the plant and on abiotic environmental conditions. For some crops, the storage of pollen grains is possible in appropriate conditions, allowing their subsequent use for crossing with living plant material. It is also possible to regenerate haploid plants from pollen culture for some crops. By controlling the storage temperature and relative humidity (0 to 10°C, 10 to 30% RH, depending on species), pollens of Citrus spp., Cocos nucifera, Fragaria sp., Olea europea, Pinus silvestris, Pistachio altantica, Pyrus malus and Vitis vinifera could maintain their viability for more than 1 year.

For long-term conservation, cryopreservation seems to be the most efficient method. For example, maize pollen could be dried to 50% of its original water content in an air current for 1 h and then stored at -196°C in liquid nitrogen. Deep-frozen maize pollen can be used for fertilization after 10 years storage. Successful cryopreservation of pollen from various 24 crops has been reported (Barnabas and Kovacs, 1997).

In situ Conservation

In situ conservation refers to conservation of genetic resources within their ecosystem and natural habitats. These techniques involve maintenance of genetic variation at location where it is encountered, either in wild or traditional farming systems.

- **Genetic reserves:** in this type of conservation location, management, and monitoring of genetic diversity is carried in natural wild populations within defined areas designated for active, long-term conservation.

- **On-farm conservation:** This refers to the sustainable management of genetic diversity of locally developed traditional crop varieties with associated wild and weedy species or forms by farmers within traditional agricultural, horticultural or agrisilvicultural cultivation systems.

- **Home gardens:** This type of conservation is similar to on-farm conservation, involves smaller scale but more species-diverse genetic conservation in home, kitchen, backyard or door-yard gardens.

- Complementary conservation

For ex situ conservation of PGR in a crop or crop group, a gene pool approach has to be followed for safe and effective conservation. Following this approach, it is very likely that a range of ex situ conservation methods would be applicable to satisfy the needs of a gene pool. For example, the rice gene pool consists of self-pollinated cultigens and a range of wild Oryza spp. habitat to range of climatic conditions with breeding ranging from obligate vegetative to facultative and obligate self-pollination. In a situation, it is quite logical to have an approach, which is appropriate and has balanced application of both in situ and ex situ conservation methods. This will lead to the adoption of a more "holistic" approach to conservation. Even withex situ, a balance has to be struck as per the need. For example, in case of wild Oryza species, it has to be assessed, whether they would be best conserved in a field gene bank or in vitro as cell, tissue, organ, pollen or perhaps as DNA or in combination thereof. Therefore, a network of complementary and comprehensive strategy is needed to ensure effective conservation and sustainable use of PGR for food and agriculture by present and future generations.

Svalbard Global Seed Vault

A major achievement for the conservation of the germplasm have been the creation of the of the Svalbard Global Seed Vault (SGSV) in 2008, established to serve as the ultimate safety net for seed samples from the world's most important collections (GCDT, 2010). The Svalbard Global Seed Vault is a secure seedbank located on the Norwegian Island of Spitsbergen near the town of Longyearbyen in the remote Arctic Svalbard archipelago, about 1,300 km (810 miles) from the North Pole. The facility preserves a wide variety of plant seeds in an underground cavern. The seeds are duplicate samples, or «spare» copies, of seeds held in genebanks worldwide. The seed vault will provide insurance against the loss of seeds in genebanks, as well as a refuge for seeds in the case of large scale regional or global crises. The seed vault is managed under terms spelled out in a tripartite agreement between the Norwegian government, the Global Crop Diversity Trust (GCDT) and the Nordic Genetic Resource Center (also known as NordGen and previously named the Nordic Gene Bank, a cooperative effort of the Nordic countries under the Nordic Council of Ministers). The Prime Ministers of Norway, Sweden, Finland, Denmark, and Iceland participated in a ceremonial «laying of the first stone» on 19 June, 2006. The Svalbard Global Seed Vault opened officially on February 26, 2008. The first seeds arrived in January 2008. Five percent of the seeds in the Vault, about 18,000 samples with 500 seeds each, come from the Centre for Genetic Resources of the Netherlands (CGN), part of Wageningen University, Netherlands.

Construction of SGSV

The seedbank is constructed 120 m (390 ft) inside a sandstone mountain at Svalbard on Spitsbergen Island. The bank employs a number of robust security systems. Seeds are packaged in special four-ply packets and heat sealed to exclude moisture. The facility is managed by the Nordic Genetic Resource Center, though there is no permanent staff on-site.

Spitsbergen was considered ideal due to its lack of tectonic activity and its permafrost, which will aid preservation. The location 130 m (430 ft) above sea level will ensure that the site remains dry even if the icecaps melt. Locally mined coal provides power for refrigeration units that further cool the seeds to the internationally recommended standard -18°C (0°F). Even if the equipment fails, at least several weeks will elapse before the temperature rises to the -3°C (27°F) of the surrounding sandstone bedrock. Prior to construction, a feasibility study determined that the vault could preserve seeds from most major food crops for hundreds of years. Some seeds, including those of important grains, could survive far longer, possibly thousands of years.

Mission and Seed Storage

The Svalbard Global Seed Vault's mission is to provide a safety net against accidental loss of diversity in traditional genebanks. While the popular press has emphasized its possible utility in the event of a major regional or global catastrophe, it will certainly be more frequently accessed when gene banks lose samples due to mismanagement, accident, equipment failures, funding cuts and natural disasters. Such events occur with some regularity. In recent years, some national genebanks have also been destroyed by war and civil strife. There are some 1,400 crop diversity *collections* around the world, but many are in politically unstable or environmentally threatened nations. The seeds are stored in four-ply sealed envelopes, then placed into plastic tote containers on metal shelving racks. The storage rooms are kept at -18°C (-0°F). The low temperature and limited access to oxygen will ensure low metabolic activity and delay seed aging. The permafrost surrounding the facility will help maintain the low temperature of the seeds if the electricity supply should fail. Approximately 1.5 million distinct seed samples of agricultural crops are thought to exist. The variety and volume of seeds stored will depend on the number of countries participating – the facility has a capacity to conserve 4.5 million.

Gene Bank Standards

Research on seed storage has indicated that the potential of seeds to store, that is, retaining genetic integrity and seed viability, is influenced by storage seed moisture content and temperature. Germplasm is generally conserved as a base collection or an active collection. Base collections are those that are being conserved on a long-term basis for posterity. These are unique accessions that are closest to the original samples and are not to be disturbed except for regeneration of active collections. Active or working collections are those that are immediately available for multiplication and distribution for use in research and crop improvement. To minimise the alteration in genetic structure and loss of viability in germplasm accessions during storage, the seed genebanks (that are part of the national network) preferably follow the genebank standards as recommended by FAO/IPGRI (Anonymous, 2001) in relation to various factors important to the good maintenance of active and base collections. The base collections are being stored in modules maintained at -20°C. Such a low temperature minimises metabolic activities and is expected to enable the seed to retain viability for 50 to 100 years without any change in genetic structure. Active collections are stored in modules maintained at 4°C and 35 to 40% relative humidity, under which seeds are expected to remain viable for 15 to 50 years without substantial change in viability and genetic integrity. For both types of collections, seed is processed after validating physical and genetic purity of seed, assessment of seed viability and seed moisture content. In most crops, seed samples with more than 85% seed viability are conserved. However, recognising inherent problems, such as indeterminate nature, which limits the harvest of physiologically mature seed of the same age in certain crops like cotton, several forages and vegetable crop species, the initial viability standards have been lowered down to between 50 to 75%. For long-term storage, the seed moisture content is brought down to 3 to 7%, while for medium-term storage the seed moisture content is brought down to 8 to 10%. For base collections to be put under long-term conservation, the preferable size of accession is 2,000 seeds in the case of self-pollinated and 4,000 in the case of cross-pollinated crops. However, in many cases, such as groundnut and castor, because of large seed size and low multiplication rates, the sample-size of the accessions has been reduced to between 1,000 to 1,500 seeds. The base and active collections are regularly monitored for seed viability, seed quantity, seed health, etc., at recommended intervals of 10 and 5 years, respectively. However, the monitoring of accessions at the National Seed Genebank (NSGB) in the Germplasm Conservation Division, NBPGR has generated valuable information on storability in a number of crop species, such as wheat, minor millets, cotton, grain legumes etc. (Anonymous, 2001). These results suggest a revision of the exact period of monitoring intervals. This information will be useful in revising the seed genebank standards in relation to other components and make seed conservation more cost effective. Seed storage problems are more common in India, because a large part of the country has a predominantly hot and humid, tropical and sub-tropical climate with great variation in temperature, rainfall and relative humidity across the year.

NATIONAL NETWORK ON CONSERVATION OF PGR

Efficient conservation of PGR in a country of the size and dimension of India, one of the 12 mega-centres of plant biodiversity and where 384 crop plants are reported to be cultivated (of which 168 species were earlier reported under the Hindustani centre, one of the eight Vavilovian centres of origin and diversity (Paroda et al., 1999), essentially requires a network approach. Network facilitates short-, medium-, and long-term conservation requirements, the division of responsibilities, application of complementary conservation strategies, and access for the use of these genetic resources in crop improvement programmes. The national network consists of the NSGB at NBPGR headquarters, New Delhi, 11 NBPGR Regional Stations situated in different agro-climatic zones of the country, and 40 crop-based National Active Germplasm Sites (NAGS), located generally at various ICAR institutes. The network is linked with the All India Co-ordinated Crop Improvement Projects, various research institutes (crop-based institutes, project directorates and national research centres; multi-crop based institutes) in the ICAR, SAUs, etc. All network components operate in close collaboration to ensure the efficient conservation and sustainable use of germplasm in crop improvement, in which the National Seed Genebank plays a pivotal role in conservation.

The National Seed Genebank

The NSGB is responsible for conservation of seeds of unique accessions on a long-term basis, as base collections for posterity. In addition, it provides technical support to the network in the planning, development and operation of medium-term genebank facilities, in human resource development, and in provision of accessions for the regeneration of active collections. The Indian NSGB has 12 modules with a capacity to hold around 1 million accessions.

NBPGR Regional Stations

The NBPGR has 11 regional stations/base centres/satellite stations located in different agroecological and phytogeographical zones of the country. They are responsible for the collection, characterisation, evaluation and/or conservation of germplasm in the region. The regional stations also coordinate various PGR activities in the region with other partners. Seven of the regional stations have medium-term seed storage modules for the conservation of active collections to meet the requirement of the region for various crops. The regional stations hold around 98,498 active collections. In addition, plant quarantine is looked after at the NBPGR headquarters, New Delhi and at the NBPGR regional station, Hyderabad.

National Active Germplasm Sites (NAGS)

The NAGS are based at ICAR institutes, at All India Co-ordinated Crop Improvement Projects and at SAUs. They are entrusted with the responsibility of crop specific collection, multiplication, evaluation, maintenance and conservation of active collections and their distribution to users at a national level. Large multiplications of active collections are preferred to reduce the number of regeneration cycles that can cause possible genetic changes and to meet the demand of seed distribution. The NAGS have a multidisciplinary team of scientists to study all the aspects of crop improvement, production and management. Therefore, the NAGS, in addition to their conservation role, are well equipped for the evaluation of germplasm and the generation of information on the potential value of accessions. This information forms the basis for use of accessions in research and crop improvement. Eleven of the NAGS have been provided with medium-term seed storage modules, to facilitate the use of active collections in research and breeding programmes.

Safety Duplicates of Crop Species

There is a built-in duplicity of accessions in the system, wherein the accessions conserved at NAGS and the crop-based institutes as active collection are conserved as base collection in the National Genebank. The active collections are used in research and crop improvement and the National Genebank helps in restoration of lost accessions to the active sites. This also serves as safety mechanism. There exists medium to high capability for research and use of improved methodologies for ex situ conservation. Nevertheless, strengthening of technical and infrastructure capabilities is required in some cases.

The capacity building in genebank management and information systems has been carried out satisfactorily, though there is a need for extension of medium-term facilities to more crop based institutes to cover larger number of crops. In last ten years 196,745 accessions were collected under 166 projects involving 599 professional and of these 104 084 accessions have been conserved. The maximum number of accessions conserved in ex situ is in the category of traditional cultivars and landraces. A significant number of collections belonging to wild and weedy relatives and advanced and improved cultivars developed using various genetic resources is also being conserved.

Major Achievements Through Germplasm Conservation

Plant Genetic Resources for Food and Agriculture (PGRFA) are vital to the development and welfare of human society. They contribute enormously towards achieving the global objectives of food security and poverty alleviation, environment protection and sustainable development. The local communities and farmers in India have sustained and enriched the diversity of these resources which they domesticated, used, conserved and made available to meet the ever increasing needs of the present and future generations. Characterization and evaluation of germplasm is required to know its worth or usefulness and availability of information on characterization and evaluation of conserved genetic resources is the key to utilization. Plant breeding provides many examples of the use of genetic resources for the improvement of the varieties of crop plants. There are examples that range from highly specific improvement to one major factor such as susceptibility to a pest or disease to all round improvement in yield, agronomical traits, disease resistance and to changes in the form and structure of the plant type.

FUTURE THRUST

- Endangered germplasm from the threatened areas of diversity to be salvaged and conserved for future use.
- Morphological and molecular characterization of germplasm to enhance their utilization in crop improvement.
- Conservation, management and protection of bio-resources especially

plant resources, through the participation of the people.

- Conservation and use of diversity needed to be addressed in a holistic manner and to meet the demands of the users of germplasm.

- Research on core and mini-core collections and identification of new diverse sources.

- Public awareness of the importance of CWR and neglected and underutilized species.

- Need to maximize synergy through appropriate collaboration between various national, sub-regional and international levels.

REFERENCES

1. Ahmad M, Zaffar G., Razvi SM, Mir SD, Bukhari SA, Dar ZA, Habib M (2014). Resilience of cereal crops to abiotic stress: A review. Afr. J. Biotechnol. 13(29):2908-2921.

2. Allard RW (1970). Populations structure and sampling methods. in Genetic Resources in Plants (O.H. Frankel and E. Bennett, eds.) 97–107. Blackwell, Oxford. Council, National Research. 1993. Managing Global Genetic Resources, Agricultural Crop Issues and Policies. National Academy Press, Washington D.C.

3. Andersson MS, Fuquen EM, de Vicente MC (2006). State of the art of DNA storage: results of a worldwide survey, 6-10. In: M.C. de Vicente Eds. DNA Banks- providing novel options for genebanks. Topical reviews in agricultural biodiversity. International Plant Genetic Resources Institute, Rome, Italy.

4. Anonymous (2001). Annual report 2000. National Bureau of Plant Genetic Resources, New Delhi, India.

5. Arora RK (1991). Plant Diversity in Indian Gene Centre. In: Paroda, R.S. and R.K. Arora (eds) Plant Genetic Resources - Conservation and Management. IPGRI, Reginoal Office for South Asia, New Delhi, India, pp. 25-54.

6. Arora RK, Pandey A (1996). Wild Edible Plants of India: Diversity, Conservation and Use. NBPGR, New Delhi, India.

7. Burke MB, Lobell DB Guarino L (2014). Shifts in African crop climates by 2050 and the implications for crop improvement and genetic resources conservation. Global Environmental Change. 19(3):317-325.

8. Damania AB (2002).the Hindustan centre of origin of important plants. Asian Agri-history 6:333-341.

9. Dulloo ME, Thormann I, Jorge AM, Hanson J (2013) Crop specific

regeneration guidelines [CD-ROM]. CGIAR System-wide Genetic Resource Programme (SGRP), Rome, Italy. (CD-ROM)

10. Ellis RH, Hong TD. Roberts EH (1990). An intermediate category of seed storage behaviour. J. Exp. Bot. 41:1167-1174.

11. Ellis RH, Hong TD, Roberts EH, Soetisna U (1991). Seed storage ehaviour in Elaeis guineensis. Seed Sci. Res. 1:99-104'.

12. Engelmann F (2000). Importance of cryopreservation for the conservation of plant genetic resources. In Engelmann F, Takagi H (eds) Cryopreservation of tropical plant germplasm. Current research progress and application. IPGRI, Rome, Italy, pp. 8-20.

13. Engelmann F, Engels JMM (2002). Technologies and strategies for ex situ Conservation, 89-103 In: Engels, J.M.M., V. Rao Ramanatha, A.H.D. Brown, M.T. Jackson, Eds. Managing Plant Genetic Diversity. CABI Publishing, UK.

14. Engelmann F, Panis B (2009). Strategy of Bioversity International cryopreservation research - a discussion paper. Bioversity international, Rome, Italy. Unpublished. 8.

15. Escobar R, Mafla G, Roca WM (1997). Methodology for recovering cassava plants from shoot tips maintained in liquid nitrogen. Plant Cell Rep. 16:474-478.

16. FAO (1996). Report on the state of the World's Plant Genetic Resources, Rome Italy.

17. FAO (1998). The state of the world's plant genetic resources for food and agriculture. Food and Agriculture Organization of the United Nations, Rome, Italy.

18. FAO (2010). The second report on the state of the world's plant genetic resources for food and agriculture. Food and Agriculture Organization of the United Nations, Rome, Italy.

19. Frankel OH, Hawkes JG (1975). Crop Genetic Resources for Today and Tomorrow. Cambridge University Press, Cambridge. P. 492.

20. Frankham R (2010). Challenges and opportunities of genetic approaches to biological conservation. Biological conservation 143:1919-1927.

21. GCDT (2010) http://www.croptrust.org/main/arcticseedvault. php?itemid=211, (last accessed 20 August 2010)

22. Hodkinson TR, Waldren S, Parnell JAN, Kelleher CT, Salamin K, Salamin N (2007). DNA banking for plant breeding, biotechnology and biodiversity evaluation. J. Plant Res. 120:17-29.

23. Jarvis DI, Hodgkin T (2008). The maintenance of crop genetic diversity

24. Kartha KK, Engelmann F (1994). Cryopreservation and germplasm storage. In: (Vasil IK, Thorpe TA (eds) Plant Cell and Tissue Culture. Kluwer Academic Publishres, Netherlands, pp. 195-230.

25. Laikre L, Allendorf FD, Aroner LC, Baker CS, Gregovich DP, Hansen MH, Jackson J, Kendal AKC, Mckelvey K, Neel MC, Olivieri I, Ryman NM, Schwartz K, Bull R S, Jeffrey B Stetz JB, Tallmon DA, Taylor BL, Vojta CD, Waller DM, Waples RS (2010). Neglect of genetic diversity in implementation of the convention of biological diversity. Conservation Biology 24(1):86-88.

26. Laikre L, Larsson LC, Palme A, Charlier J, Josefsson M, Ryman N (2008). Potentials for monitoring gene level biodiversity: using Sweden as an example. Biodivers. Conserv. 17:893-910.

27. Lane A, Jarvis A (2007). Changes in climate will modify the geography of crop suitability: agricultural biodiversity can help with adaptation. J. SAT Agric. Res. 4(1):1-12.

28. Maxted N, Kell SP (2009). Establishment of a Global Network for the In Situ Conservation of Crop Wild Relatives: Status and Needs. FAO Commission on Genetic Resources for Food and Agriculture, Rome, Italy.

29. Millennium Ecosystem Assessment (2005). Ecosystems and Human Well-being: Biodiversity Synthesis. World Resources Institute, Washington, DC.

30. MoEF (1994). Ethnobotany in India – A status report. All India Coordinated Research Project on Ethnobotany. Ministry of Environment and Forest, Government of India, New Delhi, India.

31. NAS (1975). Under-exploited Tropical Plants with Promising Economic Value. National Academy of Sciences, Washington DS, USA.

32. Paroda RS, Rai M, Gautam PL, Kochher S, Singh AK (1999). National action plan on agrobiodiversity in India. National Academy of Agricultural Sciences, Indian

33. Rao VR (1980). Groundnut genetic resources at ICRISAT. in Proceedings of the International Workshop on Groundnuts, 13–17 October, 1980. 47–57. ICRISAT, Patancheru, 502324, A.P., India.

34. Rice N, Henry R, Rossetto M (2006). DNA Banks: a primary resources for conservation research, 41-48. In: M.C. de Vicente Eds. DNA Banks-providing novel options for genebanks. Topical reviews in agricultural biodiversity. International Plant Genetic Resources Institute, Rome, Italy.

35. Roberts EH, King MW (1986). Storage of recalcitrant seeds. in Crop

Genetic Resources – The Conservation of Difficult Material (L.A. Withers and J.T. Williams, eds.) 39–48. IUBS/IGF/IBPGR, Paris.

36. Ryder OA, McLaren A, Brenner S, Zhang Y, Benirschke K (2000). DNA Banks for Endangered Animal Species. Science 288 (5464):275-277.

37. SBSTTA (2010). Proposals for a consolidated update of the global strategy for plant conservation. UNEP/CBD/SBSTTA/14/9,-http://www. cbd.int/doc/?meeting=SBSTTA-14.

38. Thormann I, Hunter D, Danielian A, Djataev S, Ramelison J, Wijesekara A, Zapata Ferrufino B, Borelli T (2010). Meeting the information challenge for crop wild relatives in situ conservation: A global portal hosting national and international data on CWR. Presented at the IAALD XIIIth World Congress - 26-29 April 2010, Montpellier, France.

39. Thormann I, Dulloo ME, Engels JMM (2006). Techniques of ex situ plant conservation, 7-36. In: Robert Henry (ed.) "Plant Conservation Genetics". Centre for Plant Conservation Genetics, Southern Cross University, Lismore, Australia. The Haworth Press inc.

40. UNCED (1992). Convention on Biological Diversity. United Nations Conference on Environment and Development, Geneva.

41. United Nations (2010). The Millennium Development Goals Report 2010. New York. USA.

42. Van de Wouw M, Kik C, Hintum T, Van Treuren R, Visser B, (2009).

43. Plant genetic resources; Characterization and utilization 2009:1-15.

44. Vavilov NI (1951). Phyto-geographical basis of plant breeding. In: (selected Writings of N.I. Vavilov and translated by K.S. Chester) The Origin, Variation, Immunity and Breeding of Cultivated Plants, Chronica Botanica 13:364.Waltham Mass., USA.

45. Walpole Matt, Rosamunde EAA, Besançon C, Butchart SHM, Campbell-Lendrum D, Carr GM, Collen B, Collette L, Davidson NC, Dulloo E, Fazel AM; Galloway JN, Gill M, Goverse T, Hockings M, Leaman DJ, Morgan DHW, Revenga C, Rickwood C J, Schutyser F, Simons S, Stattersfield AJ, Tyrrell TD, Vié JC and Zimsky M (2009). Tracking progress towards 2010 biodiversity target and beyond. Science 135:1503-1504.

46. Withers LA, Engelmann L (1993). In vitro conservation of plant genetic Resources. In biotechnology in Agriculture, (A. Altman, ed.), Marcel Dekker Inc., New York, in press.

47. Zheng GH, Jing XM, Tao KL (1998). Ultradry seed storage cuts costs of gene bank. Nature 393:223-224.

48. Zhou M, Engels JMM, Ramanatha RaoV 1(995). Effect of low seed

moisture content on seed storage. 14 (ab) in International Symposium on Research and Utilization of Crop Germplasm Resources, 1–3 June, Beijing, China.

Chapter 3

CLIMATE-SMART AGRICULTURE GLOBAL RESEARCH AGENDA: SCIENTIFIC BASIS FOR ACTION

Kerri L Steenwerth[1], Amanda K Hodson[2], Arnold J Bloom[3], Michael R Carter[r4], Andrea Cattaneo[5], Colin J Chartres[6], Jerry L Hatfield[7], Kevin Henry[8,9], Jan W Hopmans[2], William R Horwath[2] Bryan M Jenkins[10], Ermias Kebreab[11], Rik Leemans[12], Leslie Lipper[13], Mark N Lubell[14], Siwa Msangi[15], Ravi Prabhu[16], Matthew P Reynolds[17], Samuel Sandoval Solis[2], William M Sischo[18], Michael Springborn[19], Pablo Tittonell[20], Stephen M Wheeler[21], Sonja J Vermeulen[22], Eva K Wollenberg[23], and Louise E Jackson[2]

[1]Crops Pathology and Genetics Research Unit, c/o Department of Viticulture and Enology, United States Department of Agriculture (ARS/USDA), Agricultural Research Service

[2]Department of Land, Air and Water Resources, University of California at Davis

[3]Department of Plant Sciences, University of California at Davis

[4]Department of Agricultural and Resource Economics, University of California at Davis

[5]Climate Smart Agriculture Project, Food and Agriculture Organization of the U.N., Viale delle Terme di Caracalla

[6]eWater, University of Canberra Innovation Centre, University Drive South

[7]National Laboratory for Agriculture and the Environment, ARS/USDA

[8]Where the Rain Falls, CARE France

[9]School of Global Envirfonmental Sustainability, Colorado State University

[10]Department of Biological and Agricultural Engineering, University of California at Davis

[11]Department of Animal Science, University of California at Davis

[12]Environmental Sciences, Wageningen University

[13]Agricultural and Development Economic Analysis Division, Food and Agriculture Organization of the U.N., Viale delle Terme di Caracalla

[14]Department of Environmental Science and Policy, Center for Environmental Policy and Behavior, University of California at Davis

[15]Environment and Production Technology Division, International Food Policy Research Institute (IFPRI)

[16]World Agroforestry Center (ICRAF)

[17]Plant, International Maize and Wheat Improvement Center, Consultative Group on International Agricultural Research (CGIAR) Apdo, Postal

[18]Food- and Water-borne Disease Research Program, College of Veterinary Medicine, Washington State University

[19]Department of Environmental Science and Policy, University of California at Davis

[20]Plant Sciences, Wageningen University

[21]Department of Landscape Architecture, University of California at Davis

[22]Department of Plant and Environmental Sciences, Climate Change, Agriculture and Food Security, Consultative Group on International Agricultural Research (CGIAR), University of Copenhagen

[23]Gund Institute for Ecological Economics and Rubenstein School of Environment and Natural Resources, University of Vermont

ABSTRACT

Background

Climate-smart agriculture (CSA) addresses the challenge of meeting the growing demand for food, fibre and fuel, despite the changing climate and fewer opportunities for agricultural expansion on additional lands. CSA focuses on contributing to economic development, poverty reduction and food security; maintaining and enhancing the productivity and resilience of natural and agricultural ecosystem functions, thus building natural capital; and reducing trade-offs involved in meeting these goals. Current gaps in knowledge, work within CSA, and agendas for interdisciplinary research and science-based actions identified at the 2013 Global Science Conference on Climate-Smart Agriculture (Davis, CA, USA) are described here within three themes: (1) farm and food systems, (2) landscape and regional issues and (3) institutional and policy aspects. The first two themes comprise crop physiology and genetics, mitigation and adaptation for livestock and agriculture, barriers to adoption of CSA practices, climate risk management and energy and biofuels (theme 1); and modelling adaptation and uncertainty, achieving multifunctionality, food and fishery systems, forest biodiversity and ecosystem services, rural migration from climate change and metrics (theme 2). Theme 3 comprises designing research that bridges disciplines, integrating stakeholder input to directly link science, action and governance.

OUTCOMES

In addition to interdisciplinary research among these themes, imperatives include developing (1) models that include adaptation and transformation at either the farm or landscape level; (2) capacity approaches to examine multifunctional solutions for agronomic, ecological and socioeconomic challenges; (3) scenarios that are validated by direct evidence and metrics to support behaviours that foster resilience and natural capital; (4) reductions in the risk that can present formidable barriers for farmers during adoption of new technology and practices; and (5) an understanding of how climate affects the rural labour force, land tenure and cultural integrity, and thus the stability of food production. Effective work in CSA will involve stakeholders, address governance issues, examine uncertainties, incorporate social benefits with technological change, and establish climate finance within a green development framework. Here, the socioecological approach is intended to reduce development controversies associated with CSA and to identify technologies, policies and approaches leading to sustainable food production and consumption patterns in a changing climate.

Introduction

Globally, agricultural and forestry systems are expected to change significantly in response to future climate change, manifesting as major transitions in livelihoods and landscapes [1–4]. During the few past decades, crop yields have been reduced because of warming [5], and the results of modelling studies suggest that climate change will reduce food crop yield potential, particularly in many tropical and midlatitude countries [6–9]. Rising atmospheric CO_2 concentrations will decrease food and forage quality [10]. Price and yield volatility likely will continue to rise as extreme weather continues, further harming livelihoods and putting food security at risk [11]. Global demand for agricultural products, be they food, fibre or fuel, continues to increase because of population growth, changes in diet related to increases in per capita income and the need for alternative energy sources while there is less and less additional land available for agricultural expansion. Agriculture thus needs to produce more on the same amount of land while adapting to a changing climate and must become more resilient to risk derived from extreme weather events such as droughts and floods.

The term *climate-smart agriculture* (CSA) has developed to represent a set of strategies that can help to meet these challenges by increasing resilience to weather extremes, adapting to climate change and decreasing agriculture's greenhouse gas (GHG) emissions that contribute to global warming (Figures 1 and 2). CSA also aims to support sustainable and equitable transitions for agricultural systems and livelihoods across scales, ranging from smallholders to transnational coalitions. Forming a core part of the broader green development agenda for agriculture [12–14], CSA focuses on meeting the needs of people for food, fuel, timber and fibre through science-based actions; contributing to economic development, poverty reduction and food security; maintaining and enhancing the productivity and resilience of both natural and agricultural ecosystem functions, thus building natural capital; and reducing the trade-offs involved in meeting these goals. It invokes a continuous, iterative process for stakeholders, researchers and policymakers to meet the challenges presented by climate change and collectively transform agricultural and food systems towards sustainability goals [15]. Increased awareness and adaptive management are essential components of the CSA strategy. Yet, CSA is controversial. Such a broad agenda can be appropriated to support conflicting agendas or promote specific ecosystem services [16]. GHG emission mitigation by resource-poor farmers raises equity as an issue in developing countries because it may bring farmers little benefit unless it directly provides them with adaptive capacity.

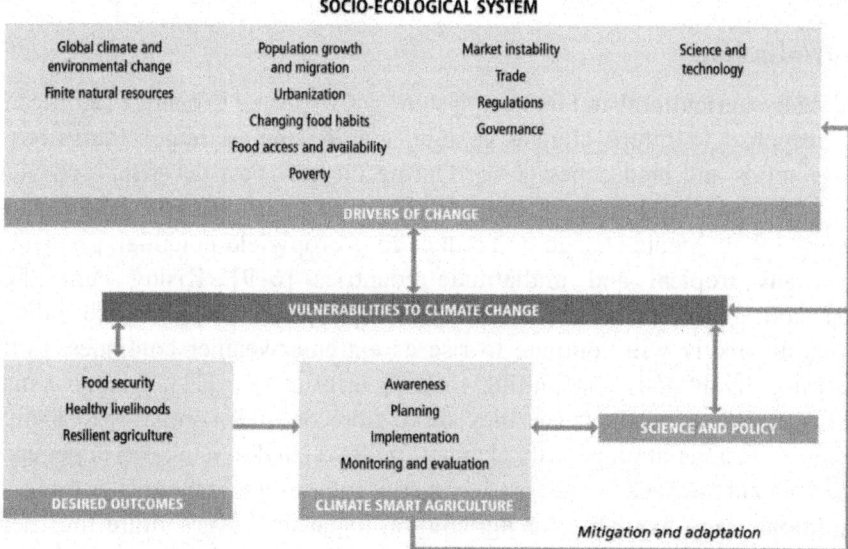

Figure 1: Diagram illustrating how climate-smart agriculture can be utilised as an agent for developing resilience, mitigation and adaptation within the socioecological system. Although not exclusively within the purview of climate-smart agriculture (CSA), 'adaptation and mitigation' in this figure are implied to be derived from an

iterative CSA strategy. Adaptation and mitigation affect 'drivers of change' to diminish existing 'vulnerabilities to climate change' in the socioecological system, leading to the long-term goals of CSA in 'desired outcomes'. The arrow between CSA and 'science and policy' indicates the vital role of novel science–policymaking partnerships and science-based actions in CSA.

Setting CSA in the context of a safe operating space for humanity with socioecological systems that support adaptive management and governance will require scientific metrics and science–policymaking dialogues [16] that depend on strong engagement of the scientific community.

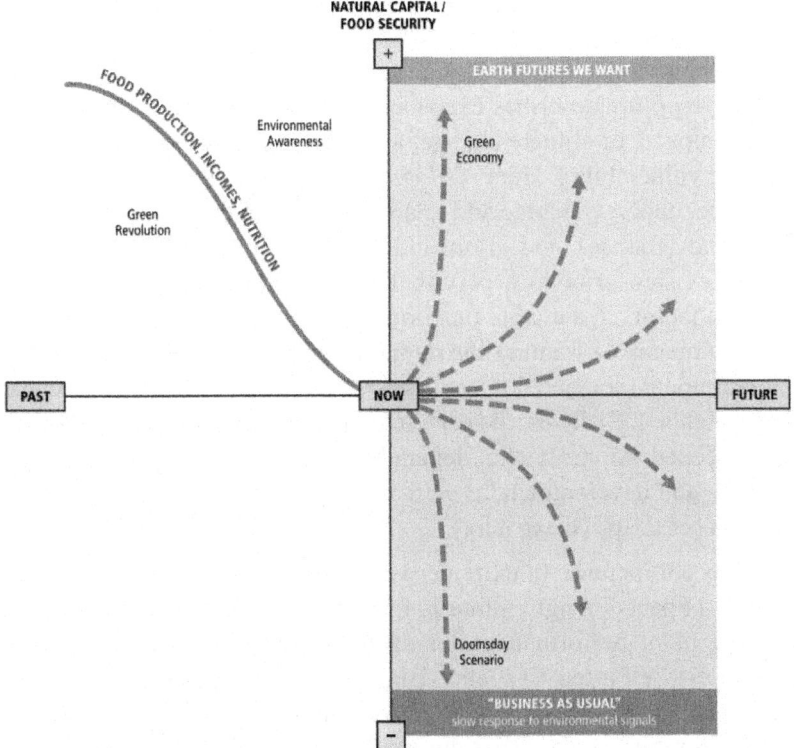

Figure 2: Schematic of the historical trajectory of food production, incomes and nutrition within the socioecological system with respect to food security and natural capital over time juxtaposed with potential future outcomes. Placement of 'environmental awareness' indicates that this factor gained prominence in the socioecological system as food security and natural capital diminished with respect to the trajectory of food production, income and nutrition. Positive outcomes are shaded in green, and the 'business as usual' and 'doomsday' outcomes are shaded in orange. Climate-smart agriculture plays a role in building positive outcomes such as the 'green economy'. Adapted with permission from M van Noordwijk (personal communication, 2014).

At the 2013 Global Science Conference on Climate-Smart Agriculture (Davis, CA, USA), participants examined the state of global science and best practices concerning climate and agriculture worldwide. Participants built on the consensus achieved at the 2011 Global Science Conference on Climate-Smart Agriculture conference (Wageningen, the Netherlands), agreeing on a broad strategy for science and policymaking to strengthen food security, mitigation and adaptation [17]. Participants further examined current gaps in knowledge, identified existing and promising work within CSA and formulated agendas for interdisciplinary research and science-based actions to support CSA.

The relationship between vulnerability, resilience and adaptation was an overarching theme echoed across the conference and is crucial to CSA. *Vulnerability* describes exposure, sensitivity and capacity to respond to negative impacts of climate change, and *adaptation* is the means by which to reduce the vulnerability. Here *resilience* is regarded as the capacity to tolerate disturbance, undergo change and retain the same essential functions, structure, identity and feedback and is not indicative solely of returning to the same state that existed prior to a perturbation or disturbance [18–20]. Resilience focuses on factors that enable functioning despite adverse conditions [21, 22], provides a means of framing the dynamic relationships between humans and the environment (socioecological systems) and considers society's capacity to manage change [23]. Thus, the principle of resilience can guide transformative change needed to meet the demands of food security, natural resource protection, and development, as well as to diminish vulnerability and promote adaptation (or adaptive capacity).

The recent increase in extreme weather events (climate shocks) threatens disruptive impacts on agriculture [24, 25]. Projected adaptive actions include improving plant performance (for example, nutrition, yields, food quality) in response to elevated CO_2 and rising temperatures [26–28]; avoiding pest damage and food waste [28, 29]; developing forecasting, management and insurance options to decrease the risk due to unexpected rainfall patterns, higher temperatures and shifting length in growth seasons [14, 28, 30]; and managing natural resources at the landscape and regional levels to assure the environmental quality and ecosystem services upon which agriculture depends [31–33]. Solutions involve trade-offs. For instance, planning now for higher temperatures and declining precipitation in arid zones may reduce water deficits for agriculture, but it will require institutional investment to support both the intensified demand for ground and surface waters [34, 35] and the necessary improvements in irrigation efficiencies [36]. Along with these adaptive actions, CSA seeks to contribute to the mitigation and reduction of GHG, mainly nitrous oxide (N_2O) and methane (CH_4) emissions, and to balance trade-offs with food security and livelihoods [7, 37, 38]. For example, combining agroforestry, afforestation and conservation efforts with agriculture to meet global food demand will help to mitigate GHG emissions, support biodiversity and concomitantly preserve ecosystem services [39, 40]. Other trade-offs that occur when abrupt environmental changes stress agricultural systems include changes in rural and urban human migration patterns, as well as loss of cultural resources, which reduces the ability to manage land use effectively [41–43]. Without doubt, the development status of a country or region will influence the approach to mitigating and adapting to climate uncertainty and will affect the implementation and focus of the CSA strategy. For example, industrialized nations focus more strongly on mitigation of climate change through reduction of agriculture's environmental impacts, whereas developing countries' approaches to climate uncertainty emphasize stabilizing and boosting food production, improving incomes and building adaptive capacity [7, 15, 44]. Gender can also influence decisions and capacity for mitigation and adaptation. Women in some regions in Africa have experienced greater exposure and vulnerability, especially to extreme events, than men, but they also have demonstrated greater collective action in farming decisions linked to social networking [45, 46].

Crucial science questions and challenges for food systems in the face of climate change and uncertainty require comprehensive, collaborative investments and science-based actions. In the past few years, policies and programmes have included landscape-scale research on food security and natural resources, policy and governance to achieve agricultural resilience to climate change and capacity building [47]. Under CSA, transformative changes to achieve food security, poverty relief, mitigation and adaptation target novel types of science–policymaking partnerships and involve stakeholders and decision-makers in the public and private sectors to gain long-term commitment and investment to carry the new actions to fruition. CSA emphasizes the involvement of scientists with farmers, land managers, agroforesters, livestock keepers, fishers, resource managers and policymakers (stakeholders) to empower them in the formation of palatable choices to enact adaptive capacity and resilience 'on the ground' and within broader policies [14, 15]. Farmer-led innovative approaches and social learning are crucial parts of this process, where social learning represents a 'change in understanding that goes beyond the individual to become situated within wide social units or communities of practice through social interactions between actors within social networks' [48, 49]. In this article, we summarize and synthesize the discussions and ideas presented at the 2013 CSA conference by an international community of scientists, growers, policymakers, research scientists, government officials, nonprofit entities and students who are working to achieve food security, poverty reduction, mitigation and adaptation within the CSA context. The three sections of this article reflect the scientific themes presented at the conference: (1) farm and food systems, (2) landscape and regional issues and (3) institutional and policy-related aspects. Within the first and second themes, parallel sessions at the conference charged participants to identify knowledge gaps, research initiatives and transformative actions required to address these specific issues. We provide a summary of the 12 sessions and highlights of the oral presentations by subject experts, and we conclude with recommendations offered during discussions as well as a consensus agenda for future actions [50]. Finally, broad outcomes and messages are presented, largely adhering to the actual proceedings to reflect the spirit and outcomes of this conference. Thus, the emphasis is on structuring disciplinary and interdisciplinary science in a CSA context rather than mechanisms for implementing science in action. This article is intended to serve as a benchmark and guide for future CSA research activities.

THEME 1

Farm and Food System Issues: Sustainable Intensification, Agro-ecosystem Management and Food Systems

Considerable research on climate change and agriculture exists at the farm and food system levels, including topics such as farming practices for mitigation of agricultural GHG emissions, choice and adaptation of crops and livestock to new climate regimes, decision-making by farmers and life-cycle assessments [51–55]. The tendency has been to apply disciplinary science that informs particular problems and solutions for agriculture, as demonstrated by the topics of the six sessions in theme 1. Sustainable intensification, focused initially on increased agricultural production and food security, has now moved to a broader set of goals with multiple social, ethical and environmental dimensions [56, 57]. The integrative challenge for CSA is to better understand the trade-offs and choices farmers must make for greater multifunctionality and resilience to climate change. Because planning for climate change can be highly farm-, commodity- and context-specific, especially in response to extreme events, CSA is committed to new ways of engaging in participatory research and partnerships with producers [14].

Crop Physiology and Genetics under Climate Change

Responding to effects of climate change (for example, changes in nutrient availability and plant nutrient acquisition, higher CO_2 concentrations and temperatures, water deficits and flooding) that influence the closure of the yield gap between potential and actual production will require continuation of existing 'best management practices' coupled with improvements in agronomic management practices and crop-breeding [58, 59]. Uncertain is the degree to which advances in crop physiology and genetics will continue to support higher agricultural production in response to more frequent climate shocks. Whereas successful crop adaptation to new production locations may be a good predictor of future outcomes, much higher CO_2 concentrations and temperatures are conditions beyond our current set of experiences [21,60]. Molecular approaches and genetic engineering will foster better understanding and manipulation of physiological mechanisms responsible for crop growth and development, as well as the breeding of stress-adapted genotypes [61–63], but there are social controversies surrounding the use of some of these technologies. High-throughput phenotyping platforms and comprehensive crop models will lead to more rapid exploration of genetic resources, enabling both gene discovery and better physiological understanding of how crop improvement can increase tolerance to environmental stress [64–68]. Development of new crop genotypes to meet the need to thrive under future management and climate conditions, the expected increases in the frequency of climate shocks and the uncertainty of rates of climate change presents a challenge. The specific examples set forth in the following paragraphs demonstrate how greater understanding of biochemical pathways, plant traits and phenotypes and germplasm evaluation could help overcome bottlenecks in both yield and development of physiological resilience to environmental stresses.

Molecular approaches provide opportunities to establish linkages between biochemical pathways and physiological responses. In cereals such as rice, grain yield is highly dependent on the carbohydrate source (top leaves) and sink (florets) relationship, which is strongly influenced by the plant hormone cytokinin [69]. Cytokinin production also affects drought tolerance and senescence, and isopentenyl transferase (IPT) expression controls upregulation of pathways for cytokinin degradation. Therefore, it follows that tolerance of abiotic stress by delaying stress-induced senescence through manipulation of IPT expression in transgenic lines could maintain optimal levels of cytokinin, resulting in greater fitness and more seed and grain production [62]. When exposed to varying drought intensities pre- and postflowering, transgenic rice with higher IPT expression maintained consistently higher grain yields and concentrations of sucrose and starch compared to the wild-type genotype. The delayed onset of drought-related symptoms in the transgenic lines caused positive source–sink relationships for a relatively longer period with higher photosynthetic rates than the wild type.

Combinations of multiple plant traits to survive stress, however, may produce more resilient crop production in the face of climate change [64]. Survival strategies employed by plants include early flowering to escape drought periods, stomatal control to prevent water loss, enhanced root growth in deeper soil layers to access water [70] and reduced leaf growth to minimize the transpiring surface [71]. These adaptations come at a cost, where reductions in the growth cycle, light interception and carbon (C) assimilation by photosynthesis are often accompanied by a higher C requirement to build additional plant roots, especially under nutrient stress [72]. Thus, the trade-offs of introducing new plant traits must be considered for specific types of environmental stress [65].

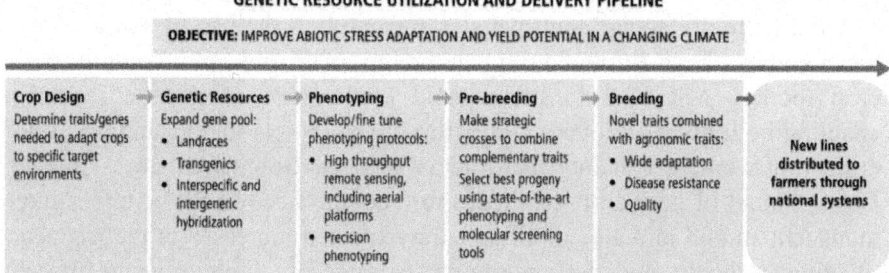

Figure 3: The genetic resource and utilization pipeline reflects the combination of physiological, molecular and traditional breeding approaches. Adapted with permission from M Reynolds (personal communication, 2014).

By examining the genetic basis of physiological mechanisms and environmentally induced stress responses, crops such as maize, wheat and other cereals can be bred to produce better yields and tolerances through targeted accumulation of alleles that confer robust responses to environmental stressors such as drought [73] (Figure 3). This approach is used by the International Maize and Wheat Improvement Center for the discovery and accumulation of drought-adaptive traits in wheat and maize germplasm from wild-type crop relatives and cultivars from a wide range of climates and growing conditions [65, 67, 74]. Screening for physiological traits can be highly effective in selecting such lines for a breeding programme. Canopy temperature (CT) is an example of a widely used, high-throughput germplasm screening tool. CT is linked to stomatal conductance, an indirect indicator of water uptake by roots, especially under drought and heat stress [75, 76]. In one study, researchers found that 60% of variation in yield from recombinant inbred lines grown under drought conditions was explained by CT [77]. Screening for physiological traits in candidate genotypes as an initial step may thus accelerate the search for novel genes [75] and genotypes that will be needed to deal with rapid changes in climate, such as the greater intensity and frequency of drought. Trait-based breeding programmes will be most effective when approaches are developed to simultaneously screen a broad array of genotypes for phenotypic responses to environmental stresses quickly (for example see, [78]).

Complementary approaches are necessary for solving complex physiological plant responses to climate and management. Changes in temperature, precipitation, water delivery, salinity and CO_2 concentrations will occur simultaneously. Direct experimentation, high-throughput screening platforms using molecular-based techniques and predictive modelling are a set of tools for achieving multiple goals [79–81], which include exploration of genetic resources for broader use and dissemination, gene pool expansion and yield stability in the face of interannual weather variation. In addition, these tools can help with other crop selection criteria, including quality of food and feed, source–sink relationships, pest and disease resistance, plant–microbe interactions that reduce CH_4 and N_2O emissions, and postharvest storage [60, 81]. Regional networks that examine environmental and physiological tolerances and yield potentials, as well as their coalescence into global crop improvement networks [82], will provide large-scale screening approaches to assessing both germplasm and phenotypic responses of crop plants. These networks already exist in representative target environments, such as the Network for the Genetic Improvement of Cowpea for Africa, Sorghum and Millet Networks, International Wheat Improvement Network, International Maize Improvement Network; and other regional networks linked to CGIAR that focus on grains and legumes in Africa, Latin America, the Caribbean and Asia. They also include networks for research and extension supported by Association for Strengthening Agricultural Research in Eastern and Central Africa. Participatory breeding by farmers and other stakeholders will eventually be an essential way to advance this agenda [83, 84].

Livestock Management and Animal Health

Livestock production not only contributes to climate change via GHG emissions (see [85]) but also suffers due to extreme weather events and disease related to climate change. Direct and indirect challenges in both mitigation and adaptation include fluctuating feed prices, habitat changes, expansion of vector-borne diseases in warm climates, impaired reproduction, pasture quality and availability and physiological heat stress [86, 87]. Opportunities for mitigating emissions include dietary manipulation, genetic improvement and mortality reduction to enhance overall production potential; manure management; and reduction of deforestation and pasture burning through payments for ecosystem services [88,89]. Adaptation strategies include income and livelihood diversification by mixing crop and livestock production; sustainable intensification through pasture regeneration or destocking; diversifying livestock feeds; manipulation of rumen microbial composition; matching animal breeds to local environments and moving animals to other sites; and better risk management and transformative change (for example, exit from or entry into animal agriculture) [88, 90–92]. These strategies rely heavily on sustainable intensification, as in the improvement of productivity and efficiency that exists in conjunction with incentives and investments that allow systems to intensify and in the development of regulations and limits on intensifying systems, among other aspects [93]. Access to credit or savings, land and resource inputs, and livelihood diversification are other potential pathways towards adaptation and food security [94, 95]. Technology, supporting policies and investments will require varied mixtures of strategies, as shown by the examples described in the following paragraphs.

Flexibility in livelihood options for pastoralist, agropastoralist and ranching communities can increase a household's capacity to manage risk and adapt in the face of burgeoning external stress [96]. Adaptation options depend on household objectives and attitudes; local access to natural resources, inputs and output markets; and sustainable intensification. Nutrient management is fundamental to maintaining a livelihood in livestock production. In Madagascar, external nitrogen (N) inputs are not commonly used to replenish the N losses that occur through erosion, leaching, GHG emissions and harvest. Hence, Alvarez et al. [97] examined N flows through crop-livestock systems to determine management scenarios leading to improvement in their N use efficiency, productivity and economic viability. They evaluated four intensification scenarios for system productivity, food self-sufficiency and gross margins: (1) using supplementary feed (N inputs) to increase dairy production; (2) applying mineral N fertilizer to increase crop production; (3) improving conservation of manure N during storage and soil application; and (4) combining scenarios 1 and 3. They found that gross margin increased in response to improved retention of manure N and that increased N supply through supplementary feeding (scenario 4) across farm types led to increases in whole-farm N use efficiencies from 2% to 50%, in N cycling from 9% to 68% and in food self-sufficiency from 12% to 37%. An example of adaptation to manage risk in East Africa is pastoralists who have shifted from cows to camels, which are better-adapted to survive periods of water scarcity and able to consistently provide more milk [98]. Risk adaptation by farmers may also involve changing from cultivated crops to livestock, as crops may be more environmentally and spatially constrained in the pastoralists' home regions [99].

Mitigation options at farm to regional scales form a large part of Brazil's multifaceted approach to managing direct and indirect GHG emissions from livestock. Brazil's commercial cattle industry is the largest in the world (more than 170 million head in 2006), and emissions from raising cattle are responsible for about half of the country's total emissions [100]. The principal targets for mitigating GHG emissions associated with cattle production in Brazil are reduction in deforestation and enteric fermentation, regeneration of secondary forest, recuperation of degraded pasture and soils and elimination of fire in pasture management. Maintenance of grazing productivity and high stocking rates through pasture reclamation and adoption of integrated crop-livestock systems, such as rotational grazing and introduction of legumes in pastures, buffers pressure on deforestation. Such pasture regeneration creates a potential for increasing soil C storage, with increases of up to 0.72 Mg of $C\cdot ha^{-1}\cdot yr^{-1}$ reported under improved management [101]. However, other pasture maintenance practices increase emissions. For example, burning accounted for 1.69 CO_2eq (Mt from total biome) in the Cerrado ecosystem from 2003 to 2008. Key mitigation efforts include reduction in enteric CH_4 emissions by genetic stock improvement and dietary manipulation [91]. This dietary manipulation through grain supplementation increases forage digestibility and reduces enteric fermentation, but it leads to greater emissions of N_2O through the use of fertilizers to grow the grain [100]. Several other promising technologies include grass and legume species with lower GHG emission potential, additives (for example, ionophores and secondary plant compounds such as tannins) and use of propionate precursors in feed to reduce methanogenesis [102]. To complement farm-based efforts, uniform and fair economic procurement and incentivized policies must be in place and enforced across the supply chain in order to establish supply and trade chains with low C footprints. Regional and national policies must contain mechanisms that balance market pressure to convert from low-impact land uses (for example, forests) to relatively more intensive uses (for example, ranching). The Norwegian Agency for Development Cooperation (NORAD) and the Brazilian organization Aliança da Terra, which includes farmers, researchers and agribusiness entrepreneurs, are partnering to increase contributions by private landowners throughout the Brazilian Amazon and Cerrado to the goals of the Reducing Emissions from Deforestation and Forest Degradation (REDD+) programme by combining sustainable rangeland economic development with forest and water protection [103].

Effectively managing emerging zoonotic diseases and outbreaks due to climate change is a strong component of maintaining agropastoralist livelihoods. Increased temperatures due to climate change will affect the survival of pests in the winter and thus distribution of pests and diseases (for example, zoonotic, endemic, emerging, foodborne and noninfectious diseases), though some regions may find relief from these existing pressures in a changing climate. The Emerging Pandemic Threats Program (PREDICT) addresses the broad geographic issues in disease emergence from farm to national to global scales related to a shifting climate [104, 105]. The programme, which is run within the US Agency for International Development, leverages existing networks within national and local governments as well as networks of scientists and specialists involved in outbreak reporting, microbial characterization and pathogen discovery [104]. The programme includes 20 developing countries where hotspots of emerging infectious diseases exist and is focused on surveillance of human–animal interfaces where transmission is more likely. PREDICT is focused on the prevention of pandemics by addressing underlying ecological, economic and social drivers of change, such as shifts in land use from forested systems to livestock. It is used to develop and deliver new technologies (such as information management and communication tools) to improve control efforts close to the pathogen's source. This type of interdisciplinary effort that moves from farm-level to broad spatial scales is considered necessary for creating comprehensive strategies for the control and prevention of emerging zoonotic diseases in a changing climate [104].

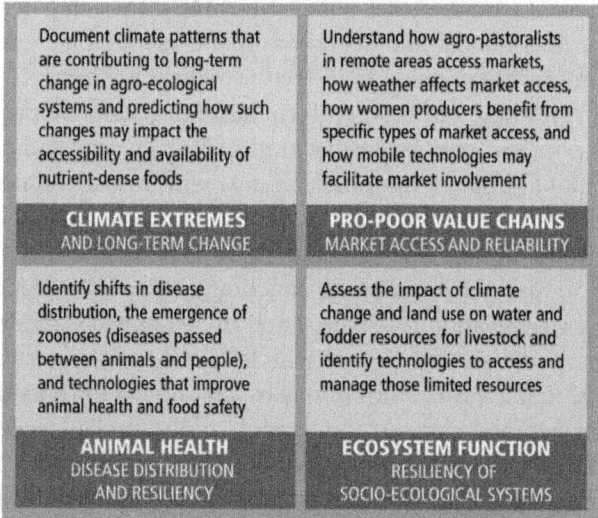

Figure 4: Four aspects of livestock management in climate-smart agriculture. Adapted from [111] with permission from R Bowen (personal communication, 2014). CSA, Climate-smart agriculture.

In support of agropastoral farming systems, models must integrate mitigation options, alternative intensification pathways, zoonotic disease and vector ecology (for example, genetic shifts, patterns of emergence); mechanisms of effecting behavioural change; and adaptation to future climate change scenarios. Some existing models for predicting regional GHG emissions from livestock production include BEEFGEM (Ireland; [106]), IFSM (USA; [107]) and SIMS$_{DAIRY}$ (UK; [108]). Reisinger*et al.* [109] recently evaluated different metrics on the integrated assessment model, MESSAGE, and the land-use model, the Global Biosphere Management Model (GLOBIOM), to examine the global costs of abatement strategies used to reduce the magnitude of climate change and subsequent effects on regional food production and supply prices for livestock products and other agricultural commodities. Other transformative approaches to livestock production include identifying the value of a blend of market-orientated smallholders vs. large-scale farms, evaluating ecosystem services payments as a means of income diversification, forming institutional and market mechanisms for reaching smallholders to foster technological change, finding the best locations for both livestock production and marginal land rehabilitation, and creating new capacity of the livestock sector for mitigation and adaptation in the face of climate change [90, 110] (Figure 4). The social and economic impacts (for example, income stability, human nutrition, product value chains, transaction and opportunity costs of other alternatives) of land-sparing and reducing livestock consumption, two recently suggested mitigation options, merit further investigation, especially with respect to gender, region and income differentiation [92].

Nitrogen Management: Agricultural Production, Greenhouse Gas Mitigation, and Adaptation

Future food security will continue to rely on N fertilizer inputs, but cropping systems must achieve yield potential (that is, close the yield gap) while minimizing trade-offs in air, water and soil quality [58, 59]. The long-term ramifications of N-related GHG emissions; off-site movement of N on eutrophication, acidification and pollution of aquatic and terrestrial ecosystems; and human health problems have led to a recommendation that anthropogenic inputs of reactive N to terrestrial ecosystems be reduced by up to one-fourth of present quantities, or about 35 million tonnes of N per year [112]. Even if this reactive, anthropogenic N entering agroecosystems is emitted as N_2 rather than N_2O, the energy associated with the Haber-Bosch process and transport of fertilizers will still contribute to GHG emissions [113]. Cropping system diversification, careful selection of crop rotations to reduce nutrient loss, and improved soil organic matter content are means by which to promote sustainable intensification. Yet, this often involves a set of complex trade-offs for producers and their livelihoods [114], emphasizing the need for a CSA strategy that involves stakeholders from the beginning to develop viable scenarios that include both mitigation and adaptation to climate change. The examples presented here demonstrate how strategies for N fertilization practices provide both mitigation and adaptation benefits by decreasing GHG emissions, reducing reliance on synthetic mineral fertilizer and enhancing food security.

Enhanced-efficiency fertilizers (EEFs), such as slow-release fertilizers or those containing nitrification inhibitors and urease inhibitors, hold potential to mitigate GHG emissions. According to the Intergovernmental Panel on Climate Change (IPCC) Fourth Assessment Report [114], the mean mitigation potential of N_2O by nutrient management using nitrification inhibitors and slow-release fertilizers has been estimated to be 0.07 t CO_2-eq·ha^{-1}·yr^{-1} (as a reference, agriculture accounted for an estimated 5.1 to 6.1 $GtCO_2$eq·yr^{-1} in 2005, which amounts to 10% to 12% of total global anthropogenic emissions of GHGs). In practice, N_2O emissions decreased by 54% from a no-till corn–dry bean rotation receiving urea, urease and nitrification inhibitors in comparison to a urea-only application in Colorado (USA) [115]. According to a recent global meta-analysis of enhanced-efficiency fertilizers, nitrification inhibitors can reduce N_2O emissions by 38% and polymer-coated fertilizers by 35%, on average, compared to conventional fertilizer, but urease inhibitors alone are not as effective in reducing N_2O emissions [116]. Nitrification inhibitors are compatible with both chemical and organic fertilizers, making them a seemingly attractive mitigation option, but their efficacy varies with edaphic factors. For example, EEF materials were applied to rainfed corn in the central Corn Belt (Midwest region, USA), a more humid region than Colorado [117]. Although all EEF treatments had lower cumulative emissions than the treatment that did not include EEFs, episodic N_2O emissions from EEF treatments corresponded to rainfall patterns, and the relative effectiveness among EEF materials was similar. Together, these findings suggest that the impact of EEF materials may be diminished in rainfed agriculture systems compared to irrigated systems with regulated water availability. Although yield responses to EEF materials may also vary with respect to crops and location, consistent yield increases in corn (central Corn Belt) grown with EEF materials were reported to occur as a result of the increased duration of photosynthetic leaf area during grain-filling [118]. In microirrigation systems, the results are less impressive, likely due to the increased efficiency of EEF where fertilization by fertigation matches crop needs more precisely, leaves less residual fertilizer and avoids its loss [119]. With this emerging evidence that use of EEF materials could have a positive impact on crop production and limit N_2O emissions, the results of research on understanding the conditions for which these materials are useful could underpin both the development of risk assessment tools and the feasibility of grower adoption of these technologies.

Mitigating GHG emissions through C sequestration depends on the stability of soil C pools. Declining productivity in the rice–wheat cropping systems of India's Indo-Gangetic plains has been attributed to reductions in soil C [120]. Mandal et al. [121] found that, to combat this, addition of NPK fertilizer during double-rotations of rice led to increases in soil organic C stocks compared to adding just N or NP alone [121]. When compost was applied during rice production, as much as 29% of compost-derived C was stabilized [121]. This was attributed to high lignin and polyphenol content in crop residue and compost and also to the diminished soil C decomposition stemming from anaerobic conditions due to soil submergence under rice cultivation. Crop residue management improves poor soil fertility through soil organic matter accumulation, leading to reductions in soil N loss by leaching and gaseous emissions; in many situations in developing countries, however, crop residues are used to feed animals, to provide fuel for cooking or are turned into biochar [122, 123]. Developers of mitigation strategies for increasing soil C and decreasing N_2O emissions have to take into account the dynamics of crop residue, tillage and nutrient management, along with climate, in order to evaluate the efficacy of different practices across locations [124].

Legumes, a form of ecological intensification, offer both mitigation and adaptation options, especially to smallholder farms susceptible to deficits in soil fertility, climatic uncertainty and reduced economic access to agricultural inputs such as mineral fertilizer. The biologically fixed N from legumes is often tightly synchronized with plant N demand and has a much lower C footprint than industrially produced synthetic N fertilizers [125]. For instance, intercropping with N-fixing trees in Sub-Saharan Africa were found not only to reduce reliance on fertilizers but also to enhance soil C sequestration and reduced N_2O emissions [126]. In this intercropping system, 10.9 Mg C·ha^{-1}·yr^{-1} were sequestered in the soil. The potential for N_2O mitigation was only 0.12 to 1.97 kg N_2O·ha^{-1}·yr^{-1}[126]. However, the authors of a review of 71 site-years of pasture, cropping and agroforestry systems indicated that providing N additions via legumes can increase accumulation of soil C at rates greater than can be achieved with other crops, such as cereals or grasses, even when they are supplied with N fertilizer [125]. Furthermore, intercropped mixtures of peas and barley (Hordeum vulgare L.), compared to the respective sole crops, were found to lead to effective weed suppression in weed communities across sites in Western Europe [127]. Adaptation options that include legumes to reduce dependence on fossil-fuel derived fertilizers include integration of intercropped or rotational legumes into management regimes, development and facilitation of access to new legume cultivars with broader stress tolerance and removal of barriers to legume use and consumption in the food system (for example, competing uses, seed availability, labour).

The design of more efficient N management strategies will only be conducive to climate change solutions if based on knowledge systems and participatory research with stakeholders to ensure viable action and adaptive management. Although decision-making support tools and metrics are being developed to aid producers in tempering N inputs for the desired outcomes of higher crop production (for example, quantity and nutritional quality) and lower environmental impacts [128], adoption is a major obstacle. When extension agents are involved in troubleshooting with and training of participants, the new knowledge systems that are created begin to delineate clear pathways that benefit farmers' livelihoods. In regions dominated by smallholder farmers who are already experiencing climate impacts such as increased drought, flooding or heat waves, the priority is on adaptive measures for reliable N availability to support food security and minimize vulnerability. Combining low inputs of synthetic N fertilizers with practices that increase soil quality through organic matter management and acquisition of N from biological N fixation allows adaptation measures to contribute to GHG mitigation. However, synthetic N sources are fraught with constraints such as high cost, price fluctuations and availability, whereas biological N sources are affected by constraints of labour, time and physiological tolerance. Future food security also will depend on a substantial rate of yield gains for major cereal crops. Maintaining these yield increases above a 1% annual growth rate will require constant improvement in crop yields, stress avoidance and agronomic management to achieve physiological yield potential [129]. However, maintaining a compounding rate of yield increases is not consistent with historical trends and likely is not achievable without great effort [130]. Therefore, the limits of current crop productivity need to be estimated using potential yield and water-limited yield levels as benchmarks. Determining and closing the yield gap, especially in developing countries, is fundamental to achieving food security because variety improvement through breeding and genetic modification might be insufficient [129, 131, 132].

Farmer Decision-Making and Barriers to the Adoption of Climate-Smart Agriculture Practices

Climate change challenges farmers' decisions by altering risks and uncertainty and incorporating new information into their traditional knowledge-processing systems. The unfolding of the decision-making process and its translation into action depends on the socioecological context in which farmers are embedded. How well innovation models apply to all climate-related behaviours is a major question, especially given that governance regimes at the national and international levels strongly influence farmers' actions [133]. The massive literature on innovation systems has established the basic hypothesis that farmers evaluate the costs and benefits of different practices in light of information accessed through social networks and other communication channels. The diffusion of innovation model can provide critical insights into adoption decisions. In this model, adoption of innovations follows a sequence of stages: knowledge, persuasion, decision, implementation and confirmation [134]. Innovations generated by agricultural research are communicated by extension agents to farmers. This approach may place too much emphasis on traditional socioeconomic variables and ignore how other social factors (for example, networks, gender, social norms, values, climate-change attitudes), and uncertainty may be implicated by practices that are ostensibly consistent with CSA priorities (for example, adoption of new crops and cultivars or changes in N fertilization) [135–137]. Effective outreach strategies will manifest with greater understanding of farmers' beliefs about climate change and their readiness to respond to climate change through mitigation and adaptation. Little is known about farmers' and their advisors' willingness to use outreach tools, their information needs with respect to climate change or their ability to incorporate this knowledge into existing decision-making processes. A survey of almost 5,000 farmers in 22 top corn-producing watersheds across the United States showed that farmers' climate change beliefs correlated with both their perceptions of climate risk and their willingness to respond and adapt to changing conditions [138]. Farmers who believed that climate change is occurring, and is due in large part to human activity, were significantly more likely to support both mitigation and adaptation actions and also more likely to support government- and farm-level GHG reduction efforts. Most farmers supported adaptive strategies, with two-thirds agreeing that they should take efforts to protect land from increased weather variability. Many (59%) expressed lower levels of support, however, for mitigation through GHG reduction. These farmers obtained much of their information through social networks that included professional advisors. A survey of corn grower advisors, including government, nonprofit, for-profit and agricultural extension personnel, found that advisors are more influenced by current weather conditions and 1- to 7-day forecasts than by longer-term climate outlooks [139]. The advice given to farmers has been based predominately on historical weather trends and focused on short-term operational decisions rather than on long-term strategies. For climatic data to be useful to such populations, designing outreach strategies that target extension agents and other professional advisors will increase the potential to influence beliefs and practices of farmers. Furthermore, though mitigation policies alone might not resonate with farmers, those that combine mitigation with adaptation could be effective. In general, adoption of best management practices can be promoted by focusing on implementation among farmers most likely to adopt them, followed by leveraging social networks to inform other farmers about the benefits of adoption [140].

The constraints that farmers face when making decisions, such as whether to use conservation agricultural techniques, may create barriers to practices that could improve resilience to climate change. Conservation agriculture includes practices such as minimum mechanical soil disturbance, permanent organic soil cover and crop rotation, all of which typically increase soil C storage, especially when applied in concert [141, 142]. Cited benefits of conservation agriculture in Sub-Saharan Africa include increased yields, reduced labour, improved soil fertility, reduced erosion and land-saving [141–143]. Reports of conservation agriculture's widespread adoption may be overrated, though, because many farmers seem to adopt technologies only while incentives are offered and the project is actively supported, and then they quickly return to their former crop management practices once project support ceases [144]. Constraints to adoption include strong competition for mulched crop residues for livestock feeding, increased labour demand for weeding (which often changes cultural gender divisions of agricultural work) and lack of access to and/or use of herbicides and other inputs [143,144]. Although there are some recognized factors that influence adoption (for example, larger farm size and more education), no universal variables seem to explain adoption [145], leading some to suggest that conservation farming may be successful only under certain agroecological conditions [144, 146].

Recent work in Zambia may help to explain regional variation in farmer adoption and rejection of conservation agriculture practices. Analysis of surveys of rural incomes and livelihoods revealed that rates of rejection in Zambia were high (approximately 95%), and practice dropped from 13% to 5% of farmers between 2004 and 2008 [145]. Rainfall data reveal that, during the past 10 years, the onset of the first rains needed for planting have been progressively delayed. Although adoption decisions are not strongly or explicitly based on labour constraints, farmer age or education level, farmers in districts that experience more rainfall variability are more likely to adopt conservation agriculture practices and to implement those practices with greater intensity [147]. Because conservation agriculture allows planting to occur as soon as the rains begin, it offers an adaptive response to changing rainfall regimes [148].

Fundamentally, an existing lack of food security and farmers' concerns about poor health will counteract incentives to their adoption of new farming technology [149, 150]. Although many farmers believe climate risk is real, they are less likely to believe it is caused by human behaviour. They have paid the most attention to climate variables that have traditionally constrained their operations and have relied on an existing suite of adaptive behaviours [53]. Thus, knowledge networks are especially critical to their understanding of trade-offs between the short-term costs and longer-term benefits of adopting new farming technology and practices that will help them mitigate and adapt to the effects of climate change as well as to increases in climate variability. Adaptation to climate change and the idea of climate change itself define and change human cultures. Indeed, cultural factors (for example, place attachment, value systems, individual and collective identities) shape how people support and respond to adaptation interventions and must be woven into climate change policies and programmes [151]. Key to this effort is linking science, technology and decision-making to the context of socioecological systems to better achieve balance between economic, cultural and social needs [152]. Systems that effectively leverage science and technology in support of sustainability efforts create salience, credibility and legitimacy across boundaries where boundaries exist between science and policy, disciplines, public and private sectors, and/or organizational hierarchies. Actions employed within these systems include convening (bringing all stakeholders in the CSA context together to foster communication and build trust), translation (defining a shared ontology and language), collaboration (actors working together to produce applied knowledge and specific outcomes, with specific mechanisms in place to facilitate interactions across multiple boundaries) and mediation. Specifically, mediation is 'a process by which different interests are represented and evaluated so that mutual gains can be crafted and value created in a way that leads to perceptions of fairness and procedural justice by multiple parties' [152], p. 470. These components, as well as broad stakeholder engagement from the initiation of a project, are keys for linking science with action, developing knowledge networks and forming critical capacity to reach desired outcomes also see [135–137, 152]. Other approaches for forming new knowledge networks and adaptive capacity in the socioecological system combine both back-casting and explorative scenarios [137]. Interactions between climate change and culture, as well as ideas regarding the ethics and morality involved with climate change and the role of these constructs in stakeholders' and the larger society's adoption of actions related to mitigation and adaptation, are outside the scope of this article, but they are discussed by Hayward [153] and Markowitz and Shariff [154].

Climate Risk Management: Financial Mechanisms, Insurance and Climate Services for Farmers

An alternative to emergency aid in the face of climate shocks is reliable programmes developed to minimize farmer risk, which could prove to be more effective by preventing the slide into poverty traps [155]. The uncertainty of climate change, especially extreme events, makes it difficult for individual farmers to incorporate risk into their decision-making [156, 157]. Vulnerabilities to climate effects on production, pests, disease and price volatility depend on farmers' assets and natural resource base [158]. Appropriate risk management tools, such as improved forecasts and extension support, and appropriately designed safety nets or insurance instruments must revolve around the vulnerabilities in specific farming situations. Rural households in developing countries, limited in both resources and access to information, could be disproportionately affected unless appropriate measures are introduced to manage the additional risk and uncertainty related to climate change [159–161]. Innovative management of risk and uncertainty employs financial mechanisms (for example risk transfer or insurance contracts) that use several types of methods to understand investment decisions, technology choices, and risk perceptions. These methods include remote-sensing technology, micro-level household data, analysis of diversification, and farm surveys. Implementation of such insurance instruments requires appropriate technical innovation, building awareness and trust, ensuring viable market demand, and enhancing local capacity building among local financial institutions [162, 163].

Index insurance is one such instrument that effectively reduces farmers' risk under a changing climate and generally has many advantages. With index insurance, indemnity payments are decoupled from actual crop losses, instead of linking payments to changes in attributes that impact or reflect crop growth or survival over a given spatial extent. This then reduces transaction costs associated with verifying ownership and losses, removes the opportunity for individuals to change their risk behaviours to increase the likelihood of receiving a payout, and allays the problem of adverse selection, in which high risk individuals are disproportionately represented in the insured pool. Most vitally, the rural poor are no longer widely excluded from insurance by the need to demonstrate assets as a prerequisite to purchasing a policy [161]. For example, the Index Based Livestock Insurance (IBLI) programme recently launched by The Index Insurance Innovation Initiative seeks to accurately represent the insured's loss experience through the use of landscape-level data derived from measures such as the Normalized Difference Vegetation Index (NDVI) (Figure 5). NDVI is a satellite-derived indicator of photosynthetic activity or a proxy for plant production to feed livestock, which is available in real time every 10 days [165]. Livestock in Northern Kenya's arid and semiarid lands account for more than two-thirds of average income, with most livestock mortality associated with severe drought [164]. Herd losses that push a household below a certain threshold tend to result in long-term consequences, including destitution, which can trap the household in poverty. The data derived from the developed index showed that the NDVI performed well when tested against other herd mortality data from the same region, and, when compared to drought experiences over the past 27 years, removed 25% to 40% of total livestock mortality risk in simulations. The IBLI programme has been implemented, with initial payouts issued to households in October 2011 [166]. Actions needed to facilitate establishment of the IBLI include identification of systematic criteria for end users to evaluate whether they need to purchase this insurance product [167] and development of programmes for client recruitment, low-cost marketing, and claim settlements.

To provide long-term farm and community security in support of CSA, bundling agronomic breeding programmes for drought tolerance and financial programmes with index-based drought insurance will maximize farms' resilience to financial shocks due to drought, especially as the drought tolerance of crops diminishes with more severe drought stress.

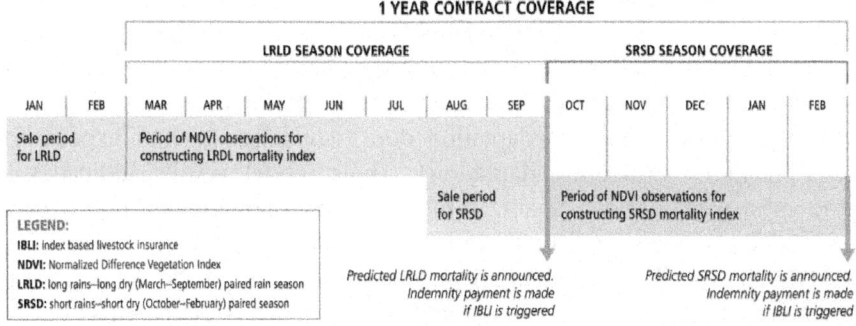

Figure 5: Depiction of a 1-year contract for index-based livestock insurance and its implementation. Adapted from Chantarat *et al.* [164] with permission from John Wiley & Sons.

Developing and planting crops with drought tolerance is primarily a more cost-effective risk management tool than index insurance in the face of less extreme climatic events; however, index insurance could complement both private and public crop improvement programmes by providing assistance when even drought-tolerant varieties fail during extreme climatic events. Demand for bundled strategies seems likely to be high [168, 169], thus creating a sustainable market for both drought-tolerant varieties and index insurance. To assess how bundled strategies affect household welfare and operate in practice in a drought-prone region of Ecuador, Carter and Lybbert [170] estimated the underlying probability structure for traditional maize yields from yield data collected annually by the Ecuadorian government from random samples of producers in different regions of the country. The certainty equivalent of the drought-tolerant technology was 6% higher than that of traditional technology. Incomes were most stable under drought pressure when drought-tolerant and insurance index technologies were combined, but interactions of such bundled strategies with other risk management and safety net programmes remain to be determined.

Uncertainty influences individual farmers' expectations of yield and dramatically impacts their adaptation behaviour. For example, government policies to protect farmers against climate-change risks, such as insurance programmes and direct*ex post facto* payments after extreme climate shocks, may reduce farmers' incentives to diversify farm production away from more climate-sensitive crops. Antón *et al.* [30, 171] examined farmers' responses to agricultural risk management policies under conditions of climate change using a stochastic microeconomic simulation model calibrated with data derived from farming in Australia, Canada and Spain. They distinguished between farming risk and uncertainty with regard to climate and farmers' beliefs. They examined the impacts of *ex post facto* disaster payments and three types of crop insurance (individual yields, area-based yield and weather index) utilizing a combination of climate-change scenarios (no change, marginal change, change with an increase in extreme events) and farmers' behavioural response options (lack of adaptation due to misalignment of expectations, diversification, structural adaptation). Their model results indicated that farmers in Australia and Spain, in the absence of government policy, would respond by increasing diversification, assuming they correctly anticipated climate change. The introduction of risk management policies in these two countries tended to crowd out diversification, and this effect increased with climate change. The relative cost-effectiveness of policies depended strongly on the extent of extreme events and farmers' misperceptions of climate (that is, misalignment), which can greatly inflate a policy's budget. Reducing the uncertainty that farmers face, with regard to how climate change will affect them, by developing information strategies will aid in the design of robust risk management policies and will limit the excessive financial costs brought on by misperceptions [30].

The goal in using the risk management instruments described here is to promote resilience of rural households to weather shocks and climatic variability, a key premise of CSA. Although not addressed here, other index insurance products promote the integration of rural households into market production and often are used in concert with programmes aimed at promoting agricultural value chains and supply chain risk management [162]. These kinds of programmes consolidate and facilitate the participation in the agricultural value chain by specific populations in discrete regions, and they are intended to help increase farmers' access to credit and to encourage investment in appropriate technology to increase productivity.

Energy and Biofuels: Development of Production Methods and Technologies to Cut Emissions without Interfering With Food Production

Bioenergy is the native energy resource embedded within agriculture, but, more fundamentally, agriculture is itself an energy conversion process with the capacity to develop a rich portfolio of products for diverse markets, including markets for food and energy. The role of biofuels in achieving reduction goals (that is, mitigation) for GHG emissions and meeting future energy needs (that is, adaptation), as well as their impact on food commodity prices, remains a principally global issue [172, 173]. Estimates of increases in food and commodity prices suggest that between 3% and 70% of retail food price increases can be attributed to biofuels; however, this wide range stems from differences in time periods, data sets using different price series (export, import, wholesale, retail) and different food products [174–176]. Global models used to predict mid- to long-term effects of biofuel production growth on prospective prices, production of feedstocks (for example, maize, sugar cane, oilseeds), mitigation and adaptation measures, and land-use change are general or partial equilibrium (PE) models. General equilibrium models encompass supply, demand and prices in the entire economy and take into account multiple markets and associated inputs; PE models are focused on equilibrium conditions in an individual market or sector of a national economy, in which prices, quantities under demand and product supply remain constant. Along with models used to assess land-use change in response to bioenergy production [177] are models such as the Asia-Pacific integrated model, which is used for analysis of global and national CO_2 emissions, mitigation costs and C taxes [178]; the Modular Applied General Equilibrium Tool which is used to examines links between agricultural markets, the general economy and agricultural policy issues [179]; the Global Change Assessment Model, which is an integrated assessment model of energy, agriculture and climate used extensively by IPCC and others [180, 181]; GLOBIOM, which is used in analysis of mid- to long-term land-use change scenarios in agriculture, forestry and bioenergy [182]; and the Model of Agricultural Production and its Impact on the Environment which is utilized in evaluating spatially explicit patterns of production, land-use change and water use in different global regions and linking economic development with food and energy demand [183, 184]. These models can provide information regarding uncertainties, costs and trade-offs crucial to CSA for (1) climate policymaking, GHG mitigation and sustainable energy futures and (2) projections regarding agriculture, agricultural markets and the future of the world's food and feed supplies. The case studies described here are used to assess costs and trade-offs of biofuel expansion at the farm and global scales as well as the impacts of enacted policies in the European Union (EU) and the State of California in the United States.

Increased future demands for food, fibre and fuels from biomass can only be met if the available land and water resources on a global scale are used and managed much more efficiently than they are now. Therefore, developers of an integrated bioenergy framework must incorporate not only bioenergy's mitigation potential but also its costs and trade-offs with food, water security and land use. To assess the cost-effectiveness of bioenergy for climate change mitigation, Popp *et al.* [184] coupled global models of vegetation and hydrology [185, 186], land-use optimization (MAgPIE) and the energy–economy–climate interface [187]. If all suitable land for agricultural production was made available, bioenergy from specialized grassy and woody bioenergy crops, such as *Miscanthus* (poplar), could produce 100 EJ globally by 2055 and up to 300 EJ by 2095. However, bioenergy cropland would grow from 1.52 billion ha to 1.83 billion ha, thereby increasing CO_2 emissions due to deforestation. Meeting bioenergy needs while preserving intact and frontier forests would require higher rates of technological change in agriculture (by 0.9% per year until 2095), thus leading to additional costs. The potential trade-offs of conserving forests and cultivating bioenergy crops on a large scale include conflicts with respect to food supply, food prices (especially in the tropics) and water resource management [188].

In the EU, market demand for biofuels and biomass will likely increase as the region becomes less reliant on fossil fuels and the EU implements targets for renewable energy, such as the Renewable Energy Directive and the ensuing national renewable energy action plans. This demand was first met with imported biomass sources from residue streams, such as palm kernel shells and wood pellets, and industrially produced biomass, such as palm oil and ethanol [189]. In an analysis conducted for the International Energy Agency, Hoefnagels *et al.* estimated future intra and inter European trade of solid bioenergy biomass by combining geographic information system models of transport routes with models of supply and demand for energy crops, forestry products and/or residues and agricultural residues [189]. They estimated that intra European biomass trade could increase to 6,560 kilotonnes of oil equivalent (ktoe) by 2020 in the low-import scenario and to 5,640 ktoe in the high-import scenario. Transportation costs could contribute substantially to these totals (for example, up to 75% (9 €/GJ) of the total cost (12 €/GJ) in the case of forestry residues). However, they determined that the lower transportation costs of pelletized biomass would not make up for its high production costs. In both scenarios, the chief future exporting regions for inter European biomass trade included Poland, Estonia, Hungary and Slovakia and the major importing regions included Germany, Italy, the United Kingdom and the Netherlands. Within the CSA strategy, these modelled outcomes can help in the identification of the issues and stakeholders that should be involved in the development of future energy use and policy.

Newly enacted low carbon fuel standard (LCFS) policies in California and the EU offer promising approaches to reducing the C footprint of transportation fuels. The LCFS applies to itself a direct life-cycle C intensity analysis that captures all GHGs emitted per unit of fuel energy during extraction, cultivation, land-use conversion, processing, transport and fuel use [190]. Both California's LCFS and the European Parliament's revised fuel-quality directive require a 10% reduction in GHG emissions by 2020, and both allow credit-trading. These standards differ from previous policies aimed at reducing petroleum fuels, which comprised volumetric mandates and only indirectly required reductions in GHG emissions. As a case in point, the US renewable fuels standard requires annual sales of 36 billion gallons of biofuels by 2022, 21 billion gallons of which must derive from advanced biofuels and achieve a 50% reduction from baseline life-cycle GHG emissions. The other 15 billion gallons must come from corn ethanol [190]. With this focus on total GHG emissions rather than on volume, biofuels under LCFS will not be forced into a small number of categories, and transformative innovation, a key part of the CSA strategy, will be promoted. The flexibility and performance-based nature of the LCFS allows industry, rather than government, to pick the likely biofuel winners [190]. If implemented on a global scale, such changes in biofuel policies will heavily influence agricultural markets and environmental outcomes. Tokgoz *et al.* [191] simulated a reduction in maize ethanol production of the magnitude suggested by the LCFS analysis by utilizing a modified version of the International Food Policy Research Institute's (IFPRI) PE model, or the International Model for Policy Analysis of Agricultural Commodities and Trade (IMPACT). IMPACT was developed to project future global food supply, demand and security in 115 country regions. Holding biodiesel production levels constant at 2010 levels in this model dramatically decreased rapeseed and soybean oil prices and increased the availability of food calories. Building future international policies upon the LCFS policies implemented in Europe and California will further the demonstrated benefits of reducing fuel C intensity rather than promoting policies that benefit biofuel producers who pursue ongoing profit-driven growth.

Policymakers and financial institutions have been hesitant to invest in bioenergy, owing to negative press and the resultant uncertainty about its long-term sustainability. In response, the scientific community must present a balanced perspective of how bioenergy can (or cannot) be managed as part of CSA (for example, see [172, 173]). Models comprising the global impacts of bioenergy, along with agricultural productivity at local, regional and country scales, can be utilized to effectively assess the realization of environmental and economic objectives via policy and technology [192]. Separate consideration of bioenergy in the agricultural context will lead to suboptimization of the system with the likelihood of realizing lower environmental and economic benefits [193]. The viability of biofuels will be achieved when their cost is competitive with those of fossil fuels when it includes both the cost of the feedstock seed and the value of coproducts derived from the biofuel by-products, which can provide additional revenue. In some cases, large subsidies are required to make biofuels competitive with fossil fuels (for example, *Jatropha*-based oil in Senegal) and/or feedstock seeds must be imported to satisfy demand, suggesting that alternative feedstocks should be adopted [193]. A stable supply of feedstock, determination whether other industries strongly compete for the same feedstock and access to a well-functioning value chain for the product are all crucial to facilitating vertical integration of production, conversion and processing, as observed in Brazil's biofuel sector. Msangi and Evans [194] suggested that growing a biofuel feedstock that can serve as a food product with coproducts will create greater stability for the farmer and that solving problems of food security in developing countries will lead to a flourishing biofuel sector. Furthermore, increases in food crop production and efficiency underpin the success of increased reliance on bioenergy and the conservation of forested lands in lieu of expansion of agricultural lands [188]. Lignocellulosic biofuels also can be a strong component of GHG mitigation with small impacts on global food prices, especially if sufficient land for feedstock production exists and does not compete with land devoted to food production, as indicated by modelled outcomes [173]. It is imperative to engage producers and affiliated industries in research to better understand how markets for new development of bioenergy and nontraditional biological products can become an integral part of energy-efficient agriculture.

THEME 2

Landscape and Regional Issues: Land Use, Ecosystem Services and Regional Resilience

Recently, extensive research on climate impacts on landscape and regional scales has been stimulated in part by policies that require institutional action to mitigate and adapt to climate change [14, 195]. Such research includes use of remote sensing to analyse land-use mosaics, inventory approaches to assessing C stocks and water resources, and models to examine the potential of land-use change in different climate scenarios [196–198]. These techniques are being combined with farm- and field-scale data on crop performance, soil biogeochemistry and irrigation use to analyse if and how mitigation and/or adaptation strategies build food security and ecosystem services [34, 199–201]. Interdisciplinary science underpins an integrated landscape approach, along with involvement of stakeholders who hold key information for developing climate-change scenarios and innovation pathways [202, 203]. Landscape approaches that expand beyond agriculture itself are needed to understand how extreme events trigger rural outmigration and create new types of rural–urban connections. The development of metrics and indicators to track responses of climate change and ecosystem services is accelerating with broader recognition of the need for greater accessibility of data, formation of more types of socioecological assessments [203–205] and charting of the progress of climate-change policies.

Climate Change and Food Security: Modelling Adaptation and Uncertainty

Determining the adaptive capacity of mitigation and adaptation scenarios that will evolve with CSA's participatory processes rely, in part, on biophysical models. Models that will be used to examine the limits to crop adaptation as well as the impacts of climate change on biodiversity, land use and ecosystem services are now available [2, 206]. They still contain much uncertainty due to (1) the ability of process models to accurately simulate the growth and development of crops when exposed to very high temperatures and elevated CO_2 levels, (2) the rate and degree to which agricultural productivity and development can progress in concert with reductions in GHG emissions and (3) the ramifications of successful agricultural adaptation to climate change for land-use change and associated ecosystem services [207–209]. Despite these uncertainties, the use of models and scenario-building has led to the exploration of potential synergies and obstacles to coping strategies in agricultural that would not have been possible with empirical data alone [210, 211]. Here we present modelling approaches to evaluating adaptation scenarios across the EU, the Mediterranean region and the United States.

Modelling can be used to identify climate-change impacts and sensitivities as well as possible adaptation strategies. Rather than being focused solely on climate-change constructs, such vulnerability assessments also include changes in CO_2 concentrations, GHG emission management, N deposition, land use, and socioeconomic trends to manage vulnerability. The Advanced Terrestrial Ecosystem Analysis and Modelling (ATEAM) program produced a new set of climate scenarios for Europe in multiple global change scenarios and ecosystem models [212]. A dialogue among relevant stakeholders from the private sector, governmental and nongovernmental organizations and policymakers was conducted. Unlike global trends, European trends included moderate or no population increase, little urbanization, increased forest area and decreasing demand for agricultural land. The modelled outcomes allowed for changes in land management that could decrease vulnerability, such as C sequestration due to reforestation. Modelled outcomes indicated that the Mediterranean region could face increased risks of forest fires, water shortages, changes in tree species distribution and losses of agricultural potential. Under the different scenarios, which ranged from business as usual to greatly reduced GHG emissions, 20% to 38% of the population in the Mediterranean would live in watersheds under stress and experience water scarcity exacerbated by increased tourism and demand for irrigation. Mountain regions would be especially vulnerable because of less snow cover and subsequent changes in river runoff. These modelled outcomes provide opportunities for back-casting and identification of sensitivities where mitigation and adaptation efforts should be focused, as well as how subsequent research could inform policies around such efforts.

The participants in the EU SmartSOIL project [213] employ a CSA-like strategy that includes stakeholder involvement and is used to examine the implications of findings for economics and policy implementation. As of 2012, consultation with policymakers and advisors had begun in six case study regions [214]. The creators of SmartSOIL developed a framework of C flows and stocks informed by new data and meta-analysis of long-term European experiments that are relevant to short- and long-term CSA management decisions. This framework will be used to improve existing soil and crop simulation models out of which a simplified model will be derived to predict scenarios for future management systems to improve productivity and enhance C sequestration. As an example of modelling for C sequestration, Lugato *et al.* [197] used the CENTURY model to inform proposed European policies on the mitigation potential of agricultural soils through C sequestration and to assist in the evaluation of the agricultural sector's deployment of 'greening' measures for agriculture that benefit climate and environment as required in the EU's post-2013 Common Agricultural Policy. Nearly 16 soil–climate–land combinations in the EU and neighbouring countries (Serbia, Bosnia and Herzegovina, Croatia, Montenegro, Albania, former Yugoslav Republic of Macedonia and Norway) were used in calculations, including the main arable crops, orchards and pastures as well as management practices (for example, irrigation, mineral and organic fertilization, tillage) (nearly 164,000 scenarios). Testing modelled results against soil inventories collected using comprehensive and standardized approaches (the European Environment and Observation Network and the Land Use/Cover Statistical Area Frame Survey) strengthened the examination of the uncertainty of modelled outcomes. Consideration of a broad spatial extent (pan-EU scale) allowed for better evaluation of C sequestration, in which an estimated current stock of 17.63 Gt C is predicted to increase through 2100. Within the pan-EU region, stocks will diminish in the southern and eastern parts because of higher soil respiration, whereas these losses will be offset by increases in the central and northern regions due to increased CO_2 atmospheric concentration and favourable crop-growing conditions. Such survey and monitoring programmes support the need for further spatiotemporal analysis of climate trends and stakeholder dialogue in modelling efforts so that proposed adaptation strategies are relevant to economic and socioecological contexts such as local, national and EU-wide policies and regulations.

Many climate modelling studies are focused on yield variations in response to changes in mean climate conditions [215]; yet, this approach overlooks several key factors, such as the occurrence of extreme events in which variance is changing [216]. Empirical approaches that capture the effects of extreme temperatures can be used to more efficiently assess climate impacts and adaptation. For example, in Mediterranean sunflower and wheat, an increase in both mean temperatures and climate extremes modelled under A2 and B2 scenarios (year 2100 business-as-usual and reduced GHG emissions scenarios, respectively) would cause severe yield reductions by shortening growing seasons and intensifying heat stress [217]. In the United States, yield patterns of rainfed maize have been explained by accounting for extreme events using the process-based Agricultural Production Systems Simulator (APSIM). With APSIM, observed negative yield responses to extreme heat shocks (measured as accumulated extreme degree days) were best explained by increased vapour pressure deficit (VPD). VPD contributed to water stress by increasing plant demand for soil water and reducing future water supply as a consequence of higher plant transpiration rates [218]. The ratio of water supply to demand, as modelled with APSIM, was three times more responsive to a 2°C mean warming than to a 20% reduction in rain. The results of these studies direct researchers, policymakers and extension agents to take science-based actions that rely on climate scenarios and predicted outcomes that are not based solely on the change in climatic means but include climate extremes. Despite incongruences between actual biological patterns and model simulations, model outputs provide an evolving information base for planning strategies and new research directions.

Quantitative assessments of adaptation to consider the effects of extreme events on agriculture can inform policymaking by providing a much wider set of outcomes than is possible with perceptions or projected impacts. Modelled outcomes evaluated in a socioecological context allow investigation into the limits of adaptation and related consequences for agricultural productivity, other economic sectors and land use (for example, an indicator-based, spatially explicit and scenario-driven adaptive capacity model [211]). Coordinated cycles of model improvement and projection across multiple spatial scales (global, regional, local) will facilitate model validation and calibration as well as effective use of studies with different geographical domains [219]. Challinor *et al.* [219] recommended that different model intercomparisons and improvement programmes (MIPs) form separate but linked strategies, that detailed modelling studies of response mechanisms (for example biophysical processes, crop yields) and robust experimental data (for example, see [208]) underpin the models and that systematic comparisons of impact studies and their outcomes be used to address sources of models' uncertainties. Involvement of stakeholders at the outset of model development also aids in development of relevant scenarios and tools [152]. Modelled outcomes form a key part of the climate policy and governance process necessary to attain the Copenhagen 2°C target, which requires 70% to 90% worldwide emission reduction targets and which has been questioned as being too weak [220]. A wider set of options for targets will be facilitated by examining options for climate governance (that is, institutional mechanisms to guide and direct societal policymaking) and their societal implications, as well as assessing the potential for success in achieving the target within existing political structures (for example, democracy, autocracy) [221, 222]. For example, the 'mitigate for 2°C but adapt for 4°C' option implies that society will take steps to adapt to existing in a warmer world, but will maintain the goal of reaching the current 2°C target. This approach will diminish conflicts and trade-offs associated with the water–food–energy nexus. Yet, if it is perceived that the 2°C target is unattainable and investment is strongly supportive of adaptation, then acceptance of even higher target values and a greater need for adaptation-related burden-sharing will be a consequence [221].

Soil Carbon and Achieving Multifunctionality through Mitigation and Adaptation

Soil resource degradation has led to loss of functions and ecosystem services, such as water availability, water-holding capacity, C storage, mitigation of GHG emissions and sustained agricultural productivity [223, 224]. Soil degradation limits resilience to climate change and extreme events, such as drought, and therefore impacts food security and augments susceptibility to poverty, especially in vulnerable regions such as Sub-Saharan Africa. Better understanding of the biophysical capacity of agricultural landscapes to act as C sinks through capture and storage of atmospheric CO_2 in soils and perennial vegetation leads to strategic design and operational management for both mitigation and adaptation actions [122, 225]. Improving biophysical capacity for desired functions such as GHG mitigation, food production and maintenance of soil and ecosystem biodiversity is a form of ecological intensification and is enhanced within a multifunctional landscape. Ecological intensification builds resilience by leveraging ecological processes to increase outputs from agricultural lands to promote (provisioning supporting, and regulatory ecosystem services) and decrease dependence on external inputs [93]. Balancing trade-offs between the different types of services can be facilitated by assessing indicators such as soil organic C (SOC). Trade-off analysis can employ simulation methods and modelling tools (for example, the Agricultural Model Intercomparison and Improvement Project, known as AgMIP; see [9]) to evaluate existing and alternative agricultural systems, changes in market conditions affecting supply and demand, and related policies in relation to climate change. The negative trade-offs can be minimized when landscapes are managed to achieve multifunctionality objectives, such as by a diverse set of land-use types, each providing a different combination of services [31]. The case studies below are focused on tools for accounting for GHG emissions and soil C storage, processes to enhance soil C storage and use of a paired economic-biophysical model to assess impacts of mitigation efforts within multifunctional landscapes.

Climate-change mitigation and adaptation within multifunctional landscapes depends on the multiple roles of SOC, which include a reservoir for plant nutrients (N and P) to support crop production and reduce external inputs, a substrate for soil organisms affecting their activity and diversity, and a promoter of soil physical structure leading to enhanced water quality and reduced erosion [223]. To maximize mitigation efforts, accurate GHG calculation can engage stakeholders and other end users to form a database with which to understand the C budget of their practices, such as SOC sequestration and storage and CO_2 emissions from fossil fuel combustion. Many models used to calculate GHG emissions and SOC are designed for specific geographical areas to meet distinct needs. Colomb *et al.* [226] provided information on the features of 18 available calculators and created a framework for choosing the most suitable GHG and C calculators for a given situation. They found that major sources of GHG emissions were usually well-identified, but that the calculators used failed to account for landscape effects due to land-use change. Few calculators accounted well for emissions from the loss of previous biomass, which is especially crucial in cases of deforestation–reforestation or rehabilitating and restoring grazed and ungrazed grasslands. To illustrate this point, Colomb *et al.* [226] used seven calculators to assess the GHG balance of replacing grassland by wheat, a case where the average emissions due to land-use change were greater than those that occurred during the production of wheat itself. Owing to differences in reporting units, measurement of emissions and scope, the results obtained with different calculators could not be directly compared and uncertainty levels were very high. Minimizing uncertainty in C and GHG accounting methods will provide reliable data to aid global markets and agencies for use in developing GHG- and C-footprinting and life-cycle assessment criteria. Greater standardization of metrics will also help in the enumeration of trade-offs in balancing between crop management and land use.

The design of multifunctional, ecologically intensive landscapes when providing ecosystem services of local and global interest is informed by analysing synergies between agricultural practices and landscape attributes [58]. For example, an analysis of carbon stocks and flows in smallholder farms in Kenya revealed positive synergies between agricultural production, on-farm biodiversity and above-ground C storage [227]. Dominant land-use types considered included home gardens, food-crop plots, cash-crop plots, pasture plots and woodlots. Close to the homestead, home gardens received the most organic nutrients in the form of compost, kitchen waste and manure, and downslope and farthest away maize, vegetables and eucalyptus woodlots were planted. Tree species diversity was highest in home gardens and near crop fields. Although such trees contributed up to 39% of total aboveground C storage, the greatest contribution came from monospecific woodlots dominated by *Eucalyptus saligna* (which contributed up to 81% of total aboveground farm C). In a landscape survey of 250 farms across 6 regions in Kenya, SOC, available P and exchangeable K^+ varied widely but generally varied by management practice and reflected diminished soil fertility with greater distance from the homestead [58]. Thus, a combination of land-use practices contributed to C storage below and above ground as well as to multiple functions on the farm (Figure 6). Including the diverse agricultural landscapes in such studies leads to understanding of how management practices support ecological processes for C storage, and farmer participation supports identification of economically viable options for smallholder farmers [58].

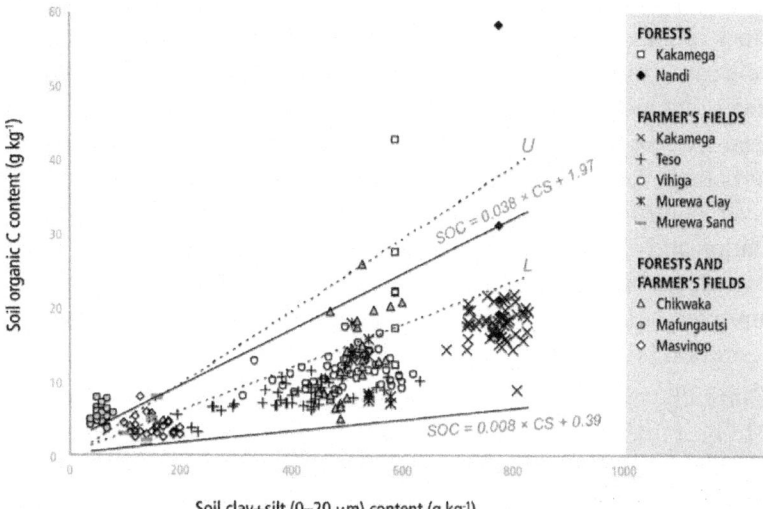

Soil clay+silt (0–20 μm) content (g kg⁻¹)

Figure 6: Soil organic C from the upper 0 to 20 cm as a function of clay plus silt. The tropical forests (Forests) and cultivated fields converted from the forests (Farmer's Fields) represented here were in Zimbabwe (Chikwaka, Mafungautsi, Masvingo, Murewa) and Kenya (Kakemega, Nandi, Teso, Vihiga). The Zimbabwean forests were the Miombo woodlands with unimodal rainfall of 800 mm. The Kenyan forests were rain forests with bimodal rainfall of 1,800 mm. Upper (U) and lower (L) boundary lines were fitted to the 95th and 5th quantiles, respectively. Samples collected at Chikwaka, Mafungautsi and Masvinga (Forests and Farmer's Fields) were collected along a temporal gradient from 0 to 60 years after conversion from tropical forests, but were not differentiated by land use in this figure. Adapted from Tittonell [228] with permission from the author.

Trade-offs between mitigation and adaptation occur often in agricultural systems, notably in the allocation of scarce resources between competing activities. The Trade-off Analysis model for Multi-Dimensional Impact Assessment is used to evaluate climate-change impacts and the viability of adaptation strategies by combining survey, experimental and modelling data [229]. Its next step is calculation of future land use, output, output price, cost of production and farm and household sizes for different climate-change and socioeconomic scenarios. The authors applied the model to the Vihiga and Machakos districts in Kenya to simulate changes in crop and livestock productivity and the effects of climate change to 2030. Climate change was projected to have a negative economic impact for 62% of farmers in Machakos and 76% in Vihiga, but these modelled effects could be partially offset by specific adaptation strategies. The most viable adaptation strategies included introduction of an improved maize variety or low-yielding, dual-purpose sweet potatoes in Machakos and improved feed quantity and quality combined with livestock breeds adapted to increased drought and high temperatures in Vihiga. In some cases, mitigation activities result in negative trade-offs, such as organic practices that increase SOC offset net GHG emissions, leading to competition for feed for livestock or fuel, or even to decreases in average yields, thereby exacerbating forest conversion to agricultural land [122]. Agroforestry, however, contributes to multifunctional landscapes that support mitigation and adaptation and can lead to improvements in livelihoods, whereby provision of fuel wood, timber, fruits and/or fodder is often associated with the cobenefits of improved soil fertility, water infiltration and below- and above-ground C sequestration [40, 150].

Currently, agricultural decision-makers and policymakers rarely consider SOC to be a major factor in agricultural management or land-use change, and the concept of multifunctional landscapes is an emerging idea in the science-based policymaking realm. Yet, the study of SOC formation, its functions, its physical and chemical protection and identification of those fractions most susceptible to degradation is an area of active research. Through various international conventions, this scientific knowledge is slowly becoming part of the science–policymaking interface relevant to climate-change mitigation and adaptation (for example, the United Nations Framework Convention on Climate Change, the United Nations Convention to Combat Desertification, and the Global Soil Partnership). However, the complex trade-offs in land-use decision-making regarding provision of multiple ecosystem services in a given landscape are usually local, so new interdisciplinary and socioecological research approaches are needed in order to downscale information and options regarding how to best manage soil C in relation to other ecosystem services and farmers' livelihoods [40, 122, 150,230–232].

Water Management for Food and Fishery Systems

The effects of climate change on hydrology are far more uncertain than temperature change, and yet, global irrigation water demand will likely increase by approximately 10% by midcentury [233]. IFPRI models indicate that calorie availability in developing countries could potentially reach almost 85% of that in developed countries by 2050, but in more pessimistic scenarios, calorie availability will decline in all regions, due in part to lesser water availability [36]. IPCC models for irrigated areas within this same time frame indicate that the gap between potential evapotranspiration and effective rainfall will be about 17% by 2050 under a high-emission scenario, placing extra stress on demand for irrigation water [234]. Taylor *et al.* [235] asserted that land-use change may have even more noticeable impacts on the hydrological cycle than climate change itself, but that, given the strong focus of mitigation and adaptation planning on land-use change, the two will remain intimately linked. For example, following conversion of forests and grasslands to agriculture in the West African Sahel [236], Southeastern Australia [237], New Zealand [238] and Southwest USA [239], runoff and/or groundwater recharge increased up to two orders of magnitude. Such increases are not always sustained, owing to a range of vegetation cover and hydrological response factors [240]. Forests and woodland cover can also support water quality and, in some cases, can assist in reducing dryland salinization and water-quality decline in semiarid environments [241–243]. Massive abstraction of groundwater and redistribution to agricultural land (nearly 70% of global freshwater withdrawal and 90% of consumptive water use for irrigation) has led to groundwater depletion in regions with primarily groundwater-fed irrigation (for example, regions of China and in the Ogallala Aquifer region in the United States). With projected increases in drought incidence and severity, changes in rainfall patterns and intensification, and decreases in snowpack, agricultural areas that are currently irrigated with surface water will become heavily reliant on groundwater. In Mediterranean-type climate regions in California's Central Valley and in southern Europe, groundwater recharge will be highly dependent on uncertain changes in precipitation patterns [235]. Aquifer salinization is also predicted to increase, at least in California's Central Valley [244]. Sea-level rise also threatens groundwater and surface water with saltwater inundation [245]. The case studies here depict adaptation measures that have been employed to meet the challenge of water management in the face of climate change across a range of spatial scales.

In the Central Valley of Chile, multidisciplinary teams have enacted a CSA-like strategy to address climate-related changes in water [245]. In Chile, farmers' permanent water rights are determined by estimates of minimum stream flow. In a high-emissions scenario, the Central Valley may experience temperature increases of 4°C by the end of this century [4], which would lead to decreases in water supply and thus challenge the existing system of determining water rights and their allocation. In the Maipo basin of Chile, snowmelt from the mountains will be reduced, affecting both river discharge and water demand. In a moderate climate-change scenario (B2), modelled reference evapotranspiration, an indicator metric of irrigation demand, was discovered to potentially increase by 10% to 15%, whereas under the high-emissions scenario (A2), increases ranged from 14% to almost 20% [31]. Permanent water rights vulnerability under the two scenarios, on the basis of data for monthly mean river flow and an agricultural census, indicated that water demands would be inadequately met in 40% to 50% of years under the more severe climate-change scenario. In response, farmers could change crops and/or cultivars, increase irrigation or sell their land and water rights. Even under current climatic conditions, farmers' existing water rights have been questioned because of increasing demand by urban users [245]. To address this issue of failing water rights and limited availability in future climate scenarios, a 'science-policy' strategy has been employed that involves civil society, scientists and policymakers in an iterative dialogue to identify the challenge and its solutions (Figure 7). Since 2008, annual meetings have been conducted with researchers and stakeholders from the national water service, irrigation commission, and environment ministry in Chile). The result has been increased inclusivity and quality of overall participation in topics such as climate-change impact assessment, water-allocation system reliability and water-sector adaptation evaluation, leading to improvements in decision-makers' support of studies on uncertainty in evaluating irrigation projects and future reservoir operations. The science-policy approach supports dissemination of information and projects to strengthen vulnerability assessment tools and coping strategies for irrigated agriculture.

In the Mekong River Delta in Vietnam, more than 700,000 ha of coastal habitats used for aquaculture are threatened by rising sea levels due to climate change. Kam *et al*. [246] analysed the farm-level economic costs and benefits of several alternatives: (1) autonomous adaptation, that is, spontaneous adoption or response, to climate change; (2) no climate change; and (3) planned, or policy-driven, adaptive strategies in which costs are distributed more equitably across the supply chain or are borne by government and other entities. Here 'autonomous adaptation' includes farmers' responses to changes in land and water availability, commodity prices, market incentives, and climate variability. Such responses incur incremental capital costs and include using different levels and combinations of inputs, altering species and production systems, adjusting the height of pond dikes, and increasing water volumes pumped into ponds. Shrimp farmers will be better able to bear the cost of autonomous adaptation than catfish farmers because they sustain relatively higher profit margins and require lower capital investments than catfish farmers. However, without government intervention to prevent flooding and salinity intrusion, the shrimp industry in aggregate will likely experience higher adaptation costs, as it covers more area. Planned adaptive strategies include genetic improvement of breeding stock and pathogen control. Although constructing dikes would reduce river and coastal flooding and salinity intrusion in support of fish production (a provisioning service), opportunities for expansion in both brackish-water and mangrove aquaculture systems that are key to coastal preservation (supporting service) will be lost. In general, evaluating adaptive planning with many types of metrics, including those for ecosystem services through restoration of coastal and intertidal vegetation, were found to provide more data to inform the final choices made by stakeholders [247].

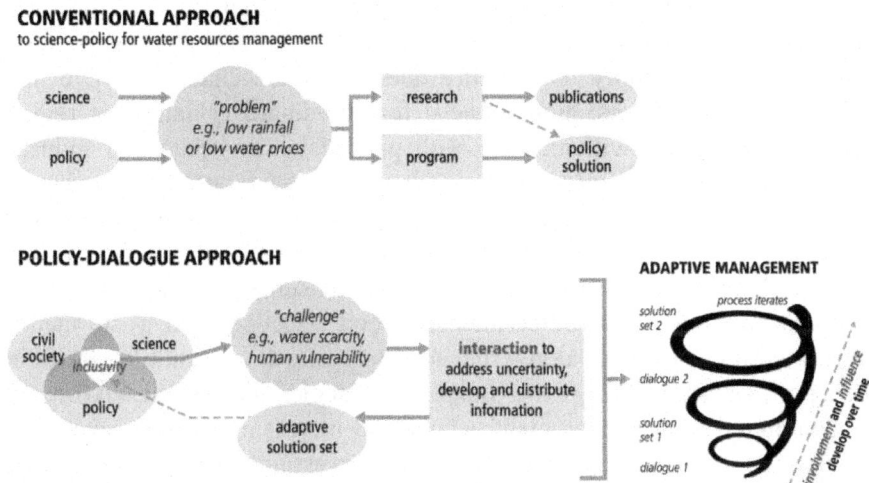

Figure 7: A comparison of the conventional approach and the policy-dialogue approach. The policy-dialogue approach led to the development of greater adaptive capacity and stakeholder engagement described by Scott *et al.* [245] and is also being employed in CSA. From Scott *et al.* [245]. Reproduced with permission from Taylor & Francis.

Recently, the concept of rainbow water, or terrestrial and oceanic evaporation as a source of atmospheric moisture and subsequent precipitation, has emerged. This conceptualization frames how to harmonize the interests of all users of the hydrologic cycle [248]. Available blue water sources—water used for irrigation, industrial or domestic use—and grey water sources cannot support the rate of agricultural intensification, so interest in green water—rainfall used by forests and other vegetation—has grown. Although controversial, passage of air over vegetation with a specific leaf index of 1 in the 10 days preceding rainfall was observed to lead to increased precipitation in Africa [249]. It follows that assessments of climate must take into consideration whether, where and how landscape changes alter large-scale atmospheric circulation patterns of water far from where the land use and cover changes occur to avoid misalignment of investment in climate mitigation and adaptation [248].

Given that climate change is likely to reduce water availability across many agricultural regions, it is critical that water policy and management practices focus on efficient and equitable water rights and allocation policies; increasing water productivity via more and better irrigation storage, conveyance and delivery systems that reduce evaporative losses; in-field water-use efficiency improvements; and technologies that reduce seawater intrusion in coastal environments. These challenges are equally important in the quest to increase agricultural productivity to feed a growing global population, irrespective of the degree of climate-change impact. Responses to the spatial and temporal shifts in water quantity and quality due to climate change involve many scales and stakeholders, and the need for coordinated planning at regional and national scales will increase with growth in the urban and industrial sectors. Approaches to increasing the efficiency of water used for food supply must employ drought-tolerant crops and irrigation technology (for example, water-conserving irrigation systems, crop coefficients and surface renewal [250, 251]). They also need to address both consumptive behaviour (that is, overconsumption and resource-intensive food selection) and waste incurred during postharvest and along the supply chain (for example, threshing, transport, storage) [252]. Other adaptive strategies include the involvement of communities and government agencies in increasing storage capacity via small-scale reservoir projects, rainwater harvesting, groundwater banking through artificial and/or natural aquifer recharge and flood harvesting (that is, directed capture of floods in floodplains) and restoration of coastal vegetation to promote opportunities for aquaculture [242, 244,252]. Additional adaptation options include reduction in end-user demand, deengineering and reoperation of water systems to create adequate supply and distribution, improved wastewater treatment plants to facilitate wastewater reuse, desalination plants and targeted water-conservation projects [253].

Managing Forest Biodiversity to Increase Ecosystem Services and Resilience

Forest loss and degradation cause GHG emissions and loss of C stocks, biodiversity and ecosystem services. Trees and forests buffer microclimates, regulate water quality and flows, store C and provide habitat for plants and animals in protected areas and corridors [248, 254, 255]. When landscapes are managed to contain a mosaic of forestry and agroforestry ecosystems, the diversification of food, feed and timber production, income sources, and markets promotes greater resilience to environmental uncertainty [149, 256]. REDD+programmes to pay developing countries for conservation and sustainable use of forests have evolved over the past decade toward greater attention on (1) increased interactions between institutional networks and (2) achieving reduced GHG emissions along with improvement of livelihoods of local communities and biodiversity conservation [257]. A systems approach involving biophysical and social sciences, as well as indigenous knowledge, is fundamental to demonstrating that REDD+projects are performance-based, fair and equitable [33]. Although afforestation and reforestation are often considered in REDD+projects, trees on farms are usually not included, owing to strict 'forest' definitions. Yet, agroforestry systems offer many REDD+-related benefits. Intentional integration of trees on farms and in agricultural landscapes increases C sequestration, along with greater food security and resilience [40, 229] (for example, see Figure 8). Assessing such multifaceted trade-offs across an agricultural landscape is relevant to the CSA strategy, but will require greater coordination on local, regional and international levels to be incorporated into REDD+.

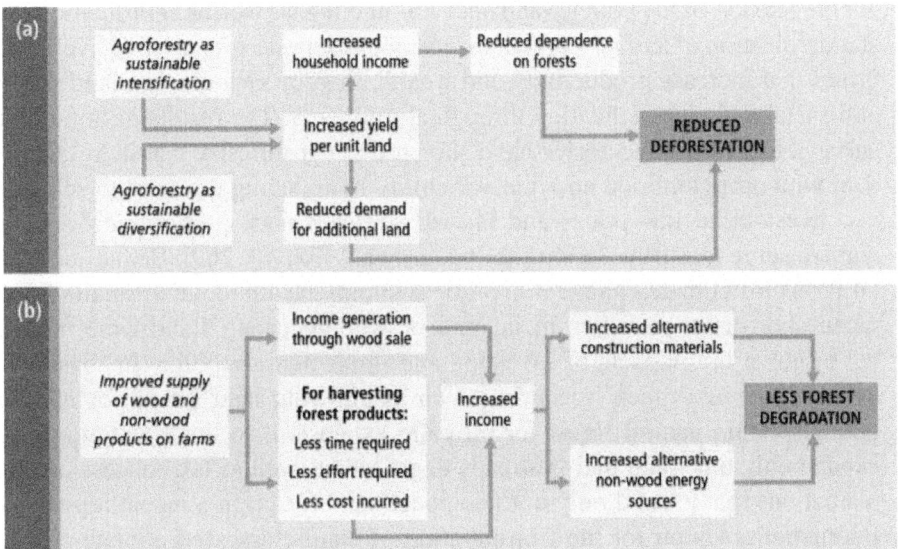

Figure 8: Contribution of agroforestry to the REDD+mechanism. (a) Sustainable intensification and diversification pathway. (b) Source of wood and nontimber forest products. Deforestation is avoided through sustainable intensification and diversification. Reproduced from Minang *et al.* [33] with permission from P Minang (personal communication, 2014).

Examples of agroforestry types in agricultural landscapes include remnant forest or savanna, agroforests, tree crops, home gardens and boundary plantings [258]. Tree species and densities for each type are selected by desired ecological processes, farmers' criteria and land-use policies. An integrated landscape approach allows valuation of the ecosystem services derived from these management options and can be used to determine potential trajectories of tree-cover transitions [31, 149]. It permits the nesting and spanning of spatial scales of different agroforestry types, the confrontation of biases for C benefits versus livelihood choices, and the optimization of tree-diversity exploration. It also opens opportunities to identify synergies and trade-offs and helps sidestep definitional challenges that result in negotiation platforms for proactive actions that reduce vulnerability and increase benefits (for example, see [259]). The landscape perspective is useful for scenario-building, such as comparing financial incentives that emphasize economic efficiency for agricultural and timber purposes versus socially 'green' and 'rights-based' approaches that support resilient livelihoods and broader sustainable development goals. The current scientific literature does not adequately detail these socioecological and community-based processes or how they underpin decision-making.

Examining trade-offs in REDD+can provide scientific information to enable science-based policies and decision-making, as well as coordination and standardization of REDD+practices. Many of the trade-offs involve livelihood issues that increase productivity and wealth, thereby encouraging land tenure and sustainable intensification through agroforestry. The results of household surveys and farm inventories have shown that agroforestry can help farmers deal with drought, flood and rain variability by reducing the need to sell land and livestock at low prices and instead sell seedlings, timber and firewood and consume tree fruit during the 'hunger gap' [33, 40, 260]. Sequestering C on farms for climate-change mitigation will only be attractive to smallholders when short-term increases in income or welfare occur. Landscape models have shown the impacts of investing and implementing policy in 'business-as-usual' versus 'green' scenarios, such as allowing land swaps for permits granted within natural forest for oil palm expansion, so that plantations can expand only onto land that is already degraded, as well as tax concessions for plantations that expand only onto degraded land [261]. In a recent report, the International Union for the Conservation of Nature assessed climate-change mitigation activities across many regions of the world where REDD+policies likely would be implemented [262]. Examination of the social, economic and environmental trade-offs and potential synergies revealed that clear tenure and property rights, including rights of access, use and ownership, are essential for effective REDD+implementation To benefit local communities, including the most vulnerable, REDD+policies must enhance the ecosystem services upon which the rural poor are most dependent and leverage new financial resources to reward local communities for management. These opportunities can easily be lost if the vulnerable are explicitly excluded as beneficiaries (for example, because of unclear tenure) or high barriers to entry (for example, forest certification) [263].

Participatory, transparent, accountable governance can help achieve benefits of implementing REDD+policy by creating synergy between parties at multiple scales. A governance approach that facilitates harmonized goals and policies between civil society and engaged stakeholders focuses on the relationships among organizations rather than on new organizational structures and financing mechanisms. Public–private partnerships can improve the effectiveness of the biodiversity governance system and complement regional and multinational efforts [263]. In Cameroon, for example, nongovernmental organizations are implementing REDD+pilot projects and acting as bridges between the public and the state, both to create awareness among local communities and to voice concerns about social safeguards [264]. Such partnerships have helped government institutions organize international biodiversity governance around an ecosystem approach, largely by changing the scale and nature of the dialogue through a community of practice with institutions outside the immediate REDD+network [257].

Although REDD+will benefit from institutional interactions that build trust and reach eventual consensus on forming, coordinating and integrating policies that support livelihoods and resilience while sequestering C in forests, the definition of appropriate ecosystems for payments still is a major issue. As pointed out by Visseren-Hamakers and Verkooijen [257], it remains to be seen whether CSA, with its integrated planning of land, agriculture, forests, fisheries and water, will be included in policymaking steps towards broadening of the REDD+agenda.

Rural Migration Due To Climate Change

A worldwide transition toward urbanization is occurring, partly in response to climate change, although rural outmigration due to climate shocks, such as hurricanes, is better documented than gradual changes, such as lower rainfall in arid areas [43]. Migration within countries is complex, having both positive and negative impacts on adaptation and household resilience. Climate shocks and disasters can propel people living under vulnerable conditions into poverty traps that force migration out of rural areas [265], where men most often migrate, leaving the women and children with increased household and farming burdens [45]. Migration can be a beneficial strategy that spreads risks through resource diversification, such as remittances that bring money back to the household [266]. Livelihood and food security, as well as culture, affect who migrates, when, for what reasons and to which destinations [267] (Figure 9). Despite the material benefits that can result from mobility and migration, displacement of people from places that they value reduces culturally based activities, such as preplanning for specific climate-change events [42]. Migration can lead to inhabiting vulnerable urban locations, such as flood-prone areas [268], and increase inequities due to poverty and lack of social networks. Opportunities exist to improve structural and institutional frameworks to reduce migration from rural areas, including greater diversification of rural livelihood systems [149, 269]; opportunities for public health, social equity and environmental welfare [270]; and connection of urban populations with local or regional food sources to support rural incomes [3, 11, 28].

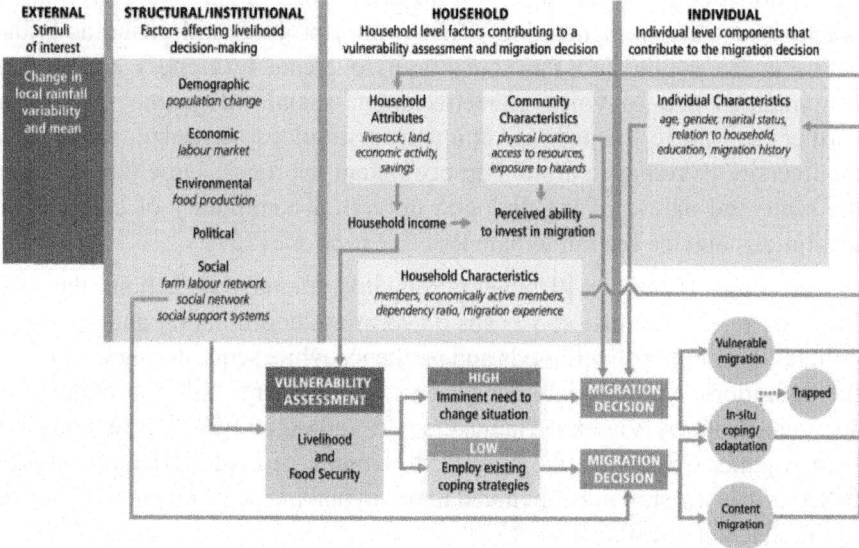

Figure 9: Decision pathway for rural migration in response to external stimuli, often related to climate change. Factors that affect decisions occur at institutional, household and individual levels. Adapted from Warner *et al.* [267] with permission from K Henry.

Land scarcity and degradation are conducive to outmigration. In Guatemala, people from households affected by flooding or soil degradation were found to be more likely to leave settled rural areas for the forest frontier to engage in clearing of forests for agriculture [271]. Surprisingly, on the basis of employing a remote-sensing approach across Central and South America over a 10-year period, rural–urban migration was not observed to strongly affect the recovery of forest vegetation [272]. The researchers in that study found that a significant increase in woody vegetation occurred in only about half of the municipalities that lost population. Thus, depopulation does not necessarily imply land-use change. In their analysis of annual satellite land cover maps, they found that 180,000 km^2 of forest was lost between 2001 and 2010, with the majority of deforestation occurring in South America (92%), particularly in Argentina, Brazil, Bolivia and Paraguay. Much of this land is in soybean production and cattle-grazing to meet the increasing global demand for meat. DeFries *et al.* [273] recently demonstrated that increases in rates of deforestation are closely linked to increases in urban populations and their demand for agricultural products rather than changes in rural populations. In Central America, temporary international migration of members of smallholder households has been indirectly associated with a lack of reforestation; remittances are spent on owning more land, and less household labour favours a transition to cattle production. This is relatively safe and risk-averse compared to row crop production, but it increases forest loss and land degradation and thus decreases the mitigation and adaptation potential [274].

Rural–rural migration offers a livelihood adaptation strategy for rural people facing stresses and shocks due to climate change, but it can also increase migrants' vulnerability. In Vietnam, migrants to the fertile Central Highlands aim to increase their economic livelihoods by producing coffee destined for international markets. Instead of settling permanently, many circulate between their new and origin communities because their social networks that remain at home allow them to avoid some of the risks of permanent relocation [275]. For example, family members in the community of origin may look after the migrants' children, take care of land and assets and provide access to loans. The lack of formal credit institutions at the new destination means that the community of origin may provide continual financial support instead of successful migrants' sending remittances home. Such social networks expose remaining household members to risk if ventures fail because of economic, social and environmental conditions. Both the migrants and origin households may then require loans to take further livelihood risks. In these cases, migration may drive both households into further poverty. Reforming Vietnam's household registration system to allow migrants access to banking, lending and other public services at their new locations could reduce the risks of such outcomes [275].

In the project 'Where the Rain Falls: Climate Change, Food and Livelihood Security, and Migration', researchers have examined rainfall, food security and human migration in eight countries in Asia, Africa and Latin America [267], mainly in agricultural areas. Four distinct household migration profiles were identified, varying along a spectrum from resilience, where migration is one of a variety of adaptation measures that progressively reduce climate sensitivity, to vulnerability, where migration either is difficult or exacerbates sensitivity to climatic stressors. Although national and regional contexts affect migration, household characteristics were discovered to be most important for migration-related decisions and outcomes. For example, migration was generally erosive for the poor and those with small land holdings. Household size and composition, land ownership, asset base, degree of livelihood diversity and education levels were associated with migration strategies that increased resilience, such as nonagricultural jobs or diversified livelihoods [267]. One of the 'Where the Rain Falls' project case studies is the Mantaro Basin of Peru, where pressures to migrate stem from lower precipitation that reduced farmer and herder incomes [276]. Two livelihood and migration profiles in the Mantaro Basin were identified in response to climatic vulnerability. Lowland farmers who often commuted on a daily basis for casual urban employment used their proximity to the city to diversify their livelihoods. In contrast, herders farther from the city were forced to migrate for longer periods or permanently, in the absence of other options, and therefore were generally more vulnerable.

The act of migration has a risk dimension, whether it is a positive form of adaptation or part of erosive coping strategies. Understanding the cultural dimensions of risk-taking under climate uncertainty is crucial for determining migration decisions, especially as the necessity for climate-driven planned resettlement becomes more urgent [42]. Although outmigrants are mainly men, the outcomes of climate-change–induced migration are likely to be highly gendered because women are disproportionately affected. Women tend to be poorer and less educated and to have lower health status and limited direct access to, or ownership of, natural resources [45]. It will become more feasible to identify risk-prone agricultural areas and circumstances if models of biophysical aspects of climate change and land use also take into consideration factors that influence migration decisions, such as landlessness, land tenure and distribution issues, as well as the role of social networks that facilitate resilience and adaptation in rural areas as well as escape from poverty traps [167].

Climate-induced outmigration from rural areas involves mitigation and adaptation issues related to urban and periurban outcomes, such as increased GHG emissions due to urban sprawl on land that once supported food production [11]. Interdisciplinary work is needed to understand effective strategies for developing and preserving smallholder agriculture near cities, expanding urban and periurban agriculture, managing urban growth for farmland preservation, connecting agricultural producers with local urban markets, ensuring availability of agricultural labour and enabling diversified rural livelihood systems. Such strategies will have combined benefits for climate change mitigation and adaptation.

Metrics for Vulnerability Assessment, Food Security and Ecosystem Services in Agricultural Landscapes

Science-based actions within CSA require integrated data sets and sound metrics for testing hypotheses about feedback regarding climate, weather data products and agricultural productivity, such as the nonlinearity of temperature effects on crop yield [277], and the assessment of trade-offs and synergies that arise from different agricultural intensification strategies. Approaches range from the development of broad indicators for identifying differences in climate vulnerability over large spatial scales down to the use of finely disaggregated spatial metrics [278]. New and innovative research and policy designs, as well as cooperative arrangements among and between government agencies, research institutions and civil society, have the potential to implement monitoring and assessment systems for decision-making. Examples presented here demonstrate how biometeorological, economic and sociological indicators can be used in vulnerability assessments and show nuances that must be addressed with respect to scale.

Novel outcomes, such as nonlinear effects of climate change on agricultural productivity (for example, US maize), are emerging based on the use of large-scale data sets, indicating that environmental change may drive agricultural productivity in unexpected ways [277]. For example, Lobell *et al.* [5] examined harvest and daily weather data derived from more than 20,000 historical maize trials conducted by the International Maize and Wheat Improvement Center and private seed companies in Sub-Saharan Africa from 1999 to 2007. 'Optimal management' and 'drought stress' were the two most common scenarios under which maize was grown. Final yield was reduced to the following different extents due to warmer temperatures: by 1% under optimal rain-fed conditions and 1.7% under drought conditions for every degree day spent above 30°C. Lobell *et al.* [5] suggested that a 1°C warming would lead to negative yield where maize is presently grown under optimal management (roughly 65% of the area) in Sub-Saharan Africa, whereas all areas in this region would show decreased yield of as much as 20% under drought stress. Similarly, in the United States, which generates 40% of global maize production, predicted increases in interannual weather variability (temperature and precipitation) could result in an 18% decrease in maize yields by 2030 to 2050 in comparison to the period from 1980 to 2000, along with increasing volatility in annual yields [279]. Expansion of cropland in other regions and retention of speculative inventories (that is, holding volumes for higher price earnings) may offset the volatility. Here metrics of climate and indicators of crop productivity and other agronomic factors predicted crop response to climate warming and drought over a widespread region, setting the stage for more research on how adaptation measures, such as improving soil moisture and breeding for drought and heat tolerance, could be used to reduce vulnerability in the future [5].

Metrics that incorporate human ecology are integral to enabling the CSA strategy. Vital Signs [280] is a monitoring programme for changes in human well☐being, agriculture and ecosystem services and is designed to provide metrics in rapidly expanding and intensifying agricultural landscapes in Africa, leading to integrated approaches that support food security (Figure 10). A primary goal of Vital Signs is characterizing the uncertainty and quantifying the sampling intensity needed to achieve different levels of accuracy and statistical power to detect change. Information gathered in the initial phase will be further evaluated for its overall utility and delivery cost. Measurements collected by Vital Signs participants are based on hierarchical spatial scales to provide integrated information that can inform structural relationships and counterfactuals involved in decision-making from the global to household scale. The global perspective facilitates comparisons between different regions (250,000 km^2·region^{-1}), whereas regional measurements deliver information at the scale on which agricultural investments are made. Information collected at the landscape scale (10 to 20 units per region) measures the relationships between agricultural intensification, water availability, soil health and other ecosystem services, together with human well-being. Plot-level (1 ha) data reflect agricultural production, including seed selection, fertilizer type and application rate, as well as crop yield response. At the household level, surveys are employed to collect information on health, nutritional status, income and assets. Stakeholder planning meetings and participatory research established both at the onset and throughout the project are integral to garnering active engagement in Vital Signs.

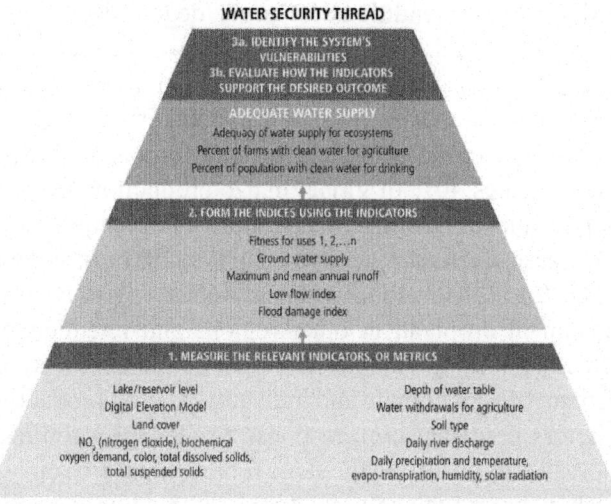

Figure 10: Water security thread from Vital Signs. The pyramid of the water security thread depicts the integration of metrics (*1*) that build the desired indices (*2*) with the outcomes of interest (*3a* and *3b*). Adapted with permission from the Vital Signs programme (S Barbour, personal communication, 2014).

Prioritizing allocation of resources and focusing policies on vulnerable regions requires metrics to assess susceptibility to a lack of food security due to climate change [281]. Biophysical climate indicators derived from global climate-change models and food insecurity indicators (that is, availability, access and utilization) can serve as such metrics. As an example of this approach, Ericksen *et al.* evaluated hotspots of vulnerability using the overlap among indicators of global climate (for example, rainfall variability, number of reliable growing degree days, and change in mean annual temperature) and food security indicators across the global tropics [281]. The latter were composed of availability (for example, crop yield and mean food production indices), access (for example, GDP per capita, transport time to markets, and monthly staple food prices) and utilization (for example, malnutrition prevalence and proportion of the population using unimproved water source). Future vulnerability was depicted by existing resource pressure (for example, annual population growth and agricultural area per capita). The resulting index of vulnerability reflected three central components: exposure of populations to the impacts of climate change, sensitivity of food systems to these impacts and coping capacity of populations to address these impacts. With this vulnerability index, it was possible to rank the most highly exposed regions, leading to the emergence of southern Africa as a highly exposed region, as well as areas within Brazil, Mexico, Pakistan, India and Afghanistan. This approach is limited by the following factors: The data represent only current food security levels; data are gathered only at the national level, which masks variability within regions and among households; and other data are needed on climate-change exposure and on food security variables other than crop yields and utilization, such as food distribution and equity.

Systems delivering real-time indicators and metrics that are tied closely to management decisions and current conditions allow science and policymaking entities to progress from using lagging indicators to finding leading indicators that can be used to identify when and where thresholds of climate-change responses will occur [112]. Indicators and metrics are often used to support public goods and services, so better standards and codified practices that support shared vocabulary and ontology will reduce the costs and streamline efforts for curating and disseminating such information. Research designed to develop metrics that inform global to local social networks for data collection, sharing and integration can also be leveraged for extension efforts. The identification of efficient and location- and situation-specific sets of indicators will complement efforts to construct human capital, social awareness and consensus regarding specific issues, leading to action strategies and policy guidelines across various temporal and spatial scales.

THEME 3

Integrative and Transformative Institutional and Policy Issues: Bridging Across Scales

Figures 11 and 12 provide an overview of some of the main points covered in each session of the 2013 Global Science Conference on Climate-Smart Agriculture. The relative emphasis on mitigation of GHG emissions versus adaptive capacity to climate change (or both) varied depending on the session topic. CSA strives for food security, adaptation, mitigation and resilience, but not all of these are achieved in the same context. The session topics often invoked multiple scientific disciplines to inform further action and problem-solving strategies in support of CSA goals in the context of the session topic, but further integration across these topics and disciplines is necessary. Scientific uncertainties are inherent in climate science, given the difficulty of forecasting climate and its interactions with other aspects of human-induced environmental change. The examples that are mentioned here require intensified scientific activity, formation of knowledge networks, and involvement of many relevant stakeholders to obtain better information to support decision-making (see also [135–137,152]). Also, there are clear social controversies challenging CSA, often derived from assumptions and questions of equity and legitimacy, such as who will implement a response to climate change and how this will occur. To obtain buy-in from vulnerable populations and countries, such issues must gain the forefront in discussions of CSA science and policy among the diverse set of stakeholders described in the Introduction section above. Many of the stakeholder-driven programmes mentioned in the conference sessions exist at regional and global levels, as climate science is often funded for large-scale initiatives. As stated previously, this article and the conference presentations do not emphasize the local knowledge-to-action processes that are essential for transformations towards climate preparedness. Nonetheless, some of the possible pathways towards such socioecological approaches to fostering greater participation and advancement of CSA objectives are shown for each of the session topics. Clearly, science must play an active and central role in developing the information base that will support food security, adaptation and mitigation in CSA and new types of inclusive, participatory decision-making as well as knowledge exchange processes [135, 152].

SESSION CONTENT FROM THEME 1 (2.0): FARM AND FOOD SYSTEM ISSUES: SUSTAINABLE INTENSIFICATION, AGRO-ECOSYSTEM MANAGEMENT, AND FOOD SYSTEMS					
Mitigation and/or adaptation	Interdisciplinary science	Scientific uncertainties	Social controversies	Stakeholder programs mentioned	Social-ecological pathways
2.1 CROP PHYSIOLOGY & GENETICS					
Mainly adaptation	Molecular biology, ecophysiology, agronomy	Plant response to high [CO2] + high temperatures	Genetic engineering	Regional crop breeding networks	Participatory plant breeding
2.2 LIVESTOCK MANAGEMENT & ANIMAL HEALTH					
Both	Animal science, range science, veterinary medicine, economics	Feasibility of scaling up of combined mitigation and adaptation strategies	Emphasis on low GHG emissions under poverty	Global programs for REDD+; for managing diseases	Management decisions based on livelihood criteria
2.3 NITROGEN MANAGEMENT					
Both	Agronomy, plant nutrition, biogeochemistry	Decision-support tools effective across agroecosystems	Emphasis on low GHG emissions under poverty	Agricultural extension programs	Knowledge systems to enhance adoption of practices
2.4 FARMER DECISION MAKING					
Both	Social & political sciences, agronomy	Overcoming barriers to adoption of new practices	Dealing with perceptions of low climate risk	Social networks in farming communities	Short- & long-term planning for uncertainty
2.5 CLIMATE RISK MANAGEMENT					
Adaptation	Economics, remote-sensing, agronomy	Design of safety nets for uncertain climate shocks	Insurance reduces incentives for adaptation	Index-based livestock insurance programs	Bundled strategies for household welfare
2.6 ENERGY & BIOFUELS					
Mainly mitigation	Biogeochemistry, life-cycle analysis, agronomy, economics, geography	Projections of costs & benefits from biofuels	Food price increase attributed to biofuels	Energy directives and action plans	Biofuel feedstock value chains that support food security

Figure 11: Conference session 2.0 content within theme 1. Farm and food system issues: sustainable intensification, agroecosystem management and food systems[a]. [a]GHG, Greenhouse gas; REDD+, Reducing Emissions from Deforestation and Forest Degradation.

Inter- and transdisciplinary scientific approaches are principal both to our understanding of how socioecological systems support the adaptive management and governance that are essential to long-term human provisioning of food and to the establishment of science–policymaking dialogues to plan for the future [47, 282–284]. These actions are keys to assessing trade-offs of mitigation in context-specific situations, such that resource-poor farmers are supported rather than undermined by CSA. To realize the CSA objectives of increased food security, resilience, mitigation and adaptation, scientific research supports awareness, analytical capacity and the evidence base to understand the impacts of climate change on agricultural growth strategies and food security, and identify climate smart options suitable to the local context [15].

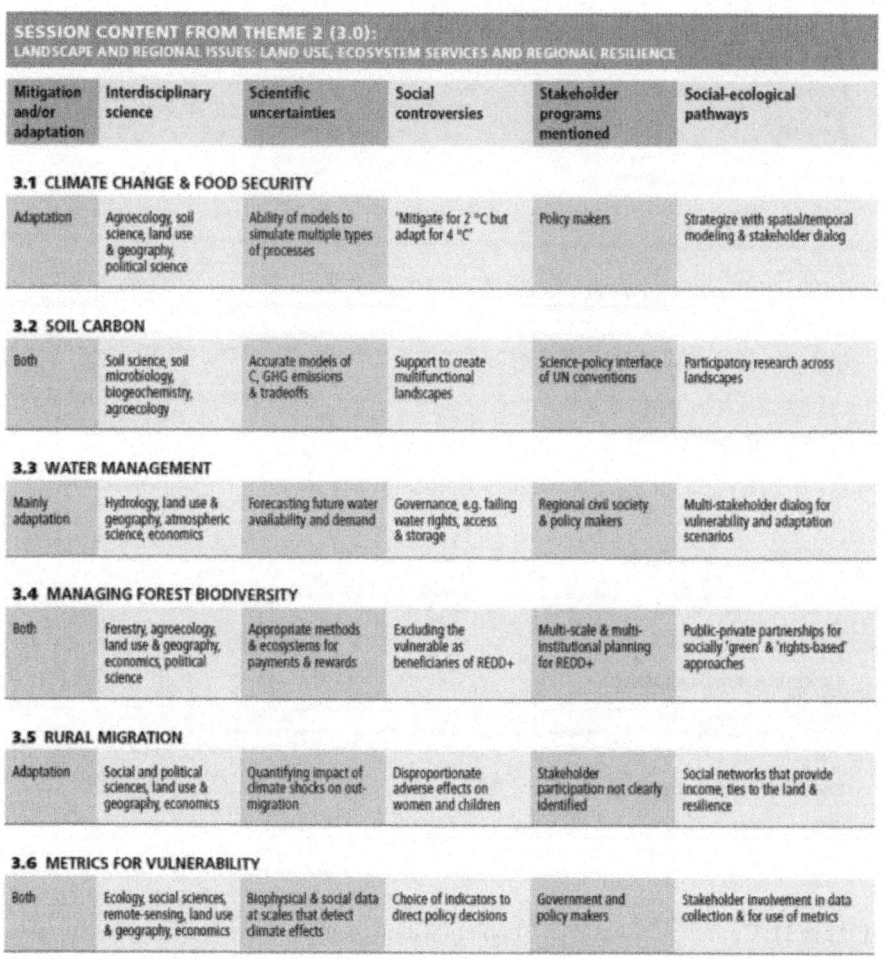

SESSION CONTENT FROM THEME 2 (3.0): LANDSCAPE AND REGIONAL ISSUES: LAND USE, ECOSYSTEM SERVICES AND REGIONAL RESILIENCE					
Mitigation and/or adaptation	Interdisciplinary science	Scientific uncertainties	Social controversies	Stakeholder programs mentioned	Social-ecological pathways
3.1 CLIMATE CHANGE & FOOD SECURITY					
Adaptation	Agroecology, soil science, land use & geography, political science	Ability of models to simulate multiple types of processes	'Mitigate for 2 °C but adapt for 4 °C'	Policy makers	Strategize with spatial/temporal modeling & stakeholder dialog
3.2 SOIL CARBON					
Both	Soil science, soil microbiology, biogeochemistry, agroecology	Accurate models of C, GHG emissions & tradeoffs	Support to create multifunctional landscapes	Science-policy interface of UN conventions	Participatory research across landscapes
3.3 WATER MANAGEMENT					
Mainly adaptation	Hydrology, land use & geography, atmospheric science, economics	Forecasting future water availability and demand	Governance, e.g. failing water rights, access & storage	Regional civil society & policy makers	Multi-stakeholder dialog for vulnerability and adaptation scenarios
3.4 MANAGING FOREST BIODIVERSITY					
Both	Forestry, agroecology, land use & geography, economics, political science	Appropriate methods & ecosystems for payments & rewards	Excluding the vulnerable as beneficiaries of REDD+	Multi-scale & multi-institutional planning for REDD+	Public-private partnerships for socially 'green' & 'rights-based' approaches
3.5 RURAL MIGRATION					
Adaptation	Social and political sciences, land use & geography, economics	Quantifying impact of climate shocks on out-migration	Disproportionate adverse effects on women and children	Stakeholder participation not clearly identified	Social networks that provide income, ties to the land & resilience
3.6 METRICS FOR VULNERABILITY					
Both	Ecology, social sciences, remote-sensing, land use & geography, economics	Biophysical & social data at scales that detect climate effects	Choice of indicators to direct policy decisions	Government and policy makers	Stakeholder involvement in data collection & for use of metrics

Figure 12: Conference session 3.0 content from theme 2. Landscape and regional issues: land use, ecosystem services and regional resilience[a]. [a]GHG, Greenhouse gas; REDD+, Reducing Emissions from Deforestation and Forest Degradation.

How does research better inform the institutional, financial and knowledge-sharing arrangements to create a sense of possibility for transformative processes that reduce vulnerability and increases climate preparedness? Truly transformative solutions tap into a sense of possibility for positive action, and, as in business value propositions, there is a promise of goods and services to be delivered and experienced [47]. Yet, a 'doomsday' attitude has permeated much of the agricultural science regarding climate change, emphasizing harsh potential impacts under business-as-usual scenarios (Figure 2). Although it is effective in stimulating awareness and action in some sectors, CSA research is potentially more conducive to achieving food security, adaptation, mitigation and resilience. Examples include models that go beyond impacts to include adaptation and transformation at either the farm or landscape scale (for example, see [211]), capacity approaches to examine multifunctional solutions within the socioecological system and direct evidence for situations, options and scenarios which increase human behaviours that build natural capital and resilience. Action-oriented research can also show how public-private partnerships can be used successfully to develop technologies, policies and approaches that may lead to sustainable food production and consumption patterns in a changing climate.Uncertainty is one of the most difficult obstacles to determining priorities for CSA research. Not only is future climate uncertain, but so also is the existence and operation of the institutions that are and will be involved in adaptation, mitigation and resilience. Uncertainty can breed scepticism about the urgency to plan for climate change, especially in agricultural communities and industries that already deal with large annual variability in production and prices. Thus, uncertainty is a barrier to mitigation and adaptation among some of the stakeholders whose investment, engagement and broad agricultural knowledge are critical for designing better research on coping strategies. CSA recognizes that the unfolding of decision-making processes, their translation into action and the formation of adaptive capacity depend on the socioecological contexts in which farmers are embedded (for example, the vital role of social networks in rural communities) (Figure 1). The ways of addressing uncertainty are likely to differ greatly among communities and socio-ecological systems, and research is needed to understand how to approach uncertainty in different contexts.

Although poverty can sometimes drive collective action, such as for improved food security in Kenya and Uganda through risk-sharing and pooling of labour and other limited assets [16], the least food-secure may be less likely to adopt new CSA practices because innovation implies additional costs before benefits can be realized [285]. Research on adoption of new farming technology and practices is needed to understand how upfront costs, lost income, worries about personal health and additional risks assumed during the conversion period can present formidable barriers to farmers [149], even if the new practices leverage ecological processes to improve sustainability and production [14]. For instance, diffusion of new germplasm with specialized traits (for example, drought tolerance) to targeted end users may suffer slow adoption even though new regional and local cultivars will likely be adapted to the range of conditions and management practices employed during climate change. To illustrate this point, modelled diffusion of a drought-tolerant variety among vulnerable (highly risk averse) farmers took four times longer than it did among those less vulnerable (less risk averse), underscoring the need for consideration of how seed prices affect the access of vulnerable farmers to new crop varieties [286]. Synthesis of information on how CSA practices have been facilitated by specific policy interventions, leading to broad community support, also is needed. A better understanding of how social benefits such as access to food and healthcare, rights to land and water, markets, and financing situations facilitate adoption of new farming practices or technologies will inform governance decisions [14].

Collective action for climate preparedness and problem-solving has already been effective in some situations. Safety nets for the poorest and most vulnerable households usually occur in the form of humanitarian relief and food aid, cash payments, agricultural inputs and public works [14], often after a critical event has occurred. Instead, communities can collectively plan safety-net strategies and resource transfers that are predictable and flexible enough to be scaled up and then scaled down when the crisis subsides [13, 14]. CSA research on learning, knowledge-sharing and social network analysis can help build awareness, early-warning indicators and criteria for benefit transfers for disaster responses and also effectively combine local collective action with national and/or international aid. Enhancing human and social capital, such as for childhood nutrition, entrepreneurship by women, and synergies between fuel use and C sequestration in trees, also rehabilitates household and community assets. Proactive planning will be more effective than reactive responses to a disastrous climate event, and research can help increase understanding of how adaptation policies must be designed accordingly [204].

Furthermore, collective action at the institutional scale is essential to avoiding conflicts that result from climate change. For example, institutional transboundary water agreements are associated with lower risk of conflict during water scarcity, but even one weak link in the communication, coordination and cooperation between coriparian nations will reduce their adaptive capacity to respond to new changes in hydrology, thus increasing the potential for risk and disputes [206]. So far, climate change has rarely been incorporated into such agreements. Collective action at the institutional scale could also address changing migration patterns of rural–urban connections that are likely due to extreme climate events and climate change and which will have potentially large ramifications on food production and food security, land tenure and cultural integrity [42] (Figure 10). At this point, research is needed to better understand how climate affects the dynamics of the rural labour force and thus on the stability of local food production for rural communities and nearby cities [267, 287].

To realize CSA, research on targeted financing is essential, especially in support of the most vulnerable. Upfront investment to plan and start implementation strategies is required, as is research to develop monitoring systems designed to track climate-related human responses by utilizing consistent metrics that demonstrate private benefits along with public goods (for example, GHG mitigation). Already existent funds, such as the Adaptation Fund established under the Kyoto Protocol [288], and the International Fund for Agricultural Development's Adaptation for Smallholder Agriculture Program [289], can improve smallholders' access to climate-smart assessments, technologies and institutions related to sustainable management of forests, providing up to 16 million additional jobs globally and increasing household income in rural areas as a result of restoring degraded forest [290]. Larger-scale investments, such as financing infrastructure for water resources and carbon capture, can potentially be provided by the Green Climate Fund [291], and private finance may also play a role. As climate finance develops, research shares a role in prioritizing investments and effective financing solutions and in monitoring outcomes.

Investment in research on food systems that are resilient to climate shocks may be more likely to occur if CSA expands beyond the agricultural sector. As examples, CSA research could more explicitly involve issues related to: (1) local, national and regional food trade, including governance and regulations, food safety, roads and infrastructure, and value chain coordination; (2) flexibility in financial arrangements, insurance and planning to cope with, and be responsive to, variability in climate and markets; and (3) integration of the interdisciplinary research to form a more holistic and service-oriented approach based on science to inform policy. For research to be utilized most effectively in policies related to CSA, pathways for communication of the latest scientific progress and research results must be established within relevant time frames. Communication must span sectors and scales in which policymakers and other stakeholders operate, crossing boundaries between scientists and local, regional and global actors such as nongovernmental organizations, governmental agencies, corporations and broad social and media networks [290].

CSA strategies support the realization of a broader green economy concept that acknowledges 'the sum total of all ecosystem services and how they collectively provide the complete life support system we need' [292], p. 9. In practice, market prices, costs, and benefits for the ecosystem services related to carbon sequestration, clean water production, flood protection and grass forage have been quantified. In Cameroon, for instance, the value (in US$·ha^{-1}·yr^{-1}) attributed to the forest's contribution to climate and flood control is 1.3- to 2.6-fold greater than that of the timber, fuel wood and nontimber products. Coordinated action resulting from CSA and green economy research not only realizes the improvement of livelihoods and food security through mitigation and adaptation to climate change but also creates cobenefits for ecosystem services and sustainable use of natural capital and enables evaluation of a broader set of trade-offs associated with a certain course of action.

Conclusions

Disciplinary, interdisciplinary and transdisciplinary scientific approaches play a fundamental and profound role in developing understanding of the processes underlying CSA and serve as partners in enumerating priorities for CSA. They form a crucial element in the knowledge base needed to implement CSA actions and manifest future transformative changes in agriculture in a changing climate. Global science conferences on CSA have already been influential in assembling scientists and other stakeholders to share knowledge [17, 49]. A third conference in Montpellier, France, is planned for 2015 with the following agenda items: discussion key scenarios in agriculture and food systems, identifying priorities for early action and designing a roadmap for moving forward with an action plan. These objectives set the stage for a much stronger emphasis on knowledge-to-action frameworks, capacity-building and the changes in human behaviour and social infrastructure that are necessary for adaptation and resilience [133, 152, 293]. The momentum that has already built among the science community for CSA forms the foundation for critical engagement by more researchers in fundamental and applied studies. To this end, establishing a more formal governance mechanism to embed science in the information base for the CSA Alliance, would be a vital step in developing priorities, scientific engagement and funding to support the knowledge needed for policymaking decisions.

ACKNOWLEDGEMENTS

We acknowledge Jill E Walker for graphic design, Ria D'Aversa for maintaining the bibliographic content and Kayla Burns for copyediting. We also acknowledge all conference participants who shared their knowledge and ideas at the Global Science Conference for Climate-Smart Agriculture held in Davis, CA, USA, in March 2013. We thank the session leaders, who also are authors on this manuscript, for composing summaries of the information presented in individual sessions at the conference and synthesizing discussions by participants. These summaries formed the basis of this paper. This article was funded by a Programmatic Initiative from the College of Agriculture and Environmental Science at the University of California, Davis (to LEJ).

AUTHORS' CONTRIBUTIONS

KLS, AKH and LEJ cowrote the manuscript. AJB, MRC, AC, CJC, JLH, KH, JWH, WRH, LSJ, BMJ, EK, RL, LL, MNL, SM, RP, MPR, SSS, WMS, MS, PT, SJV, SMW and EW contributed content. All authors read and approved the final manuscript.

REFERENCES

1. Morton JF: The impact of climate change on smallholder and subsistence agriculture. Proc Natl Acad Sci U S A. 2007, 104: 19680-19685.

2. Reidsma P, Ewert F, Lansink AO, Leemans R: Adaptation to climate change and climate variability in European agriculture: the importance of farm level responses. Eur J Agron. 2010, 32: 91-102.

3. Vermeulen SJ, Campbell BM, Ingram JSI: Climate change and food systems. Annu Rev Environ Resour. 2012, 35: 195-222.

4. Solomon S, Qin D, Manning M, Chen Z, Marquis M, Averyt KB Tignor MMB, Miller HL: Climate Change 2007: The Physical Science Basis (Contribution of Working Group I to the Fourth Assessment Report of the Intergovernmental Panel on Climate Change (IPCC)). 2007, (accessed 25 July 2014), Cambridge, UK: Cambridge University Press, accessed 25 July 2014)

5. Lobell DB, Schlenker W, Costa-Roberts J: Climate trends and global crop production since 1980. Science. 2011, 333: 616-620.

6. Cline WR: Global Warming and Agriculture: Impact Estimates by Country. 2007, Washington DC: Center for Global Development

7. Jarvis A, Lau C, Cook S, Wollenberg E, Hansen J, Bonilla O, Challinor A: An integrated adaptation and mitigation framework for developing agricultural research: synergies and trade-offs. Exp Agric. 2011, 47: 185-203.

8. Knox J, Hess T, Daccache A, Wheeler T: Climate change impacts on crop productivity in Africa and South Asia. Environ Res Lett. 2012, 7: 034032-

9. Rosenzweig C, Elliott J, Deryng D, Ruane AC, Müller C, Arneth A, Boote KJ, Folberth C, Glotter M, Khabarov N, Neumann K, Piontek F, Pugh TAM, Schmid E, Stehfest E, Yang H, Jones JW: Assessing agricultural risks of climate change in the 21st century in a global gridded crop model intercomparison. Proc Natl Acad Sci U S A. 2014, 111: 3268-3273.

10. Myers SS, Zanobetti A, Kloog I, Bloom AJ, Carlisle EA, Dietterich LH, Fitzgerald G, Hasegawa T, Holbrook NM, Huybers P, Leakey

ADB, Nelson R, Ottman MJ, Raboy V, Sakai H, Sartor KA, Schwartz J, Seneweera S, Tausz M, Usui Y: Rising CO_2 threatens food quality. Nature. 2014, 510: 139-142.

11. Wheeler T, von Braun J: Climate change impacts on global food security. Science. 2013, 341: 508-513.

12. World Bank: Climate-Smart Agriculture: Increased Productivity and Food Security, Enhancing Resilience and Reduced Carbon Emissions for Sustainable Development, Opportunities and Challenges for a Converging Agenda: Country Examples. 2011, Washington, DC: World Bank

13. World Bank: World Development Report 2014: Risk and Opportunity: Managing Risk for Development. 2013, (accessed 25 July 2014), Washington DC: World Bank, (accessed 25 July 2014)

14. Food and Agriculture Organization of the United Nations (FAO): Climate-Smart Agriculture Sourcebook. 2013, Rome: FAO

15. Climate Smart Agriculture: The 2nd Global Science Conference on Climate-Smart Agriculture. 2014, (accessed 5 August 2014), (accessed 5 August 2014)

16. Neufeldt H, Jahn M, Campbell BM, Beddington JR, DeClerck F, De Pinto A, Gulledge J, Hellin J, Herrero M, Jarvis A, LeZaks D, Meinke H, Rosenstock T, Scholes M, Scholes R, Vermeulen S, Wollenberg E, Zougmoré R: Beyond climate-smart agriculture: toward safe operating spaces for global food systems. Agric Food Secur. 2013, 2: 12-

17. Wageningen Statement: Climate-Smart Agriculture – Science for Action. The Global Science Conference on Climate-Smart Agriculture (GSCSA). 2011, (accessed 25 July 2014), (accessed 25 July 2014)

18. Carpenter S, Walker B, Anderies JM, Abel N: From metaphor to measurement: resilience of what to what?. Ecosystems. 2001, 4: 765-781.

19. Holling CS: Understanding the complexity of economic, ecological, and social systems. Ecosystems. 2001, 4: 390-405.

20. Walker B, Holling CS, Carpenter SR, Kinzig A: Resilience, adaptability, and transformability in social-ecological systems. Ecol Soc. 2004, 9: 5-

21. Obrist B: Multi-layered social resilience: a new approach in mitigation research. Progr Dev Stud. 2010, 10: 283-293.

22. Cumming GS: Spatial Resilience in Social-Ecological Systems. 2011, New York: Springer Science

23. Cabell JF, Oelofse M: An indicator framework for assessing agroecosystem resilience. Ecol Soc. 2012, 17: 18-

24. Easterling DR: Climate extremes: observations, modeling, and impacts. Science. 2000, 289: 2068-2074.

25. Battisti DS, Naylor RL: Historical warnings of future food insecurity with unprecedented seasonal heat. Science. 2009, 323: 240-244.

26. Bloom AJ, Burger M, Rubio Asensio JS, Cousins AB: Carbon dioxide enrichment inhibits nitrate assimilation in wheat and *Arabidopsis*. Science. 2010, 328: 899-903.

27. Yin X: Improving ecophysiological simulation models to predict the impact of elevated atmospheric CO_2 concentration on crop productivity. Ann Bot. 2013, 112: 465-75.

28. Vermeulen SJ, Aggarwal PK, Ainslie A, Angelone C, Campbell BM, Challinor AJ, Hansen JW, Ingram JSI, Jarvis A, Kristjanson P, Lau C, Nelson GC, Thornton PK, Wollenberg E: Options for support to agriculture and food security under climate change. Environ Sci Policy. 2012, 15: 136-144.

29. Gutierrez AP, Ponti L, d'Oultremont T, Ellis CK: Climate change effects on poikilotherm tritrophic interactions. Clim Change. 2007, 87 (1 Suppl): S167-S192.

30. Antón J, Cattaneo A, Kimura S, Lankoski J: Agricultural risk management policies under climate uncertainty. Glob Environ Change. 2013, 23: 1726-1736.

31. Jackson LE, Pulleman MM, Brussaard L, Bawa KS, Brown GG, Cardoso IM, de Ruiter PC, García-Barrios L, Hollander AD, Lavelle P, Ouédraogo E, Pascual U, Setty S, Smukler SM, Tscharntke T, van Noordwijk M: Social-ecological and regional adaptation of agrobiodiversity management across a global set of research regions. Glob Environ Change. 2012, 22: 623-639.

32. Stringer LC, Dougill AJ, Thomas AD, Spracklen DV, Chesterman S, Speranza CI, Rueff H, Riddell M, Williams M, Beedy T, Abson DJ, Klintenberg P, Syampungani S, Powell P, Palmer AR, Seely MK, Mkwambisi DD, Falcao M, Sitoe A, Ross S, Kopolo G: Challenges and opportunities in linking carbon sequestration, livelihoods and ecosystem service provision in drylands. Environ Sci Policy. 2012, 19–20: 121-135.

33. Minang PA, Duguma LA, Bernard F, Mertz O, van Noordwijk M: Prospects for agroforestry in REDD+landscapes in Africa. Curr Opin Environ Sustain. 2014, 6: 78-82.

34. Elliott J, Deryng D, Müller C, Frieler K, Konzmann M, Gerten D, Glotter M, Flörke M, Wada Y, Best N, Eisner S, Fekete BM, Folberth C, Foster I, Gosling SN, Haddeland I, Khabarov N, Ludwig F, Masaki Y, Olin S,

Rosenzweig C, Ruane AC, Satoh Y, Schmid E, Stacke T, Tang Q, Wisser D: Constraints and potentials of future irrigation water availability on agricultural production under climate change. Proc Natl Acad Sci U S A. 2013, 111: 3239-3244.

35. Meza FJ, Wilks DS, Gurovich L, Bambach N: Impacts of climate change on irrigated agriculture in the Maipo Basin, Chile: reliability of water rights and changes in the demand for irrigation. J Water Resour Plan Manag. 2012, 138: 421-430.

36. Nelson GC, Rosegrant MW, Palazzo A, Gray I, Ingersoll C, Robertson R, Tokgoz S, Zhu T, Sulser TB, Ringler C, Msangi S, You L: Food Security, Farming, And Climate Change to 2050: Scenarios, Results, Policy Options. 2010, (accessed 25 July 2014), Washington DC: International Food Policy Research Institute, (accessed 25 July 2014)

37. Thornton PK, Gerber PJ: Climate change and the growth of the livestock sector in developing countries. Mitig Adapt Strateg Glob Change. 2010, 15: 169-184.

38. Valin BH, Havlík P, Mosnier A, Herrero M, Schmid E, Obersteiner M: Agricultural productivity and greenhouse gas emissions: trade-offs or synergies between mitigation and food security?. Environ Res Lett. 2013, 8: 035019-

39. Kraxner F, Nordström EM, Havlík P, Gusti M, Mosnier A, Frank S, Valin H, Fritz S, Fuss S, Kindermann G, McCallum I, Khabarov N, Böttcher H, See L, Aoki K, Schmid E, Máthé L, Obersteiner M: Global bioenergy scenarios–future forest development, land-use implications, and trade-offs. Biomass Bioenergy. 2013, 57: 86-96.

40. Mbow C, Van Noordwijk M, Luedeling E, Neufeldt H, Minang PA, Kowero G: Agroforestry solutions to address food security and climate change challenges in Africa. Curr Opin Environ Sustain. 2014, 6: 61-67.

41. Adger WN: Social and ecological resilience: are they related?. Prog Hum Geogr. 2000, 24: 347-364.

42. Adger WN, Barnett J, Brown K, Marshall N, O'Brien K: Cultural dimensions of climate change impacts and adaptation. Nat Clim Chang. 2012, 3: 112-117.

43. Gray CL, Mueller V: Natural disasters and population mobility in Bangladesh. Proc Natl Acad Sci U S A. 2012, 109: 6000-6005.

44. Ifejika Speranza C: Drought coping and adaptation strategies: Understanding adaptations to climate change in agro-pastoral livestock production in Makueni District. Kenya. Eur J Dev Res. 2010, 22: 623-642.

45. Chindarkar N: Gender and climate change-induced migration: proposing a framework for analysis. Environ Res Lett. 2012, 7: 025601-

46. Villamor GB, van Noordwijk M, Djanibekov U, Chiong-Javier ME, Catacutan D: Gender differences in land-use decisions: shaping multifunctional landscapes?. Curr Opin Environ Sustain. 2014, 6: 128-133.

47. Vermeulen SJ, Challinor AJ, Thornton PK, Campbell BM, Eriyagama N, Vervoort JM, Kinyangi J, Jarvis A, Läderach P, Ramirez-Villegas J, Nicklin KJ, Hawkins E, Smith DR: Addressing uncertainty in adaptation planning for agriculture. Proc Natl Acad Sci U S A. 2013, 110: 8357-8362.

48. Reed MS, Evely AC, Cundill G, Fazey I, Glass J, Laing A, Newig J, Parrish B, Prell C, Raymond C, Stringer LC: What is social learning?. Ecol Soc. 2010, 15: r1-

49. Speelman EN, Groot JCJ, García-Barrios LE, Kok K, van Keulen H, Tittonell P: From coping to adaptation to economic and institutional change–trajectories of change in land-use management and social organization in a Biosphere Reserve community, Mexico. Land Use Policy. 2014, 41: 31-44.

50. The Davis Statement–Climate-Smart Agriculture Global Research Agenda: Science for Action. Climate-Smart Agriculture: Global Science Conference, 19–23 March 2013, University of California, Davis, USA. (accessed 26 July 2014), Climate-Smart Agriculture: Global Science Conference, 19–23 March 2013, University of California, Davis, USA. (accessed 26 July 2014)

51. de Graaf MA, van Groenigen KJ, Six J, Hungate B, van Kessel C: Interactions between plant growth and soil nutrient cycling under elevated CO_2: a meta-analysis. Glob Chang Biol. 2006, 12: 2077-2091.

52. Thornton PK, Jones PG, Alagarswamy G, Andresen J, Herrero M: Adapting to climate change: agricultural system and household impacts in East Africa. Agric Syst. 2010, 103: 73-82.

53. Haden VR, Niles MT, Lubell M, Perlman J, Jackson LE: Global and local concerns: what attitudes and beliefs motivate farmers to mitigate and adapt to climate change?. PLoS One. 2012, 7: e52882-

54. Stavi I, Lal R: Agroforestry and biochar to offset climate change: a review. Agron Sustain Dev. 2012, 33: 81-96.

55. Weber CL, Matthews HS: Food-miles and the relative climate impacts of food choices in the United States. Environ Sci Technol. 2008, 42: 3508-3513.

56. Garnett T, Appleby MC, Balmford A, Bateman IJ, Benton TG, Bloomer P, Burlingame B, Dawkins M, Dolan L, Fraser D, Herrero M, Hoffmann I, Smith P, Thornton PK, Toulmin C, Vermeulen SJ, Godfray HCJ: Sustainable intensification in agriculture: premises and policies. Science. 2013, 341: 33-34.

57. Godfray HC, Garnett T: Food security and sustainable intensification. Philos Trans R Soc Lond B Biol Sci. 2014, 369: 20120273-

58. Tittonell P, Giller KE: When yield gaps are poverty traps: the paradigm of ecological intensification in African smallholder agriculture. Field Crops Res. 2013, 143: 76-90.

59. George T: Why crop yields in developing countries have not kept pace with advances in agronomy. Glob Food Sec. 2014, 3: 49-58.

60. Newton AC, Johnson SN, Gregory PJ: Implications of climate change for diseases, crop yields and food security. Euphytica. 2011, 179: 3-18.

61. Mittler R, Blumwald E: Genetic engineering for modern agriculture: challenges and perspectives. Annu Rev Plant Biol. 2010, 61: 443-462.

62. Peleg Z, Reguera M, Tumimbang E, Walia H, Blumwald E: Cytokinin-mediated source/sink modifications improve drought tolerance and increase grain yield in rice under water-stress. Plant Biotechnol J. 2011, 9: 747-758.

63. Reguera M, Peleg Z, Blumwald E: Targeting metabolic pathways for genetic engineering abiotic stress-tolerance in crops. Biochim Biophys Acta. 2012, 1819: 186-194.

64. Cossani CM, Reynolds MP: Physiological traits for improving heat tolerance in wheat. Plant Physiol. 2012, 160: 1710-1718.

65. Reynolds M, Foulkes J, Furbank R, Griffiths S, King J, Murchie E, Parry M, Slafer G: Achieving yield gains in wheat. Plant Cell Environ. 2012, 35: 1799-1823.

66. Pask AJD, Reynolds MP: Breeding for yield potential has increased deep soil water extraction capacity in irrigated wheat. Crop Sci. 2013, 53: 2090-2104.

67. Pask A, Pietragalla J, Mullan D, Reynolds M: Physiological Breeding II: A Field Guide To Wheat Phenotyping. 2011, (accessed 26 July 2014), Mexico City: International Maize and Wheat Improvement Center (CIMMYT), (accessed 26 July 2014)

68. Araus JL, Cairns JE: Field high-throughput phenotyping: the new crop breeding frontier. Trends Plant Sci. 2014, 19: 52-61.

69. Peleg Z, Blumwald E: Hormone balance and abiotic stress tolerance in

crop plants. Curr Opin Plant Biol. 2011, 14: 290-295.

70. Reynolds M, Dreccer F, Trethowan R: Drought-adaptive traits derived from wheat wild relatives and landraces. J Exp Bot. 2007, 58: 177-186.

71. Sadok W, Naudin P, Boussuge B, Muller B, Welcker C, Tardieu F: Leaf growth rate per unit thermal time follows QTL-dependent daily patterns in hundreds of maize lines under naturally fluctuating conditions. Plant Cell Environ. 2007, 30: 135-146.

72. Wang X, Taub DR: Interactive effects of elevated carbon dioxide and environmental stresses on root mass fraction in plants: a meta-analytical synthesis using pairwise techniques. Oecologia. 2010, 163: 1-11.

73. Roy SJ, Tucker EJ, Tester M: Genetic analysis of abiotic stress tolerance in crops. Curr Opin Plant Biol. 2011, 14: 232-239.

74. Reynolds M, Manes Y, Izanloo A, Langridge P: Phenotyping approaches for physiological breeding and gene discovery in wheat. Ann Appl Biol. 2009, 155: 309-320.

75. Pinto RS, Reynolds MP, Mathews KL, McIntyre CL, Olivares-Villegas JJ, Chapman SC: Heat and drought adaptive QTL in a wheat population designed to minimize confounding agronomic effects. Theor Appl Genet. 2010, 121: 1001-1021.

76. Rebetzke GJ, Chenu K, Biddulph B, Moeller C, Deery DM, Rattey AR, Bennett D, Barrett-Leonard G, Mayer JE: A multisite managed environmental facility for targeted trait and germplasm phenotyping. Funct Plant Biol. 2013, 40: 1-13.

77. Olivares-Villegas JJ, Reynolds MP, McDonald GK: Drought-adaptive attributes in the Seri/Babax hexaploid wheat population. Funct Plant Biol. 2007, 34: 189-203.

78. Bouteillé M, Rolland G, Balsera C, Loudet O, Muller B: Disentangling the intertwined genetic bases of root and shoot growth in Arabidopsis. PLoS One. 2012, 7: e32319-

79. Tester M, Langridge P: Breeding technologies to increase crop production in a changing world. Science. 2010, 327: 818-822.

80. Langridge P, Fleury D: Making the most of "omics" for crop breeding. Trends Biotechnol. 2011, 29: 33-40.

81. Heslot N, Akdemir D, Sorrells ME, Jannink JL: Integrating environmental covariates and crop modeling into the genomic selection framework to predict genotype by environment interactions. Theor Appl Genet. 2014, 127: 463-480.

82. Philippot L, Hallin S: Towards food, feed and energy crops mitigating

climate change. Trends Plant Sci. 2011, 16: 476-80.

83. Reynolds MP, Hellin J, Govaerts B, Kosina P, Sonder K, Hobbs P, Braun B: Global crop improvement networks to bridge technology gaps. J Exp Bot. 2012, 63: 1-12.

84. Ceccarelli S, Galie A, Grando S: Participatory breeding for climate change-related traits. Genomics and Breeding for Climate-Resilient Crops: Concepts and Strategies, Volume 1. Edited by: Kole C. 2013, 331-376. Berlin: Springer Science & Business, 8

85. Opio C, Gerber P, Mottet A, Falcucci A, Tempio G, MacLeod M, Vellinga T, Henderson B, Steinfeld H: Greenhouse Gas Emissions From Ruminant Supply Chains: A Global Life Cycle Assessment. 2013, (accessed 26 July 2014), Rome: Food and Agriculture Organization of the United Nations (FAO), (accessed 26 July 2014)

86. Thornton PK, van de Steeg J, Notenbaert A, Herrero M: The impacts of climate change on livestock and livestock systems in developing countries: a review of what we know and what we need to know. Agric Syst. 2009, 101: 113-127.

87. Sutherst RW: Implications of global change and climate variability for vector-borne diseases: generic approaches to impact assessments. Int J Parasitol. 1998, 28: 935-945.

88. Hristov AN, Oh J, Lee C, Meinen R, Montes F, Ott T, Firkins J, Rotz A, Dell C, Adesogan A, Yang W, Tricarico J, Kebreab E, Waghorn G, Dijkstra J, Oosting S: FAO Animal Production and Health Paper No 177. Mitigation of Greenhouse Gas Emissions in Livestock Production: A Review of Technical Options for Non-CO2 Emissions. Edited by: Gerber PJ, Henderson B, Makkar HPS. 2013, (accessed 26 July 2014), Rome: Food and Agriculture Organization of the United Nations, (accessed 26 July 2014)

89. Niles MT: Achieving social sustainability in animal agriculture: challenges and opportunities to reconcile multiple sustainability goals. Sustainable Animal Agriculture. Edited by: Kebreab E. 2013, 193-211. Wallingford, UK: CABI

90. Herrero M, Thornton PK, Bernués A, Baltenweck I, Vervoort J, van de Steeg J, Makokha S, van Wijk MT, Karanja S, Rufino MC, Staal SJ: Exploring future changes in smallholder farming systems by linking socio-economic scenarios with regional and household models. Glob Environ Change. 2014, 24: 165-182.

91. Thornton PK, Herrero M: Potential for reduced methane and carbon dioxide emissions from livestock and pasture management in the tropics.

Proc Natl Acad Sci U S A. 2010, 107: 19667-19672.

92. Herrero M, Thornton PK: Livestock and global change: emerging issues for sustainable food systems. Proc Natl Acad Sci U S A. 2013, 110: 20878-20881.

93. Doré T, Makowski D, Malézieux E, Munier-Jolain N, Tchamitchian M, Tittonell P: Facing up to the paradigm of ecological intensification in agronomy: revisiting methods, concepts and knowledge. Eur J Agron. 2011, 34: 197-210.

94. Silvestri S, Osano P, de Leeuw J, Herrero M, Ericksen P, Hariuki J, Njuki J, Notenbaert A, Bedelian C: ILRI Position Paper. Greening Livestock: Assessing the Potential of Payment for Environmental Services in Livestock-Inclusive Agricultural Production Systems in Developing Countries. 2012, Nairobi: International Livestock Research Institute (ILRI)

95. Rufino MC, Thornton PK, Ng'ang'a SK, Mutie I, Jones PG, van Wijk MT, Herrero M: Transitions in agro-pastoralist systems of East Africa: impacts on food security and poverty. Agric Ecosyst Environ. 2013, 179: 215-230.

96. Thornton PK, Boone RB, Galvin KA, Burnsilver SB, Waithaka MM, Kuyiah J, Karanja S, González-Estrada E, Boone B: Coping strategies in livestock-dependent households in East and Southern Africa: a synthesis of four case studies. Hum Ecol. 2007, 35: 461-476.

97. Alvarez S, Rufino MC, Vayssières J, Salgado P, Tittonell P, Tillard E, Bocquier F: Whole-farm nitrogen cycling and intensification of crop-livestock systems in the highlands of Madagascar: an application of network analysis. Agric Syst. 2014, 126: 25-37.

98. Bengoumi M, Vias G, Faye B: Camel milk production and transformation in Sub-Saharan Africa. Desertification Combat and Food Safety: the Added Value of Camel Producers. Edited by: Faye B, Esenov P. 2005, 200-208. Amsterdam: IOS Press

99. Jones PG, Thornton PK: Croppers to livestock keepers: livelihood transitions to 2050 in Africa due to climate change. Environ Sci Policy. 2009, 12: 427-437.

100. Bustamante MMC, Nobre CA, Smeraldi R, Aguiar APD, Barioni LG, Ferreira LG, Longo K, May P, Pinto AS, Ometto JPHB: Estimating greenhouse gas emissions from cattle raising in Brazil. Clim Change. 2012, 115: 559-577.

101. Maia SMF, Ogle SM, Cerri CEP, Cerri CC: Effect of grassland management on soil carbon sequestration in Rondonia and Mato Grosso

states, Brazil. Geoderma. 2009, 149: 84-91.

102. Newbold CJ, López S, Nelson N, Ouda JO, Wallace RJ, Moss AR: Propionate precursors and other metabolic intermediates as possible alternative electron acceptors to methanogenesis in ruminal fermentation *in vitro*. Br J Nutr. 2007, 94: 27-35.

103. Norad: the Norwegian Agency for Development Cooperation. (accessed 26 July 2014), (accessed 26 July 2014)

104. Morse SS, Mazet JAK, Woolhouse M, Parrish CR, Carroll D, Karesh WB, Zambrana-Torrelio C, Lipkin WI, Daszak P: Prediction and prevention of the next pandemic zoonosis. Lancet. 2012, 380: 1956-1965.

105. US Agency for International Development (USAID): Pandemic Influenza and Other Emerging Threats. 2013 (accessed 26 July 2014), Washington, DC: USAID, (accessed 26 July 2014)

106. Foley P, Crosson P, Lovett DK, Boland TM, O'Mara FP, Kenny D: Whole-farm systems modelling of greenhouse gas emissions from pastoral suckler beef cow production systems. Agric Ecosyst Environ. 2011, 142: 222-230.

107. Rotz C, Montes F, Chianese DS: The carbon footprint of dairy production systems through partial life cycle assessment. J Dairy Sci. 2010, 93: 1266-1282.

108. Del Prado A, Misselbrook T, Chadwick D, Hopkins A, Dewhurst RJ, Davison P, Butler A, Schröder J, Scholefield D: $SIMS_{DAIRY}$: a modelling framework to identify sustainable dairy farms in the UK framework description and test for organic systems and N fertiliser optimisation. Sci Total Environ. 2011, 409: 3993-4009.

109. Reisinger A, Havlik P, Riahi K, Vliet O, Obersteiner M, Herrero M: Implications of alternative metrics for global mitigation costs and greenhouse gas emissions from agriculture. Clim Change. 2012, 117: 677-690.

110. Havlík P, Valin H, Herrero M, Obersteiner M, Schmid E, Rufino MC, Mosnier A, Thornton PK, Böttcher H, Conant RT, Frank S, Fritz S, Fuss S, Kraxner F, Notenbaert A: Climate change mitigation through livestock system transitions. Proc Natl Acad Sci U S A. 2014, 111: 3709-3714.

111. Collaborative Research on Adapting Livestock Systems to Climate Change. Feed the Future Innovation Lab Progress Report: Alignment with Feed the Future. 2013, (accessed 26 July 2014), Fort Collins, CO: Colorado State University, US Agency for International Development, (accessed 26 July 2014)

112. Rockström J, Steffen W, Noone K, Persson A, Chapin FS, Lambin EF, Lenton TM, Scheffer M, Folke C, Schellnhuber HJ, Nykyist B, de Wit CA, Hughes T, van der Leeuw S, Rodhe H, Sorlin S, Snyder PK, Costanza R, Svedin U, Falkenmark M, Karlberg L, Corell RW, Fabry VJ, Hansen J, Walker B, Liverman D, Richardson K, Crutzen P, Foley JA: A safe operating space for humanity. Nature. 2009, 461: 472-475.

113. Sutton MA, Oenema O, Erisman JW, Leip A, van Grinsven H, Winiwarter W: Too much of a good thing. Nature. 2011, 472: 159-161.

114. Smith P, Martino D, Cai Z, Gwary D, Janzen H, Kumar P, McCarl B, Ogle S, O'Mara F: Climate Change 2007: Mitigation. Edited by: Metz B, Davidson O, Bosch P, Dave R, Meyer L. 2007, 497-540. (26 July 2014), Agriculture, Cambridge, UK: Cambridge University Press,Contribution of Working Group III to the Fourth Assessment Report of the Intergovernmental Panel on Climate Change, (26 July 2014)

115. Halvorson AD, Del Grosso SJ, Alluvione F: Tillage and inorganic nitrogen source effects on nitrous oxide emissions from irrigated cropping systems. Soil Sci Soc Am J. 2010, 74: 436-445.

116. Akiyama H, Yan X, Yagi K: Evaluation of effectiveness of enhanced-efficiency fertilizers as mitigation options for N_2O and NO emissions from agricultural soils: meta-analysis. Glob Chang Biol. 2009, 16: 1837-1846.

117. Parkin TB, Hatfield JL: Enhanced efficiency fertilizers: effect on nitrous oxide emissions in Iowa. Agron J. 2014, 106: 694-702.

118. Hatfield JL, Parkin TB: Enhanced efficiency fertilizers: effect on agronomic performance of corn in Iowa. Agron J. 2014, 106: 771-780.

119. Kallenbach CM, Rolston DE, Horwath WR: Cover cropping affects soil N_2O and CO_2 emissions differently depending on type of irrigation. Agric Ecosyst Environ. 2010, 137: 251-260.

120. Singh A, Kaur J: Impact of conservation tillage on soil properties in rice-wheat cropping system. Agric Sci Res J. 2012, 2: 30-41.

121. Mandal B, Majumder B, Adhya TK, Bandyopadhyay PK, Gangopadhyay A, Sarkar D, Kundu MC, Choudhury SG, Hazra GC, Kundu S, Samantaray RN, Misra AK: Potential of double-cropped rice ecology to conserve organic carbon under subtropical climate. Glob Chang Biol. 2008, 14: 2139-2151.

122. Smith P, Haberl H, Popp A, Erb KH, Lauk C, Harper R, Tubiello FN, de Siqueira Pinto A, Jafari M, Sohi S, Masera O, Böttcher H, Berndes G, Bustamante M, Ahammad H, Clark H, Dong H, Elsiddig EA, Mbow C, Ravindranath NH, Rice CW, Robledo Abad C, Romanovskaya A, Sperling

F, Herrero M, House JI, Rose S: How much land-based greenhouse gas mitigation can be achieved without compromising food security and environmental goals?. Glob Chang Biol. 2013, 19: 2285-302.

123. Joseph S, Khoi DD, Hien NV, Anh ML, Nguyen HH, Hung TM, Yen NT, Thomsen MF, Lehmann J, Chia CH: North Vietnam Villages Lead the Way in the Use of Biochar: Building on an Indigenous Knowledge Base. (accessed 26 July 2014), International Biochar Initiative, (accessed 26 July 2014)

124. Hatfield JL, Parkin TB, Sauer TJ, Prueger JH: Mitigation opportunities from land management practices in a warming world: increasing potential sinks. Managing Agricultural Greenhouse Gases: Coordinated Agricultural Research through GRACEnet to Address Our Changing Climate. Edited by: Liebig MA, Franzluebbers AT, Follett RF. 2012, 487-504. Waltham, MA: Academic Press

125. Jensen ES, Peoples MB, Boddey RM, Gresshoff PM, Hauggaard-Nielsen H, Alves B, Morrison MJ: Legumes for mitigation of climate change and the provision of feedstock for biofuels and biorefineries: a review. Agron Sustain Dev. 2012, 32: 329-364.

126. Kim DG: Estimation of net gain of soil carbon in a nitrogen-fixing tree and crop intercropping system in Sub-Saharan Africa: results from re-examining a study. Agrofor Syst. 2012, 86: 175-184.

127. Corre-Hellou G, Dibet A, Hauggaard-Nielsen H, Crozat Y, Gooding M, Ambus P, Dahlmann C, von Fragstein P, Pristeri A, Monti M, Jensen ES: The competitive ability of pea-barley intercrops against weeds and the interactions with crop productivity and soil N availability. Field Crops Res. 2011, 122: 264-272.

128. Delgado JA, Kowalski K, Tebbe C: The first Nitrogen Index app for mobile devices: using portable technology for smart agricultural management. Comput Electron Agric. 2013, 91: 121-123.

129. Fischer RA, Edmeades GO: Breeding and cereal yield progress. Crop Sci. 2010, 50 (Suppl 1): S85-S98.

130. Grassini P, Eskridge KM, Cassman KG: Distinguishing between yield advances and yield plateaus in historical crop production trends. Nat Commun. 2013, 4: 2918-

131. van Ittersuma MK, Cassman KG, Grassini P, Wolfa J, Tittonell P, Hochman Z: Yield gap analysis with local to global relevance—a review. Field Crops Res. 2013, 143: 4-17.

132. van Warta JK, Kersebaum C, Peng S, Milner M, Cassman KG: Estimating crop yield potential at regional to national scales. Field Crops Res. 2012,

143: 34-43.

133. Reed MS, Podesta G, Fazey I, Geeson N, Hessel R, Hubacek K, Letson D, Nainggolan D, Prell C, Rickenbach MG, Ritsema C, Schwilch G, Stringer LC, Thomas AD: Combining analytical frameworks to assess livelihood vulnerability to climate change and analyse adaptation options. Ecol Econ. 2013, 94: 66-77.

134. Rogers EM: Diffusion of Innovations. 2003, New York: Free Press, 5

135. Kristjanson P, Reid RS, Dickson N, Clark WC, Romney D, Puskur R, MacMillan S, Grace D: Linking international agricultural research knowledge with action for sustainable development. Proc Natl Acad Sci U S A. 2009, 106: 5047-5052.

136. Spielman DJ, Ekboir J, Davis K: The art and science of innovation systems inquiry: applications to Sub-Saharan African agriculture. Technol Soc. 2009, 31: 399-405.

137. Vervoort JM, Thornton PK, Kristjanson P, Forch W, Ericksen PJ, Kok K, Ingram JSI, Herrero M, Palazzo A, Helfgott AES, Wilkinson A, Havlík P, Mason-D'Croz D, Jost C: Challenges to scenario-guided adaptive action on food security under climate change. Glob Environ Change. in press. doi:10.1016/j.gloenvcha.2014.03.001

138. Arbuckle JG, Prokopy LS, Haigh T, Hobbs J, Knoot T, Knutson C, Loy A, Mase AS, McGuire J, Morton LW, Tyndall J, Widhalm M: Climate change beliefs, concerns, and attitudes toward adaptation and mitigation among farmers in the Midwestern United States. Clim Change. 2013, 117: 943-950.

139. Prokopy LS, Haigh T, Mase AS, Angel J, Hart C, Knutson C, Lemos MC, Lo YJ, McGuire J, Morton LW, Perron J, Todey D, Widhalm M: Agricultural advisors: a receptive audience for weather and climate information?. Weather Clim Soc. 2013, 5: 162-167.

140. Baumgart-Getz A, Prokopy LS, Floress K: Why farmers adopt best management practice in the United States: a meta-analysis of the adoption literature. J Environ Manage. 2012, 96: 17-25.

141. Food and Agriculture Organization of the United Nations (FAO): Investing in Sustainable Agricultural Intensification: The Role of Conservation Agriculture. A Framework for Action. 2008, (accessed 26 July 2014), Rome: FAO, (accessed 26 July 2014)

142. Shaxson F, Kassam A, Friedrich T, Adekunle A: Conservation Agriculture: Looking beneath the Surface. 2008, (accessed 4 August 2014), Nakuru, Kenya: Food and Agriculture Organization of the United Nations (FAO). Conservation Agriculture Workshop, (accessed 4 August 2014)

143. Ndlovu PV, Mazvimavi K, An H, Murendo C: Productivity and efficiency analysis of maize under conservation agriculture in Zimbabwe. Agric Syst. 2014, 124: 21-31.

144. Giller KE, Witter E, Corbeels M, Tittonell P: Conservation agriculture and smallholder farming in Africa: the heretics' view. Field Crops Res. 2009, 114: 23-34.

145. Knowler D, Bradshaw B: Farmers' adoption of conservation agriculture: a review and synthesis of recent research. Food Policy. 2007, 32: 25-48.

146. Giller KE, Tittonell P, Rufino MC, van Wijk MT, Zingore S, Mapfumo P, Adjei-Nsiah S, Herrero M, Chikowo R, Corbeels M, Rowe EC, Baijukya F, Mwijage A, Smith J, Yeboah E, van der Burg WJ, Sanogo OM, Misiko M, de Ridder N, Karanja S, Kaizzi C, K'ungu J, Mwale M, Nwaga D, Pacini C, Vanlauwe B: Communicating complexity: integrated assessment of trade-offs concerning soil fertility management within African farming systems to support innovation and development. Agric Syst. 2011, 104: 191-203.

147. Arslan A, McCarthy N, Lipper L, Asfaw S, Cattaneo A: Adoption and Intensity of Adoption of Conservation Farming Practices in Zambia. ESA Working Paper No. 13-01. 2013, (accessed 26 July 2014), Rome: Food and Agriculture Organization of the United Nations, (accessed 26 July 2014)

148. Harvey CA, Chacón M, Donatti CI, Garen E, Hannah L, Andrade A, Bede L, Brown D, Calle A, Chará J, Clement C, Gray E, Hoang MH, Minang P, Rodríguez AM, Seeberg-Elverfeldt C, Semroc B, Shames S, Smukler S, Somarriba E, Torquebiau E, van Etten J, Wollenberg E: Climate-smart landscapes: opportunities and challenges for integrating adaptation and mitigation in tropical agriculture. Conserv Lett. 2013, 7: 77-90.

149. Jerneck A, Olsson L: Food first! Theorising assets and actors in agroforestry: risk evaders, opportunity seekers and "the food imperative" in Sub-Saharan Africa. Int J Agric Sustain. 2013, 12: 1-22.

150. Jerneck A, Olsson L: More than trees! Understanding the agroforestry adoption gap in subsistence agriculture: Insights from narrative walks in Kenya. J Rural Stud. 2013, 32: 114-125.

151. Adger N, Barnett J, Dabelko G: Climate and war: a call for more research. Nature. 2013, 498: 171-

152. Cash DW, Borck JC, Patt AG: Countering the loading-dock approach to linking science and decision making: comparative analysis of El Niño/ Southern Oscillation (ENSO) forecasting systems. Sci Technol Human Values. 2006, 31: 465-494.

153. Hayward T: Climate change and ethics. Nat Clim Chang. 2012, 2: 843-848.

154. Markowitz EM, Shariff AF: Climate change and moral judgment. Nat Clim Chang. 2012, 2: 243-247.

155. Barnett BJ, Barrett CB, Skees JR: Poverty traps and index-based risk transfer products. World Dev. 2008, 36: 1766-1785.

156. Lybbert TJ, McPeak J: Risk and intertemporal substitution: livestock portfolios and off-take among Kenyan pastoralists. J Dev Econ. 2012, 97: 415-426.

157. Carter MR: Designed for development impact: next generation index insurance for smallholder farmers. Protecting the Poor: A Microinsurance Compendium, Volume II: Part IV. General Insurance. Edited by: Churchill C, Matul M. 2012, 238-257. ISBN 978-92-2-125744-8 (accessed 26 July 2014), Geneva: International Labour Office, ISBN 978-92-2-125744-8 (accessed 26 July 2014)

158. McDowell JZ, Hess JJ: Accessing adaptation: multiple stressors on livelihoods in the Bolivian highlands under a changing climate. Glob Environ Change. 2012, 22: 342-352.

159. Adger WN: Vulnerability. Glob Environ Change. 2006, 16: 268-281.

160. Valdivia C, Seth A, Gilles JL, García M, Jiménez E, Cusicanqui J, Navia F, Yucra E: Adapting to climate change in Andean ecosystems: landscapes, capitals, and perceptions shaping rural livelihood strategies and linking knowledge systems. Ann Assoc Am Geogr. 2010, 100: 818-834.

161. Johnson L: Index insurance and the articulation of risk-bearing subjects. Environ Plan A. 2013, 45: 2663-2681.

162. Patt A, Suraez P, Hess U: How do small-holders understand insurance, and how much do they want it? Evidence from Africa. Glob Environ Change. 2010, 20: 153-161.

163. Traerup S: Informal networks and resilience to climate change impacts: a collective approach to index insurance. Glob Environ Change. 2012, 22: 255-267.

164. Chantarat S, Mude AG, Barrett CB, Carter MR: Designing index-based livestock insurance for managing asset risk in Northern Kenya. J Risk Insur. 2013, 80: 205-237.

165. BASIS, United States Agency for International Development, Food and Agriculture Organization of the United Nations, Micro-Insurance Innovation Facility of the International Labour Organization, Oxfam America: Feed the Future BASIS Assets and Market Innovation Lab:

Index Insurance: Feed the Future BASIS Assets and Market Innovation Lab: Index Insurance Innovation Initiative (I4). (accessed 26 July 2014), Davis, CA: BASIS Assets and Market Access CRSP, (accessed 26 July 2014)

166. International Livestock Research Institute (ILRI): Index Based Livestock Insurance.

167. apital, economic mobility and poverty traps. J Econ Inequal. 2011, 10: 299-342.

168. Karlan D, Osei RD, Osei-Akoto I, Udry C: Agricultural Decisions after Relaxing Credit and Risk Constraints (NBER Working Paper No 18463). 2012, 1-65. Cambridge, MA: National Bureau of Economic Research

169. McIntosh C, Sarris A, Papadopoulos F: Productivity, credit, risk, and the demand for weather index insurance in smallholder agriculture in Ethiopia. Agric Econ. 2013, 44: 399-417.

170. Carter MR, Lybbert TJ: Consumption versus asset smoothing: testing the implications of poverty trap theory in Burkina Faso. J Dev Econ. 2012, 99: 255-264.

171. Antón J, Kimura S, Lankoski J, Cattaneo A: A comparative study of risk management in agriculture under climate change. OECD Food, Agriculture and Fisheries Papers. 2012, 58: 1-89.

172. Klein D, Luderer G, Kriegler E, Strefler J, Bauer N, Leimbach M, Popp A, Dietrich JP, Humpenöder F, Lotze-Campen H, Edenhofer O: The value of bioenergy in low stabilization scenarios: an assessment using REMIND-MAgPIE. Clim Change. 2014, 123: 705-

173. Lotze-Campen H, von Lampe M, Kyle P, Fujimori S, Havlik P, van Meijl H, Hasegawa T, Popp A, Schmitz C, Tabeau A, Valin H, Willenbockel D, Wise M: Impacts of increased bioenergy demand on global food markets: an AgMIP economic model intercomparison. Agric Econ. 2014, 45: 103-116.

174. US Department of Agriculture (USDA): World Agricultural Supply and Demand Estimates (WASDE-459). 2008, (accessed 26 July 2014), Washington, DC: USDA, (accessed 26 July 2014)

175. Mitchell D: A Note on Rising Food Prices (Policy Research Working Paper 4682). 2008, (accessed 26 July 2014), Washington, DC: World Bank Development Prospects Group, (accessed 26 July 2014)

176. Zhang W, Yu EA, Rozelle S, Yang J, Msangi S: The impact of biofuel growth on agriculture: Why is the range of estimates so wide?. Food Policy. 2013, 38: 227-239.

177. Schmitz C, van Meijl H, Kyle P, Nelson GC, Fujimori S, Gurgel A, Havlik P, Heyhoe E, D'Croz DM, Popp A, Sands R, Tabeau A, van der Mensbrugghe D, von Lampe M, Wise M, Blanc E, Hasegawa T, Kavallari A, Valin H: Land-use change trajectories up to 2050: insights from a global agro-economic model comparison. Agric Econ. 2014, 45: 69-84.

178. Fujimori S, Matsuoka Y: Development of method for estimation of world industrial energy consumption and its application. Energy Econ. 2011, 33: 461-473.

179. van Meijl H, van Rheenen T, Tabeau A, Eickhout B: The impact of different policy environments on agricultural land use in Europe. Agric Ecosyst Environ. 2006, 114: 21-38.

180. Clarke L, Edmonds J, Krey V, Richels R, Rose S, Tavoni M: International climate policy architectures: overview of the EMF 22 International Scenarios. Energy Econ. 2009, 31 (Suppl 2): S64-S81.

181. Edmonds JA, Reilly JM: Global Energy: Assessing the Future. 1985, New York: Oxford University Press

182. Havlik P, Valin H, Mosnier A, Obersteiner M, Baker JS, Herrero M, Rufino MC, Schmid E: Crop productivity and the global livestock sector: implications for land use change and greenhouse gas emissions. Am J Agric Econ. 2012, 95: 442-448.

183. Lotze-Campen H: Improved data for integrated modeling of global environmental change. Environ Res Lett. 2011, 6: 041002-

184. Popp A, Dietrich JP, Lotze-Campen H, Klein D, Bauer N, Krause M, Beringer T, Gerten D, Edenhofer O: The economic potential of bioenergy for climate change mitigation with special attention given to implications for the land system. Environ Res Lett. 2011, 6: 034017-

185. Bondeau A, Smith P, Zaehle SON, Schaphoff S, Lucht W, Cramer W, Gerten D, Lotze-Campen H, Muller C, Reichstein M: Modelling the role of agriculture for the 20th century global terrestrial carbon balance. Glob Chang Biol. 2007, 13: 679-706.

186. Rost S, Gerten D, Bondeau A, Lucht W, Rohwer J, Schaphoff S: Agricultural green and blue water consumption and its influence on the global water system. Water Resour Res. 2008, 44: W09405

187. Leimbach M, Bauer N, Baumstark L, Edenhofer O: Mitigation costs in a globalized world: climate policy analysis with REMIND-R. Environ Model Assess. 2009, 15: 155-173.

188. Popp A, Krause M, Dietrich JP, Lotze-Campen H, Leimbach M, Beringer T, Bauer N: Additional CO_2 emissions from land use change—forest

conservation as a precondition for sustainable production of second generation bioenergy. Ecol Econ. 2012, 74: 64-70.

189. Hoefnagels R, Junginger M, Resch G, Matzenberger J, Panzer C: Report for IEA Bioenergy Task 40. Development of a Tool to Model European Biomass Trade. 2011, (accessed 26 July 2014), Utrecht, the Netherlands: Science Technology and Society, Copernicus Institute, University of Utrecht, (accessed 26 July 2014)

190. Sperling D, Yeh S: Toward a global low carbon fuel standard. Transp Policy. 2010, 17: 47-49.

191. Tokgoz S, Zhang W, Msangi S, Bhandary P: Biofuels and the future of food: competition and complementarities. Agriculture. 2012, 2: 414-435.

192. Lotze-Campen H, Popp A, Beringer T, Müller C, Bondeau A, Rost S, Lucht W: Scenarios of global bioenergy production: the trade-offs between agricultural expansion, intensification and trade. Ecol Modell. 2010, 221: 2188-2196.

193. van Dam J, Junginger M, Faaij APC: From the global efforts on certification of bioenergy towards an integrated approach based on sustainable land use planning. Renew Sustain Energy Rev. 2010, 14: 2445-2472.

194. Msangi S, Evans M: Biofuels and developing economies: is the timing right?. Agric Econ. 2013, 44: 501-510.

195. Beddington JR, Asaduzzaman M, Clark M: The role for scientists in tackling food insecurity and climate change. Agric Food Secur. 2012, 1: 10-

196. Verburg PH, Ellis EC, Letourneau A: A global assessment of market accessibility and market influence for environmental change studies. Environ Res Lett. 2011, 6: 034019-

197. Lugato E, Panagos P, Bampa F, Jones A, Montanarella L: A new baseline of organic carbon stock in European agricultural soils using a modelling approach. Glob Chang Biol. 2014, 20: 313-26.

198. Smith P, Davies CA, Ogle S, Zanchi G, Bellarby J, Bird N, Boddey RM, McNamara NP, Powlson D, Cowie A, van Noordwijk M, Davis SC, Richter DDB, Kryzanowski L, Wijk MT, Stuart J, Kirton A, Eggar D, Newton-Cross G, Adhya TK, Braimoh AK: Towards an integrated global framework to assess the impacts of land use and management change on soil carbon: current capability and future vision. Glob Chang Biol. 2012, 18: 2089-2101.

199. Ewert F, van Ittersum MK, Heckelei T, Therond O, Bezlepkina I, Andersen E: Scale changes and model linking methods for integrated assessment

of agri-environmental systems. Agric Ecosyst Environ. 2011, 142: 6-17.

200. Therond HB, Oomen R, Russell G, Ewert F: Using a cropping system model at regional scale: low-data approaches for crop management information and model calibration. Agric Ecosyst Environ. 2011, 142: 85-94.

201. Grace PR, Antle J, Aggarwal PK, Ogle S, Paustian K, Basso B: Soil carbon sequestration and associated economic costs for farming systems of the Indo-Gangetic Plain: a meta-analysis. Agric Ecosyst Environ. 2012, 146: 137-146.

202. de Groot RS, Alkemade R, Braat L, Hein L, Willemen L: Challenges in integrating the concept of ecosystem services and values in landscape planning, management and decision making. Ecol Complex. 2010, 7: 260-272.

203. Olsson L, Jerneck A: Farmers fighting climate change-from victims to agents in subsistence livelihoods. Wiley Interdiscip Rev Clim Chang. 2010, 1: 363-373.

204. You L, Wood S, Wood-Sichra U: Generating plausible crop distribution maps for Sub-Saharan Africa using a spatially disaggregated data fusion and optimization approach. Agric Syst. 2009, 99: 126-140.

205. Milman A, Bunclark L, Conway D, Adger WN: Assessment of institutional capacity to adapt to climate change in transboundary river basins. Clim Chang. 2013, 121: 755-770.

206. Rivington M, Matthews KB, Buchan K, Miller DG, Bellocchi G, Russell G: Climate change impacts and adaptation scope for agriculture indicated by agro-meteorological metrics. Agric Syst. 2013, 114: 15-31.

207. Trnka M, Brázdil R, Olesen JE, Eitzinger J, Zahradníček P, Kocmánková E, Dobrovolný P, Štěpánek P, Možný M, Bartošová L, Hlavinka P, Semerádová D, Valášek H, Havlíček M, Horáková V, Fischer M, Žalud Z: Could the changes in regional crop yields be a pointer of climatic change?. Agric For Meteorol. 2012, 166–167: 62-71.

208. Del Prado A, Crosson P, Olesen JE, Rotz CA: Whole-farm models to quantify greenhouse gas emissions and their potential use for linking climate change mitigation and adaptation in temperate grassland ruminant-based farming systems. Animal. 2013, 7: 373-385.

209. Gourdji SM, Sibley AM, Lobell DB: Global crop exposure to critical high temperatures in the reproductive period: historical trends and future projections. Environ Res Lett. 2013, 8: 024041-

210. Smith P, Olesen JE: Synergies between the mitigation of, and adaptation

to, climate change in agriculture. J Agric Sci. 2010, 148: 543-552.

211. Acosta L, Klein RJT, Reidsma P, Metzger MJ, Rounsevell MDA, Leemans R, Schröter D: A spatially explicit scenario-driven model of adaptive capacity to global change in Europe. Global Environ Change. 2013, 23: 1211-1224.

212. Schröter D, Cramer W, Leemans R, Prentice IC, Araújo MB, Arnell NW, Bondeau A, Bugmann H, Carter TR, Gracia CA, de la Vega-Leinert AC, Erhard M, Ewert F, Glendining M, House JI, Kankaanpää S, Klein RJT, Lavorel S, Lindner M, Metzger MJ, Meyer J, Mitchell TD, Reginster I, Rounsevell M, Sabaté S, Sitch S, Smith B, Smith J, Smith P, Sykes MT: Ecosystem service supply and vulnerability to global change in Europe. Science. 2005, 310: 1333-1337.

213. Sustainable farm Management Aimed at Reducing Threats to SOILs underclimate change: SmartSOIL. (accessed 26 July 2014), (accessed 26 July 2014)

214. Ingram J, Mills J, Frelih-Larsen A, Davis M: Uptake of Soil Management Practices and Experiences with Decisions Support Tools: Analysis of the Consultation with the Farming Community (Project No 289684). 2012 (accessed 26 July 2014), SmartSOIL, (accessed 26 July 2014)

215. White JW, Hoogenboom G, Kimball BA, Wall GW: Methodologies for simulating impacts of climate change on crop production. Field Crops Res. 2011, 124: 357-368.

216. Hansen J, Sato M, Ruedy R: Perception of climate change. Proc Natl Acad Sci U S A. 2012, 109: 415-423.

217. Moriondo M, Giannakopoulos C, Bindi M: Climate change impact assessment: the role of climate extremes in crop yield simulation. Clim Change. 2010, 104: 679-701.

218. Lobell DB: Errors in climate datasets and their effects on statistical crop models. Agr Forest Meteorol. 2013, 170: 58-66.

219. Challinor A, Martre P, Asseng S, Thornton P, Ewert F: Making the most of climate impacts ensembles. Nat Clim Change. 2014, 4: 77-80.

220. Peters GP, Andrew RM, Boden T, Canadell JG, Ciais P, Le Quéré C, Marland G, Raupach MR, Wilson C: The challenge to keep global warming below 2°C. Nat Clim Chang. 2012, 3: 4-6.

221. Jordan A, Rayner T, Schroeder H, Adger N, Anderson K, Bows A, Le Quéré C, Joshi M, Mander S, Vaughan N, Whitmarsh L: Going beyond two degrees? The risks and opportunities of alternative options. Clim Policy. 2013, 13: 751-769.

222. Petherick A: Seeking a fair and sustainable future. Nat Clim Change. 2014, 4: 81-83.

223. Lal R, Lorenz K, Hüttl RF, Schneider BU, von Braun J: Ecosystem Services and Carbon Sequestration in the Biosphere. 2013, New York: Springer Science

224. Qadir M, Noble AD, Chartres C: Adapting to climate change by improving water productivity of soils in dry areas. Land Degrad Develop. 2013, 21: 12-21.

225. Cochard R: Natural hazards mitigation services of carbon-rich ecosystems. Ecosystem Services and Carbon Sequestration in the Biosphere. Edited by: Lal R, Lorenz K, Hüttl RF, Schneider BU, von Braun J. 2013, 221-293. New York: Springer Science

226. Colomb V, Touchemoulin O, Bockel L, Chotte JL, Martin S, Tinlot M, Bernoux M: Selection of appropriate calculators for landscape-scale greenhouse gas assessment for agriculture and forestry. Environ Res Lett. 2013, 8: 015029-

227. Henry M, Tittonell P, Manlay RJ, Bernoux M, Albrecht A, Vanlauwe B: Agriculture, ecosystems and environment biodiversity, carbon stocks and sequestration potential in aboveground biomass in smallholder farming systems of western Kenya. Agric Ecosyst Environ. 2009, 129: 238-252.

228. Tittonell PA: 2007, Msimu wa Kupanda, targeting resources within diverse, heterogeneous and dynamic farming systems of East Africa, (accessed 26 July 2014)

229. Claessens L, Antle JM, Stoorvogel JJ, Valdivia RO, Thornton PK, Herrero M: A method for evaluating climate change adaptation strategies for small-scale farmers using survey, experimental and modeled data. Agric Syst. 2012, 111: 85-95.

230. Maskell LC, Crowe A, Dunbar MJ, Emmett B, Henrys P, Keith AM, Norton LR, Scholefield P, Clark DB, Simpson IC, Smart SM: Exploring the ecological constraints to multiple ecosystem service delivery and biodiversity. J Appl Ecol. 2013, 50: 561-571.

231. Setälä H, Bardgett RD, Birkhofer K, Brady M, Byrne L, de Ruiter PC, de Vries FT, Gardi C, Hedlund K, Hemerik L, Hotes S, Liiri M, Mortimer SR, Pavao-Zuckerman M, Pouyat R, Tsiafouli M, Putten WH: Urban and agricultural soils: conflicts and trade-offs in the optimization of ecosystem services. Urban Ecosyst. 2013, 17: 239-253.

232. Williams A, Hedlund K: Indicators and trade-offs of ecosystem services in agricultural soils along a landscape heterogeneity gradient. Appl Soil Ecol. 2014, 77: 1-8.

233. Wada Y, Wisser D, Eisner S, Flörke M, Gerten D, Haddeland I, Hanasaki N, Masaki Y, Portmann FT, Stacke T, Tessler Z, Schewe J: Multimodel projections and uncertainties of irrigation water demand under climate change. Geophys Res Lett. 2013, 40: 4626-4632.

234. Sood A, Muthuwatta L, McCartney M: A SWAT evaluation of the effect of climate change on the hydrology of the Volta River basin. Water Int. 2013, 38: 297-311.

235. Taylor RG, Scanlon B, Doll P, Rodell M, van Beek R, Wada Y, Longuevergne L, Leblanc M, Farniglietii JS, Edmunds M, Konikow L, Green TR, Chen J, Taniguchi M, Bierkens MFP, MacDonald A, Fan Y, Maxwell RM, Yechieli Y, Gurdak JL, Allen DM, Shamsudduha M, Hiscock K, Yeh PJF, Holman I, Treidel H: Ground water and climate change. Nat Clim Chan. 2013, 3: 321-330.

236. Dye PJ: Climate, forest, and streamflow relationships in South African afforested catchments. Commonw Forest Rev. 1996, 75: 31-38.

237. Khan S, Hanjra MA: Footprints of water and energy inputs in food production–global perspectives. Food Policy. 2009, 34: 130-140.

238. Duncan MJ: Hydrological impacts of converting pasture and gorse to pine plantation, and forest harvesting, Nelson, New Zealand. J Hydrol. 1995, 34: 15-41.

239. Bosch JM, Hewlett JD: A review of catchment experiments to determine the effect of vegetation changes on water yield and evapotranspiration. J Hydrol. 1982, 55: 3-23.

240. Scanlon BR, Jolly I, Sophocleous M, Zhang L: Global impacts of conversions from natural to agricultural ecosystems on water resources: quantity versus quality. Water Resour Res. 2007, 43: W034037

241. Jobbágy EG, Jackson RB: Patterns and mechanisms of soil acidification in the conversion of grasslands to forests. Biogeochemistry. 2003, 64: 205-229.

242. Jobbágy EG, Jackson RB: Groundwater use and salinization with grassland afforestation. Glob Chan Biol. 2004, 10: 1299-1312.

243. Farley KA, Kelly EF, Hofstede RGM: Soil organic carbon and water retention following conversion of grasslands to pine plantations in the Ecuadorian Andes. Ecosystems. 2004, 7: 729-739.

244. Schoups G, Hopmans JW, Young CA, Vrugt JA, Wallender WW, Tanji KK, Panday S: Sustainability of irrigated agriculture in the San Joaquin Valley, California. Proc Natl Acad Sci U S A. 2005, 102: 15352-15356.

245. Scott CA, Varady RG, Meza F, Montaña E, De Raga B, Luckman B,

Martius C: Science-policy dialogues for water security. Sci Policy Sustain Dev. 2012, 54: 37-41.

246. Kam SP, Badjeck M, Teh L, Teh L, Tran N: Working Paper 2012-24. Autonomous Adaptation to Climate Change by Shrimp and Catfish Farmers in Vietnam's Mekong River Delta. (accessed 26 July 2014), Penang, Malaysia: Worldfish

247. Duarte CM, Losada IJ, Hendriks IE, Mazarrasa I, Marbà N: The role of coastal plant communities for climate change mitigation and adaptation. Nat Clim Chang. 2013, 3: 961-968.

248. van Noordwijk M, Namirembe S, Catacutan D, Williamson D, Gebrekirstos A: Pricing rainbow, green, blue and grey water: tree cover and geopolitics of climatic teleconnections. Curr Opin Environ Sustain. 2014, 6: 41-47.

249. Spracklen DV, Arnold SR, Taylor CM: Observations of increased tropical rainfall preceded by air passage over forests. Nature. 2012, 489: 282-285.

250. McElrone AJ, Shapland TM, Calderon A, Fitzmaurice L, Paw UKT, Snyder RL: Surface renewal: an advanced micrometeorological method for measuring and processing field-scale energy flux density data. J Vis Exp. 2013, 82: e50666-

251. Shapland TM, Snyder RL, Paw UKT, McElrone AJ: Thermocouple frequency response compensation leads to convergence of the surface renewal alpha calibration. Agric For Meteorol. 2014, 189–190: 36-47.

252. Clausen J, Jägerskog A: SIWI Report No 31. Feeding a Thirsty World: Challenges and Opportunities for a Water and Food Secure Future. 2012, (accessed 26 July 2014), Stockholm: Stockholm International Water Institute (SIWI), (accessed 26 July 2014)

253. Mukherji A, Facon T, de Fraiture C, Molden D, Chartres C: Growing more food with less water: how can revitalizing Asia's irrigation help?. Water Policy. 2012, 14: 430-446.

254. Mbow C, Smith P, Skole D, Duguma L, Bustamante M: Achieving mitigation and adaptation to climate change through sustainable agroforestry practices in Africa. Curr Opin Environ Sustain. 2014, 6: 8-14.

255. Mbow C, van Noordwijk M, Prabhu R, Simons T: Knowledge gaps and research needs concerning agroforestry's contribution to sustainable development goals in Africa. Curr Opin Environ Sustain. 2014, 6: 162-170.

256. van Noordwijk M, Hoang MH, Neufeldt H, Öborn I, Yatich T: How Trees

and People Can Co-adapt to Climate Change: Reducing Vulnerability in Multifunctional Agroforestry Landscapes. 2011, (accessed 26 July 2014), Nairobi: World Agroforestry Centre, (accessed 26 July 2014)

257. Visseren-Hamakers I, Verkooijen P: The practice of interaction management: enhancing synergies among multilateral REDD+ institutions. Forest and Nature Governance: A Practice Based Approach (World Forests Series Vol 14). Edited by: Arts B, Behagel J, van Bommel S, de Koning J, Turnhout E. 2013, 133-149. New York: Springer Science

258. Ordonez JC, Luedeling E, Kindt R, Tata HL, Harja D, Jamnadass R, van Noordwijk M: Constraints and opportunities for tree diversity management along the forest transition curve to achieve multifunctional agriculture. Curr Opin Environ Sustain. 2014, 6: 54-60.

259. Geldenhuys CJ, Ham C, Ham H: Sustainable Forest Management in Africa: Some Solutions to Natural Forest Management Problems in Africa. Proceedings of the Sustainable Forest Management in Africa Symposium. 3–7 November 2008, Stellenbosch, South Africa. Edited by: Geldenhuys CJ, Ham C, Ham H. 2011, 1-538. (26 July 2014), Matieland, South Africa: Department of Forest and Wood Science, Stellenbosch University, (26 July 2014)

260. Thorlakson T, Neufeldt H: Reducing subsistence farmers' vulnerability to climate change: evaluating the potential contributions of agroforestry in western Kenya. Agric Food Secur. 2012, 1: 15-

261. van Paddenburg A, Bassi AM, Buter E, Cosslett CE, Dean A: Heart of Borneo: Investing in Nature for a Green Economy. 2012, (accessed 26 July 2014), WWF Heart of Borneo (HoB) Global Initiative, (accessed 26 July 2014)

262. Parrotta JA, Wildburger C, Mansourian S: A Global Assessment Report Prepared by the Global Forest Expert Panel on Biodiversity, Forest Management, and REDD+ (IUFRO World Series Vol 31). Understanding Relationships between Biodiversity, Carbon, Forests and People: The Key to Achieving REDD+ Objectives. 2012, (accessed 26 July 2014), Vienna: International Union of Forestry Research Organizations (IUFRO), (accessed 26 July 2014)

263. Visseren-Hamakers IJ, Gupta A, Herold M, Peña-Claros M, Vijge MJ: Will REDD+work? The need for interdisciplinary research to address key challenges. Curr Opin Environ Sustain. 2012, 4: 590-596.

264. Somorin OA, Visseren-Hamakers IJ, Arts B, Sonwa DJ, Tiani A-M: REDD+policy strategy in Cameroon: actors, institutions and governance. Environ Sci Policy. 2014, 35: 87-97.

265. Wrathall DJ: Migration amidst social-ecological regime shift: the search for stability in Garífuna villages of Northern Honduras. Hum Ecol. 2012, 40: 583-596.

266. Gibson MA, Gurmu E: Rural to urban migration is an unforeseen impact of development intervention in Ethiopia. PLoS One. 2012, 7: e48708-

267. Warner K, Afifi T, Henry K, Rawe T, Smith C, de Sherbinin A: Global Policy Report of Where the Rain Falls Project. Where the Rain Falls: Climate Change, Food and Livelihood Security, and Migration. 2012, (accessed 26 July 2014), Bonn: CARE France and United Nations University Institute for Environment and Human Security (UNU-EHS), (accessed 26 July 2014)

268. Jha AK, Bloch R, Lamond J: Cities and Flooding: A Guide to Integrated Urban Flood Risk Management for the 21st Century. 2012, (accessed 26 July 2014), Washington, DC: World Bank, (accessed 26 July 2014)

269. Laube W, Schraven B, Awo M: Smallholder adaptation to climate change: dynamics and limits in Northern Ghana. Clim Change. 2011, 111: 753-774.

270. Jankowska MM, Lopez-Carr D, Funk C, Husak GJ, Chafe ZA: Climate change and human health: Spatial modeling of water availability, malnutrition, and livelihoods in Mali, Africa. Appl Geogr. 2012, 33: 4-15.

271. López-Carr D: Agro-ecological drivers of rural out-migration to the Maya Biosphere Reserve, Guatemala. Environ Res Lett. 2012, 7: 045603-

272. Aide TM, Clark ML, Grau R, López-Carr D, Levy MA, Redo D, Bonilla-Moheno M, Riner G, Andrade-Núñez MJ, Muñiz M: Deforestation and reforestation of Latin America and the Caribbean (2001–2010). Biotropica. 2013, 45: 262-271.

273. DeFries RS, Rudel T, Uriarte M, Hansen M: Deforestation driven by urban population growth and agricultural trade in the twenty-first century. Nat Geosci. 2010, 3: 178-181.

274. Davis J, Lopez-Carr D: Migration, remittances and smallholder decision-making: implications for land use and livelihood change in Central America. Land Use Policy. 2014, 38: 319-329.

275. Winkels A: Migration, social networks and risk: the case of rural-to-rural migration in Vietnam. J Vietnam Stud. 2013, 7: 92-121.

276. Ho R, Milan A: "Where the Rain Falls" Project. Results from Huancayo Province, Junín Region (Report no 5). Rainfall, Food Security and Human Mobility. Case Study: Peru. 2012, (accessed 26 July 2014), Bonn: United Nations University Institute for Environment and Human

Security (UNU-EHS), (accessed 26 July 2014)

277. Schlenker W, Roberts MJ: Nonlinear temperature effects indicate severe damages to U.S. crop yields under climate change. Proc Natl Acad Sci U S A. 2009, 106: 15594-15598.

278. Auffhammer M, Hsiang SM, Schlenker W, Sobel A: Using weather data and climate model output in economic analyses of climate change. Rev Environ Econ Policy. 2013, 7: 181-198.

279. Urban D, Roberts MJ, Schlenker W, Lobell DB: Projected temperature changes indicate significant increase in interannual variability of U.S. maize yields. Clim Change. 2012, 112: 525-533.

280. Vital Signs. (accessed 26 July 2014), (accessed 26 July 2014)

281. Ericksen P, Thornton P, Notenbaert A, Cramer L, Jones P, Herrero M: Mapping Hotspots of Climate Change and Food Insecurity in the Global Tropics. (CCAFS Report no 5). 2011, (accessed 26 July 2014), Copenhagen: CGIAR Research Program on Climate Change, Agriculture and Food Security (CCAFS), (accessed 26 July 2014)

282. Holm P, Goodsite ME, Cloetingh S, Agnoletti M, Moldan B, Lang DJ, Leemans R, Moeller JR, Buendía MP, Pohl W, Scholz RW, Sors A, Vanheusden B, Yusoff K, Zondervan R: Collaboration between the natural, social and human sciences in global change research. Environ Sci Policy. 2013, 28: 25-35.

283. Leemans R, Solecki W: Redefining environmental sustainability. Curr Opin Environ Sustain. 2013, 5: 3-4.

284. Andersson E, Gabrielsson S: 'Because of poverty we had to come together': collective action for improved food security in rural Kenya and Uganda. Int J Agr Sustain. 2012, 10: 245-262.

285. Kristjanson P, Neufeldt H, Gassner A, Mango J, Kyazze FB, Desta S, Sayula G, Thiede B, Förch W, Thornton PK, Coe R: Are food insecure smallholder households making changes in their farming practices? Evidence from East Africa. Food Secur. 2012, 4: 381-397.

286. Lybbert TJ, Bell A: Why drought tolerance is not the new Bt. Nat Biotechnol. 2010, 28: 553-554.

287. Shah AK, Mullainthan S, Shafir E: Some consequences of having too little. Science. 2012, 338: 682-685.

288. Adaptation Fund. (accessed 26 July 2014), (accessed 26 July 2014)

289. International Fund for Agricultural Development (IFAD): Adaptation for Smallholder Agriculture Programme. (accessed 26 July 2014), Rome, (accessed 26 July 2014)

290. UNEP International Resource Panel Working Group on Reducing Emissions from Deforestation and Forest Degradation (REDD+) and a Green Economy: Building natural capital: how REDD+ can support a green economy. 2014, (accessed 26 July 2014), Nairobi: United Nations Environment Programme (UNEP), (accessed 26 July 2014)

291. Green Climate Fund. (accessed 26 July 2014), (accessed 26 July 2014)

292. United Nations Environment Programme (UNEP): UNEP Policy Series: Ecosystems Management Policy Brief 2-2010. The Role of Ecosystems in Developing a Sustainable 'Green Economy'. 2010, (26 July 2014), Nairobi: UNEP, (26 July 2014)

293. Bernard F, van Noordwijk M, Luedeling E, Villamor GB, Sileshi GW, Namirembe S: Social actors and unsustainability of agriculture. Curr Opin Environ Sustain. 2014, 6: 155-161.

Chapter 4

EX SITU CONSERVATION OF BIODIVERSITY WITH PARTICULAR EMPHASIS TO ETHIOPIA

Mohammed Kasso and Mundanthra Balakrishnan

Department of Zoological Sciences, Addis Ababa University, Addis Ababa, Ethiopia

ABSTRACT

Biodiversity encompasses variety and variability of all forms of life on earth that play a great role in human existence. Its conservation embraces maintenance, sustainable utilization, and restoration, of the lost and degraded biodiversity through two basic and complementary strategies called in situ and ex situ. Ex situconservation is the technique of conservation of all levels of biological diversity outside their natural habitats through different techniques like zoo, captive breeding, aquarium, botanical garden, and gene bank. It plays key roles in communicating the issues, raising awareness, and gaining widespread public and political support for conservation actions and for breeding endangered species in captivity for reintroduction. Limitations of ex situ conservation include maintenance of organisms in artificial habitats, deterioration of genetic diversity, inbreeding depression, adaptations to captivity, and accumulation of deleterious alleles. It has many constraints in terms of personnel, costs, and reliance on electric power sources. Ethiopia is considered to be one of the richest centers of genetic resources in the world. Currently, a number of stakeholders/actors are actively working on biodiversity conservation through ex situ conservation strategies by establishing gene banks, botanical garden, and zoo.

INTRODUCTION

According to the Convention on Biological Diversity, biodiversity refers to the variability among living organisms (animals, plants, and microorganisms) including inter alia, terrestrial, marine, and other aquatic ecosystems with their ecological complexes. In another expression, biodiversity encompasses the variety and variability of all forms of life on earth that play a great role in human existence [1, 2]. It also includes the ethnical value of biodiversity such as tradition and traditional knowledge of the indigenous and local communities [2] and the diversity within species (genetics), between species and of ecosystems [3].

Genetic diversity refers to the variation within species of any plant, animal or microbes in the functional units of heredity. Species diversity refers to the variety of species within a geographical area, which become central in the evaluation of diversity, and used as a point of reference in biodiversity conservation. Finally, ecosystem diversity refers to the variety of life forms in a given territory or area with all its functional ecological processes, which is often evaluated based on the diversity of all of its components [1].

Biodiversity is important for the maintenance of a healthy environment and used for direct human benefits like food, medicine, and energy. It is also used for recycling of different essential elements, for mitigation of pollution, for protection of watersheds, to mitigate soil erosion and to control excessive variations in climate and catastrophic events. For example, biodiversity provides different services free of charge worth of billions dollar every year for crucial well-being of the society. Some of these services are providing clean water and air, soil formation and protection, pollination, pest control, food, fuel, fibers, medicine, and construction and industry raw materials [4]. Agricultural biodiversity is another important component of biodiversity, which has a more direct link to the well-being and livelihood of mankind than other forms of biodiversity. Food plant and animal species have been collected, used, domesticated, and improved through traditional systems of selection over many generations [5].

However, today much of the lines of evidences are increasingly pointing out a significant global decline in biodiversity by numerous, varied, and interacting drivers [6]. More than half of the habitable surface of the earth has already been significantly altered by human activities. As a consequence, biodiversity of our planet is on the verge of decline and extinction despite our limited and incomplete knowledge on them [7]. Biodiversity loss and extinction processes can occur in two phases. The first phase is known as deterministic and often resulted from human threats such as habitat loss, fragmentation and degradation, direct exploitation of the species, competition from exotic and domestic species, and persecution and killing due to human animal conflicts. The second phase is known as deterministic that resulted from failures in mitigating threats that eventually result in very small, fragmented, and isolated remnant populations. Then these small remnant populations become vulnerable to a number of other, nonhuman caused threats mainly stochastic, genetic (genetic drift and inbreeding), and demographic events [8]. Thus, small, fragmented, isolated populations can find themselves being dragged into an extinction vortex whereby genetic and demographic stochastic events can cause the species to go extinct. During this second phase of the extinction process, very intensive management of populations and individuals is often necessary to prevent extinction [9].

Several human induced impacts are leading to a mass extinction process affecting global biodiversity. The major reasons for rapid diminishing of biodiversity are attributed to conversion of land for agriculture, wild fires, poor management of available land, over-exploitation for food, fuel-wood, medicine, construction, overgrazing by cattle, displacement and loss of landraces, lower yielding varieties, pests and diseases, global climate change, pollution (e.g., acid rain), and gap of scientific knowledge on some of the biological resources [1, 7, 10]. Human beings are destroying biodiversity, particularly during livelihood activities with or without knowledge of the consequences of their actions [6]. Agriculture is one of the most important land-use that results in detrimental environmental consequences from increased use of fertilizers and biocides, land draining, irrigation, and the loss of many biodiversity-rich landscape features [6]. There are many threats to biodiversity as a result of agricultural practice through changes in land-use, replacement of traditional varieties by modern cultivars, agricultural intensification, increased population, poverty, land degradation, and environmental changes (including climate change) [5]. Recent estimates indicate that humans use more than 40% of the terrestrial components and significantly modified global biodiversity [7]. As a consequence, many species of living organisms are classified as threatened today and this has become a central concern for conservation [11].

Conserving biodiversity has economic, social, and cultural values. Conservation of biodiversity is integral to the biological and cultural inheritance of many people and the critical components of healthy ecosystems that are used to support economic and social developments. Moreover, it is used to maintain the earth's genetic library from which society has derived the basis of its agriculture and medicine [5, 12]. The twenty-first century is predicted to be an era of bioeconomy driven by advances of bioscience and biotechnology. Bio-economy may become the fourth economy form after agricultural, industrial, and information technology economies, having far-reaching impacts on sustainable development in agriculture, forestry, environmental protection, industry, food supply, health care and other micro-economy aspects. Thus, a strategic vision for conservation and sustainable use of biodiversity in the 21st century is of far-reaching significance for sustainable development economy and society [13].

Biodiversity conservation refers to the management of human use of biodiversity in order to get the greatest sustainable benefit to present and future generations. Thus, conservation of biodiversity embraces the protection, maintenance, sustainable utilization, restoration, and enhancement of biodiversity [1]. Biodiversity conservation mainly focuses on genetic conservation with its diverse life-support systems (ecosystems) for the connotation of human well-being [3].

Conservation techniques can be grouped into two basic, complementary strategies: in situ and ex situ [14]. As also outlined in the articles 8 and 9 of the Convention of Biological Diversity (CBD), biodiversity is conserved by two major methods called in situ and ex situ. The conservation efforts, either in situ or ex situ, involve the establishment and management of protected areas and relevant research institutes or academic institutions, which establish and manage arboreta, botanical or zoological gardens, tissue culture, and gene banks [1]. The concept of ex situ conservation is fundamentally different from that of in situ conservation; however, both are important complementary methods for conservation of biodiversity. The principal difference (and hence the reason for the complementarities) between the two lies in the fact that ex situ conservation implies the maintenance of genetic materials outside of the "normal" environment where the species has evolved and aims to maintain the genetic integrity of the material at the time of collection, whereas in situ conservation (maintenance of viable populations in their natural surroundings) is a dynamic system, which allows the biological resources to evolve and change over time through natural or human-driven selection processes [5].

In situ conservation is defined as conservation of ecosystems and natural habitats, the maintenance of viable populations of the species in their natural surroundings and, in the case of the cultivated species, in the surroundings where they have developed their distinctive properties. In situ conservation can be done in farmers' fields, in pasture lands, and in protected areas [15]. For cultivated species, in situ conservation concerns the maintenance of the local intra- and inter-population diversity available in various ecological and geographical sites [1, 16]. Thus, it allows ongoing host-parasite coevolution, which is likely to provide material resistance to pests and diseases, and CBD recognized it as a primary approach to conserve biodiversity [4]. However, in situ conservation has certain limitations like more difficult access to breeders requiring the application of its complimentary technique. For example, some of the natural habitats or wild habitats are very risky when compared to relatively safe captive environment [9]. The second biodiversity conservation technique receiving the most attention to conserve biodiversity is ex situ. Ex situ conservation techniques are mostly used to be applied to species with one or some of the following characteristics: endangered species, species with a past, present or future local importance, species of ethno-botanical interest, species of interest for the restoration of local ecosystems, symbolic local species, taxonomically isolated species, and monotypic or oligotypic genera [17]. Intensive conservation and management of populations and individuals can come in many different forms, like translocation, breeding in a fenced wild habitat, supplementary feeding, captive hand rearing of young of wild parents to become pregnant sooner, and captive breeding [9].

EX SITU CONSERVATION

Ex situ conservation is a technique of conservation of biological diversity outside its natural habitats, targeting all levels of biodiversity such as genetic, species, and ecosystems [1, 2, 16]. Its concept was developed earlier before its official adoption under the Convention on Biological Diversity signed in 1992 in Rio de Janeiro [2]. In general, ex situ conservation is applied as an additional measure to supplement in situconservation, which refers to conservation of biological diversity in its natural habitats [16]. In some cases, ex situ management will be central to a conservation strategy and in others it will be of secondary importance [18]. Broadly, ex situ conservation includes a variety of activities, from managing captive populations, education and raising awareness, supporting research initiatives and collaborating with in situ efforts [19]. It is used as valuable tools in studying and conserving biological resources (plants, animals, and microorganisms) for different purposes [2] through different techniques such as zoos, captive breeding, aquarium, botanical gardens, and gene banks [16].

Types of Ex Situ Conservation

Zoos

Zoos or zoological gardens or zoological parks in which animals are confined within enclosures or semi-natural and open areas, displayed to the public, and in which they may also breed. They are considered by universal thinkers and environmentalists as important means of conserving biodiversity [19–21]. Zoos attract as many as 450 million visitors each year and so are uniquely placed to have very large educational and economic values [22]. Zoos not only act as places of entertainment and observing animal behavior, but are also as institutions, museums, research laboratories, and information banks of rare animals [23]. Although some people dislike zoos, many people enjoy them. Over the last several decades, zoos have made significant progress in its cooperative management of ex situ populations of a variety of biodiversity [9]. Zoos breed many endangered species to increase their numbers. Such captive breeding in zoos has helped to save several species from extinction [19]. Management of animals in zoos includes animal identification, housing, husbandry, health, nutrition as well as addressing and ways of interaction with the public [20]. There are various processes and mechanisms used to determine whether a species or taxon is included within a zoo's collection plan. The frequently used criteria include how the species is valued, according to its uniqueness, contribution to research or education, and conservation status [19]. Zoos help the animal to secure food, shelter, social contact and mates, and to be motivated by desire (appetitive behavior), which is reinforced by pleasure (consummative behavior) [21].

In the past, some zoos paid little attention to the welfare of the animals, and some zoos today have poor environments for animals [24]. They were also once reliant on harvest from the wild to populate their exhibits and reliance on continued wild collection to breeding closed populations [23]. Many zoo animals also became endangered or extinct due to visitor disturbances, unfavorable climate and due to insufficient space [20]. From this aspect, many scholars state on the negative features of keeping animals in zoo as it causes pain, stress, distress, sufferings and evolutionary impacts [21]. Animal welfare, education, conservation, research, and entertainment are major goals of modern zoos, but these can be in conflict. For example, visitors enjoy learning about and observing behavior in captive animals, but visitors often want to observe and interact with the animals in close proximity. Unfortunately, proximity to and interactions with humans induce stress for many species [25]. The same is true for Addis Ababa Lion Zoo Park.

However, progressive zoos are engaged in education, research, and conservation, with the aim of maintaining healthy animals, which behave as if they are in their natural habitats [24]. The current paradigm for managing essential populations is to minimize the rate of genetic decay, slow adaptation to the captive environment, and retain typical behaviors [23, 26]. It is widely accepted that the more generations a population spends in captive breeding, the less suitable it is for attempted restorations in the wild. Hence, population management is designed not to deplete too quickly the resource obtained from the founders [23]. Thus, for true sustainability of the species for the purpose of conservation, display, education, and research, constant refreshing of populations is required [9, 23]. Majority of the current breeding programs base on the genetic management of populations by the analysis of individual pedigrees in order to minimize kinship [9].

Captive Breeding

Captive breeding is an integral part of the overall conservation action plan for a species that helps to prevent extinction of species, subspecies, or population. It is an intensive management practice for threatened individuals, populations, and species by anthropogenic and natural factors [9]. In small and fragmented populations, even if the human caused threats could be magically reversed, the species would still have a high probability of extinction by random demographic and genetic events, environmental variations, and catastrophes. Thus, under sufficient knowledge on the biology and husbandry of the species, captive breeding helps individuals in the relative safety of captivity, under expert care and sound management by providing an insurance against extinction [9]. Stock for reintroduction or reinforcement efforts, opportunities for education, raising of awareness, scientific and husbandry research, and other contributions to conservation are also possible through captive breeding [9, 27].

Environmental enrichment strategies are used to improve both physiological and psychological welfares of captive animals, which can be achieved by increasing the expression of natural behavior and decreasing abnormal behaviors. Successful environmental enrichment includes the improvement of enclosure design and the provision of feeding devices, novel objects, appropriate social groupings, and other sensory stimuli [27]. The minimum requirement for successful ex situ management, particularly in the captive populations, is the inclusion of as much of the genetic diversity present in wild populations. Genetic sustainability (retention of 90% of the gentic diversity of the wild population for 100 years) in captive breeding is maintained if consideration is given on number of founders, population growth rate, effective population size, and duration of the captive program [19]. However, even if at least 30 founders in captive breeding are recommended to ensure the representation of large enough proportion of the genetic diversity of the wild population, for critically endangered species, actively removing individuals from the wild population to serve as founders may compromise the survival of the wild population [9]. For example, the Arabian Oryx captive breeding program was based on fewer founders and grew to a couple of thousand individuals through breeding management, which helped to reduce risks.

However, there are several challenges (biological and environmental) that are limiting factors to the attainment of the goal of captive breeding for many species [19]. One of the major challenges is a circular consequence of small-population management that has inherent genetic and demographic problems due to genetic diversity loss and demographic stochasticity [19]. In addition, individuals that are well adapted to the circumstances in captivity may also be less well adapted to the circumstances in the wild and may show lower fitness upon reintroduction [28]. Most notably, within the captive environment, housing and husbandry will also have significant impacts on birth and death rates [19].

Aquarium

An aquarium is an artificial habitat for water-dwelling animals. It can also be used to house amphibians or large marine mammals and plant species for tourist attractions. It is usually found in zoos or marine parks with different size. The 15,750 described species of freshwater fish comprise around 25% of living vertebrate species diversity and a key for global economic and nutritional resources of which more than 11% is threatened (60-extinct, 8-extinct in the wild and 1679-threatened) [22]. Fresh waters (0.3%) of available global surface water support 47–53% of all extant fish species that are threatened by overfishing, pollution, habitat loss, damming, alien invasive species, and climate change. This requires world's zoos and aquariums to identify the potential targets (species or areas) for in situ and ex situ conservation program [18, 22]. Aquarium is used to admire at home by hobbyists, to portray as public exhibits, to provide large quantities of human food and animal fodder [18].

Fishes are often overlooked within the development of conservation priorities. This leads to the low focus on meaningful conservation efforts rather than giving more attention for their importance to food supply and livelihoods. For example, it provides job opportunity for over 60 million people, as source of food for over 200 million people in Africa, for US$1.5 billion income from trade of 4000 species global ornamental fish industry, and many are displayed in the world's public zoos and aquaria to a global audience of as many as 450 million people per year [22, 29]. However, despite the clear value of freshwater fish diversity, wetland habitats and their associated freshwater-fish species continue to be lost or degraded at an alarming rate [30]. One recommendation is for aquariums to set up sustainable breeding program that prioritizes threatened species (VU, EN, and CR) and those classified as EW to support species conservation in situ and aid the recovery of species via collaborative reintroduction or translocation efforts when appropriate [22].

Botanical Gardens

Botanical gardens consist of living plants, grown out of doors or under glass in greenhouses and conservatories. They are used to grow and display plants primarily for scientific and educational purposes. They also include herbarium, lecture rooms, laboratories, libraries, museum and experimental or research plantings. It can be taxonomic collection of a particular family, genus or group of cultivars, native plants, wild relatives, medicinal, aromatic, or textile plants [4]. There are over 2,000 botanic gardens, holding 80,000 plant species in their living collections and receiving hundreds of millions of visitors per annum [14, 31]. Furthermore, they have valuable and distinctive mix of officials dedicated to plant research, systematics, conservation education, and public awareness [31]. They are now extremely well networked both among themselves and with other professionals, conservation organizations, and nongovernmental organizations (NGOs) [31]. They provide different services for sectors that utilize and conserve plant diversity like agriculture, forestry, pharmaceutical and biofuel industries, protected area management, and ecotourism. They have a unique opportunity as visitor attraction places and scientific institutions for documentation and conservation of plant diversity by shaping and mobilizing citizens to the current environmental challenges [31]. They also play a great role in attaining target of the Global Strategy for Plant Conservation for 2020 to cultivate 75% of the world's threatened plant species in ex situ [14]. Botanical gardens give opportunity for arable plants to be grown under relatively modified environmental conditions (intense cultivation, relatively high fertility, and high levels of disturbance) [14].

However, most of the cultivated taxa are held in a small number of collections and mostly only in small populations. Lack of genetic exchange and stochastic processes in small populations make them susceptible to detrimental genetic effects [14]. The low number of ex situ populations in most botanical gardens poses a fundamental problem for conservation. The total ex situ breeding collection is therefore very small with respect to the stated aim of conserving regional gene pools [14]. The striking lack of information on source populations casts doubt on the value of using such ex situ populations for potential reintroductions. They also require testing for fitness and similarity to wild populations before they are brought to the field [14]. Thus, conservation actions of botanic gardens such as training and capacity building, needs to be better understood and better coordinated [31].

Gene Banks

Genome resource banking is another management technique used for biodiversity conservation. Different types of gene banks have been established for the storage of biodiversity, depending on the type of materials conserved. These include seed banks (for seeds), field gene banks (for live plants), in vitro gene banks (for plant tissues and cells), pollen, chromosome, and deoxyribonucleic acid (DNA) banks for animals (living sperm, eggs, embryos, tissues, chromosomes, and DNA) that are held in short term or long term laboratory storage; usually cryopreserved or freeze-dried [32]. Currently, there are about 7.4 million PGRFA accessions conserved in over 1750 gene banks

[5]. The Genome 10K (G10K) project aims to sequence the genomes of 10,000 vertebrates, of which 4,000 will be fish. To date, 60 fish genomes have been sequenced in laboratories worldwide and added to the database, and a further 100 are targeted. The Frozen Ark database holds details of 28,060 frozen DNA samples. Among these 6,997 are from species listed in the IUCN Red List [18]. The principal aim of gene bank conservation is to maintain genetic diversity alive as long as possible and to reduce the frequency of regeneration that may cause the loss of genetic diversity [5].

With the rapid development in the field of molecular genetics and genomics, DNA is becoming more and more in demand for molecular studies and is one of the most requested materials from gene banks. Establishing DNA storage facility as a complementary "backup" to traditional ex situ collections has been suggested [5]. Some efforts have been made to establish DNA banks for endangered animals [33], and a few plant DNA banks in different parts of the world such as the Missouri Botanic Garden, Kew Royal Botanic Garden and Australian Plant DNA Bank. Many research groups are already developing their own archives of extracted genomic DNA. Recently, the Global Biodiversity Information Facility in Germany has established a DNA bank network, which provides DNA samples of complementary collections (microorganisms, protists, plants, algae, fungi, and animals) [5].

Seeds are usually the most convenient and easiest material to collect and to maintain in a viable state for long periods of time and that makes it preferred for conservation in gene banks [14, 34]. Seed banking techniques rely on the storage of dried seeds of threatened or other plants at low temperatures as the most important factors influencing seed longevity are temperature, seed moisture content, and relative humidity [14, 35, 36]. Seeds are typically conserved at moisture content between 3 and 7 percent and stored at 4 degrees Celsius for short-term conservation, and between −18 and −20 degrees Celsius for long-term conservation [5, 34]. Current research is showing that there exists variability in seed longevity for different species being conserved under similar conditions. In addition, it has been found that the type of seed (endospermic or non-endospermic) and intraspecific variation may also affect accessions longevity [5]. In addition, high initial quality seeds are a major prerequisite for ensuring seed longevity in seed banks [5].

Plants that cannot be conserved as seeds because of their recalcitrant nature (i.e. seeds that are desiccation and/or cold sensitive) or are clonally propagated are traditionally conserved as live plants in ex situ field gene banks. But, field gene banks present real logistical challenges; they require large areas and are costly, they are vulnerable to pests and diseases, natural disasters, political unrest, extreme weather, fire, vandalism, and theft, and they often are at risk due to policy changes on land use [5]. In vitro conservation refers to one type of gene bank known as slow-growth conservation method. It involves culturing of different parts of the plant (meristem, tissues, and cells) into pathogen-free sterile culture in a synthetic medium with growth retardants, which has been cited as a good way of complementing and providing backup to field collections [5]. The other genome conservation technique is cryopreservation, in which living tissues are conserved at very low temperatures (−196°C) in liquid nitrogen to arrest mitotic and metabolic activities [5]. It is now realized that cryopreservation method can offer greater security for long-term, cost effective conservation of plant genetic resources, including orthodox seeds [5]. The storage in liquid nitrogen clearly prolonged shelf life of lettuce seeds with half-lives projected as 500 and 3400 years for fresh lettuce seeds stored in the vapor and liquid phases of liquid nitrogen, respectively [5].

Advantages of Ex Situ Conservation

It is generally preferred to conserve threatened species in situ, because evolutionary processes are more likely to remain dynamic in natural habitats [14, 19]. However, considering the rate of habitat loss worldwide, ex situ cultivation is becoming increasingly important [14]. Further more, as many of the taxa are located outside natural parks or reserves, in situ measures are not enough to assure their conservation. Translocation, introduction, reintroduction, and assisted migrations are species conservation strategies that are attracting increasing attention, especially in the face of climate change [37].

As approximately 450 million people per year visit zoos and aquaria globally, their education and marketing services play a key role in communicating the issues, raising awareness, changing behavior, and gaining widespread public and political support for conservation actions. Zoos support conservation by educating the public, raising money for conservation programs, developing technology that can be used to track wild populations, conducting scientific research, advancing veterinary medicine, and developing animal handling techniques [22]. By studying animals in captivity and applying that knowledge to their husbandry, zoos can provide valuable and practical information that may be difficult or impossible to gather from the wild [24]. Zoos and aquaria have significant roles to play in improveing public awareness of the issue facing species and their habitats; for example, through presentation of maps and photographs of species recently extinct as a result of anthropogenic impacts. A similar display of threatened species, even if not currently in the collections of the zoo, would help convey to the public the magnitude of the threat facing the species [18, 22]. It also reaches a wide cross-section of the society, because zoo audiences are not limited to those who are already passionately interested in wildlife and because many zoo visitors are children. Some of these children may become committed conservationists. Some may grow up to be oil company tycoons, politicians, or movie stars, with great potential influence. Some may even live next door to a poacher or wildlife dealer. Thus, instilling an interest in conservation of wildlife in people from all walks of life while they are young is one vital role zoos can play [38]. It is often claimed that zoos perform valuable conservation work by breeding endangered species and returning them to the wild. Zoos can also be used for businesses that make money. This means that animals are often bred for commercial purposes because the public like to see new-born animals. Such breeding leads to a surplus of animals, and in order to keep numbers down sold to private collectors, circuses, or even research laboratories. A zoo with good and attractive entertainments encourages initial visits and subsequent returns to the zoo, which is used to get more revenue for conservation efforts, research, and general animal care and welfare and also to develop more positive perceptions of animals in zoos and become more supportive of conservation efforts [25, 39].

Disadvantages of Ex Situ Conservation

Some ex situ conserved collections showed lower resistance levels, although still others showed higher resistance levels than their in situ conserved counterparts mainly due to the high evolutionary drive and complex nature of evolutionary scenario [40].

The behavior of animals in the zoo may be affected by the frequent arrival of large number of people, who are unfamiliar to the animals [41]. Animals housed in artificial habitats are confronted by a wide range of potentially offensive environmental challenges such as artificial lighting, exposure to loud or aversive sound, arousing odors, and uncomfortable temperatures or substrates. In addition, confinement-specific stressors such as restricted movement, reduced retreat space, forced proximity to humans, reduced feeding opportunities, maintenance in abnormal social groups, and other restrictions of behavioral opportunity are considered [42]. However, over the course of the twentieth century, as knowledge of wildlife biology improved, zoo animals began to be kept in more natural surroundings and social groupings, and diets and veterinary care began to improve. Thus, survival and breeding rates of captive populations improved [38]. Evidence mainly from studies of rodents and primates strongly indicates that prenatal stress can impair stress-coping ability and is able to cause a disruption of behavior in aversive or conflict-inducing situations. Prenatally stressed animals show retarded motor development, reduced exploratory and play behavior, and impairments of learning ability, social behavior, and sexual and maternal behavior. Prenatal stress may also affect the sex ratio at birth and the reproductive success [43].

Although populations of some species managed in ex situ may have the best hope for their long-term survival, they might be challenged if not properly managed during translocation and reintroduction with the effects of climate change [22]. Some species may lose their biological integrity particularly on morphology. For example, an experimental study on black-footed ferrets (Mustela nigripes) in ex situ indicated a decrease of 5–10% body size than pre-captive, in situ animals [44]. In other words, the small cage size and environmental homogeneity inhibit mechanical stimuli necessary for long bone development. Thus, in the absence of such an environment, "unnatural" morphologies can result that may contribute to poor fitness or perhaps even for domestication and reintroduction and relocation [44]. It would be very difficult to reintroduce some zoo-reared animals to their natural habitats because, after generations of captivity, many have lost the necessary skills to survive in their original habitats [22]. For naturally out-breeding species, the high levels of inbreeding in captivity often have negative effects on life history traits related to reproduction and survival [11]. It makes the population in captivity deteriorate due to loss of genetic diversity, inbreeding depression, genetic adaptations to captivity, and accumulation of deleterious alleles [17]. For plants, ecological shifts, small population size, genetic drift, inbreeding, and gardener-induced selection may negatively affect population structure after several generations of ex situcultivation [16, 45, 46]. These factors could seriously put at risk the success of ex situ conservation [17].

Captive breeding of threatened species has used increasingly sophisticated technologies and protocols in recent years [47]. Although, this has blurred the dichotomy between in situ and ex situ species management, the value of captive breeding as a conservation tool remains controversial [48]. It is recognized that ex situconservation has many constraints in terms of personnel, costs, and reliance on electric power sources (especially in many developing countries where electricity power can be unreliable) for gene banks. It requires high facilities and financial investments. It cannot also conserve all of the thousands of plant and animal species that make up complex ecosystems such as tropical rainforests [49]. Capture of individuals from the wild for captive breeding or translocation some times can have detrimental effects on the survival prospects of the species as a whole through disease infection [50].

Even though the management of irreplaceable animal populations in zoos and aquarium has focused primarily on minimizing genetic decay with the use of advanced technologies, recent analyses have shown that as most zoo programs are not projected to meet the stated goals due to lack of achieving "sustainability" of the populations [23]. Thus, managing zoo populations as comprehensive conservation strategies for the species requires research on determinants of various kinds of genetic, physiological, behavioral, and morphological variations, and their roles in population viability, development of an array of management techniques, tools, and training of managers [23].

Challenges to Ex Situ Conservation

Ex situ conservation requires different kinds and levels of intensity of management, and a multistakeholder approach like the input from experts on aquarium and zoo husbandry, ex situ breeding, gene-banking, reintroduction, and habitat restoration [51]. Other expert input may include taxonomy, ecology and conservation, ethnography, and sociology. For outreach program, there is a need to liaise with local communities and national government fisheries and wildlife departments; with international (nongovernmental and intergovernmental) conservation bodies [18].

The most important challenges of applying ex situ conservation (captive breading) are the difficulty in recognizing the right time, identifying the precise role of the conservation efforts within the overall conservation action plan, and setting realistic targets in terms of required time span, population size, founder numbers, resources, insurance of sound management and cooperation, and the development of much needed new technical methods and tools [9]. In captive breeding to achieve the retention of 90% of the wild gentic diversity, it is necessary to incorporate sufficient number of founders, careful pair combinations and management [9]. Evidence also exists, which demonstrates that manipulation of housing and husbandry variables can also have significant positive influence on animal reproduction in captivity [19].

In many cases, ex situ populations are founded from only a few individuals, which cause genetic bottlenecks. Small populations are exposed to threats such as stochastic demographic events as well as genetic effects, including loss of genetic diversity, inbreeding depression or accumulation of new, potentially deleterious mutations [11]. More specific problems in garden populations include poorly documented or even unknown sources of material, accidental hybridization of material from various localities, and or unintended selection for traits more suited to garden conditions [14]. In every region, most of the cooperatively managed breeding programs have too few animals, too few animals in appropriate situations for breeding, too few successful breeders, too few founders, and too many animals with undocumented ancestries and/or too little cooperation with scientifically designated breeding recommendations. These deficiencies are resulting in declining populations or declining gentic diversity or both [23, 52]. Problems associated with small founder populations such as inbreeding depression, removal of natural selection, and rapid adaptation to captivity pose considerable challenges for managers of captive populations of threatened species [48]. Equally, reintroduction of captive-bred stock to the wild may require implementation of rigorous protocols that embrace acclimation, pre- and post-release training, health screening, genetic management, long-term monitoring, and involvement of local stakeholders [53, 54]. Shortfalls in implementing such protocols may jeopardize the likelihood of achieving success [47].

Inbreeding due to the mating between two related individuals is unavoidable in small, fragmented, or isolated populations typical of many threatened species, and it can lead to a significant reduction in fitness. The deleterious effects of inbreeding on individual fitness can be large and may be an important factor contributing to population extinction. Inbreeding depression has potential significance for the management and conservation of endangered species [55]. As populations get smaller, the probability increases for all offspring in a given generation are of the same sex [19].

Evaluating the long-term efficiency of ex situ conservation is important, but is complicated because of the difficulty of finding more than one sample of a documented (origin and cultivation) ex situ population and its corresponding still-existing in situ source population [46].

Animal translocations are usually risky and expensive, and a number of biological and nonbiological factors can influence success. Biological considerations include knowledge of genetics, demography, behavior, disease, and habitat requirements. It also includes legal framework, fiscal and intellectual resources, monitoring capacity, goal of the translocation, logistic challenges, and organizational structure of decision making [56].

The regeneration process is one of the most critical steps and a major challenge in gene bank management, during which there is the highest probability for genetic erosion [57]. It is equally important to understand how different conservation methods (seed, field, and cryopreservation) and their management can affect or change the genetic make up, thereby reducing the effective population size (Ne). This will also contribute to decision-making process for determining which methods to use for conservation of the wide diversity [5].

If people are discouraged or prevented from interacting with the resident animals, fewer visitors attend, decreasing public financial support. The visitors' noise and crowding become a source of stress for many species that affects both their welfare and the enjoyment of the visitor [25].

EX SITU CONSERVATION PRACTICE IN ETHIOPIA

Ethiopia is considered to be one of the richest centers of genetic resources in the world. It is believed that indigenous crops such as teff (Eragrostis tef), Noug (Guizotia abyssinica), and Enset (Ensete ventricosum) were first domesticated in Ethiopia. Numerous major crop species including durum wheat (Triticum durum), barley (Hordeum vulgare), sorghum (Sorghum bicolor), sesame (Sesamum indicum), castor (Ricinus communis), and coffee (Coffea arabica) are also known to show significant diversity in the Ethiopian region [58]. Almost 85% of the populations of Ethiopia live in rural areas and most of this population depends directly or indirectly on biodiversity. Biodiversity also plays a crucial role in the different sectors like energy, agriculture, forestry, fisheries, wildlife, industry, health, tourism, commerce, irrigation, and power [15].

The records on biodiversity conservation efforts in Ethiopia date back to the days of Emperor Zera-Yakob (1434–1468 E.C.). The Emperor brought juniper seedlings from Wof Washa of North Shewa and planted in Managesha-Suba area. Modern conservation intervention began by Emperor Menilik in 1908 E.C. and eventually evolved to the establishment of protected areas in the 1960s [4]. Currently, a number of stakeholders are actively working on biodiversity related issues at the federal government level. These include Institute of Biodiversity Conservation (IBC), Ethiopian Institute of Agricultural Research (EIAR), Ethiopian Wildlife Conservation Authority (EWCA), Ministry of Agriculture and Rural Development (MoARD), Ministry of Science and Technology (MoST), Higher Learning Institutions (HLIs), particularly Addis Ababa University (AAU), and offices in various regional states of Ethiopia [59]. Ex situ conservation as complementary to the rehabilitation and restoration of degraded ecosystems and the recovery of threatened species was started 1976 with the establishment of IBC [4].

Ex situ conservation activities mostly focus on high socioeconomic value and internationally important crop types that are considered to be facing immediate danger of genetic erosion [15]. The collections held at IBC are mostly of indigenous landraces some of which are not seen today in farmlands. The collections of root crops, medicinal plants, weedy species, and wild relatives of cultivated species are still relatively inadequate within the existing ex situ collections [60]. However, appropriate emphasis is being placed on conservation and sustainable use of all forms of plant biological resources [60]. Since the establishment of IBC, systematic crop germplasm exploration and collection operations have been undertaken in the different administrative regions of the country, covering a wide range of agroecological zones. Collection priorities were set based on factors like economic importance, degree of genetic erosion and diversity, researchers' needs, the rate of diffusion of improved varieties, clearing of natural vegetation, agricultural policies, natural disasters, and resettlement program [4].

Currently about 68,014 seed accessions of 200 plant species, 6,704 accessions of 205 species of forestry, medicinal, forage and pasture plants (in field gene bank), 290 species of microbial genetic resources, and three semen of threatened breed of domestic animals are conserved by IBC. About 90% of the total germplasm holdings in the gene bank consist of field crops. The total collection is composed of cereal seeds, pulses, oil crops, spices, and species of medicinal and industrial value. Aside from the crop collections, the gene bank also holds 650 collections of micro-organisms. Over 9,000 accessions of horticultural crops, medicinal plants, and herbs are maintained in field gene banks. The type and nature of collection missions and number and lists of plant species and landraces collected have been documented in manuals and reports. Regular monitoring activities are performed for seed viability [15].

For plant species with recalcitrant and intermediate storage behavior, there are ten field gene banks under IBC control and small sized fields in the various research stations of the Ethiopian Institute of Agricultural Research (EIAR) and at universities. The plan for the immediate future is to increase the number of field gene banks in different agroecological zones. Community gardens, backyards, and holy places are being considered for inclusion in the future plan. Spices, vegetables and medicinal plants require management on a large scale and with the full involvement of the local communities [15]. The initiative at national level is still in its infancy and there is currently no well-established national botanical garden in Ethiopia including the Gulele Botanical Garden Center [4, 59]. The Gulele Botanic Garden Center was established through the Proclamation No. 18/2005 E.C. in October 30/2002 E.C. in a 705 hectare land at Gulele and Kolfe-Kernayo subcities. It was established with a vision to see the center to be developed as an exemplary garden in terms of education, ecotourism attraction and center for originality of the Ethiopian plant species, and to be a place of research and nurturing of plant species. The center also has a mission to provide persistent ecotourism services to tourists by taking care of plant species and carrying out educational and research works [59].

Although there is no well-established zoo or zoological garden in Ethiopia, the Addis Ababa Lion Zoo Park can be dominantly cited [59]. The Addis Ababa Lion Zoo Park was established in 1948 with five founder lions presented to Emperor Haile Selassie as gifts. The park accommodates lions with cubs, tortoises, baboons, monkeys, apes, and ducks. A team of international researchers has provided the first comprehensive DNA evidence from 15 (eight males and seven females) samples of Addis Ababa lion indicating the genetically unique samples that requires immediate conservation action. Both microsatellite and mitochondrial DNA data suggest that the zoo lions are genetically distinct from all existing lion populations for which comparative data exist.

Desiccation-intolerant seeds and species that do not readily produce seeds are conserved ex situ in field gene banks. For example, accessions of coffee (Coffea arabica), root crops such as yam (Dioscorea bulbifera) and "Oromo dinich" (Coleus edulis), and spices like ginger (Zingiber officinale) and Ethiopian cardamom (Aframomum corrorima) are conserved in agro-ecological zones in field gene banks [60].

The need for action for global biodiversity conservation is now well understood, and government agencies, nongovernmental organizations, and botanic gardens have all been working in various ways to promote environmental sustainability and reduce species and habitat loss [31].

Seed banking is the major ex situ conservation method employed in Ethiopia. There are three major seed banks operating in Ethiopia. The National Tree Seed Project processes seeds from a narrow range of tree species and uses short-term storage facilities. It aims to cater for the annual seed demand from commercial and small-scale forestry enterprises. Of the 70 species regularly collected and processed, 20 are indigenous. The Forage Genetic Resources Centre maintained by the Consultative Group on International Agricultural Research at the International Livestock Research Institute maintains long-term conservation of a wide range of native and exotic forage species. The Institute for Biodiversity Conservation and Research holds active collections of seeds mainly for research and distribution and as a base collection for long-term conservation [60].

For security reasons, the collected and stored germplasm need to be conserved in duplicate gene banks. However, except for the limited samples of the Ethiopian germplasm held by the Consultative Group on International Agricultural Research, United States Agency for International Development, and the Nordic Gene Bank, majority of the Ethiopian collections are still kept in a single copy at the National Gene Bank. Greater efforts need to be made to store duplicate collections to avoid future genetic erosion [15]. For the continuing power supply, the Ethiopian Gene Bank has independent power supply in the form of a stand by generator to overcome power cuts [15].

Lack of adequate knowledge with respect to collection, handling, and treatment of seeds often impedes the planting of indigenous trees and shrubs. Inadequate work has been done on establishing the seed storage behavior of native species resulting in only limited availability of ex situ conservation seed collections especially with respect to native forest species [60] and lack of alternative storage facilities for the existing conventional cold rooms (e.g., in vitro and cryo-preservation methods) [4]. The current holdings of the IBC gene bank reach over 60,000 accessions of plant species. Some collections are in the medium-term storage mainly due to insufficient seed samples [15].

CONCLUSION

Biodiversity plays a great role in human existence and in healthy function of natural systems although it is on the way of depletion dominantly due to anthropogenic activities. This requires conservation of biodiversity either in in situ or ex situ or both methods in combination based on the conservation situation and its objective. Although in situ conservation is more encouraged to be used for biodiversity conservation, ex situconservation is recommended as it complements through different techniques like zoo, captive breeding, aquarium, botanical garden, and gene bank. Ex situ conservation has its own advantages, disadvantages, and challenges making decision on its application by evaluating advantages, disadvantages and challenges. Although, Ethiopia is rich in biodiversity resources, more people depend on it for their livelihood directly or indirectly causing a great loss. Even if the conservation of biodiversity in Ethiopia has long-time history, its progress, coverage, and enforcement of the rule for conservation seem to be weak. Despite of good progress made in gene bank conservation, it is yet to be developed. In the same way, attention should be given for developing a National Zoological Park and a Botanical Garden.

ACKNOWLEDGMENTS

The authors are extremely grateful to Dr. Habte Jebessa for his valuable comments, suggestions and corrections on the draft of this paper. Our appreciation also goes to Professor. Afework Bekele for his remarkable encouragements and critical comments on the draft of this paper.

REFERENCES

1.   T. I. Borokini, A. U. Okere, A. O. Giwa, B. O. Daramola, and W. T. Odofin, "Biodiversity and conservation of plant genetic resources in Field Gene-bank of the National Centre for Genetic Resources and Biotechnology, Ibadan, Nigeria," The International Journal of Biodiversity and Conservation, vol. 2, pp. 37–50, 2010.

2.   M. Antofie, "Current political commitments' challenges for ex situ conservation of plant genetic resources for food and

agriculture," Analele Universității din Oradea—Fascicula Biologie, vol. 18, pp. 157–163, 2011.

3.   C. A. Tisdell, "Core issues in the economics of biodiversity conservation," Annals of the New York Academy of Sciences, vol. 1219, no. 1, pp. 99–112, 2011. ·

4.   Institute of Biodiversity Conservation, National Biodiversity Strategy and Action Plan, Institute of Biodiversity Conservation, Federal Democratic Republic of Ethiopia, Addis Ababa, Ethiopia, 2005.

5.   M. E. Dulloo, D. Hunter, and T. Borelli, "Ex situ and in situ conservation of agricultural biodiversity: major advances and research needs," Notulae Botanicae Horti Agrobotanici Cluj-Napoca, vol. 38, no. 2, pp. 123–135, 2010.

6.   J. Young, C. Richards, A. Fischer et al., "Conflicts between biodiversity conservation and human activities in the central and eastern European countries," Ambio, vol. 36, no. 7, pp. 545–550, 2007.

7.   H. Debela, "Human influence and threat to biodiversity and sustainable living," Ethiopian Journal of Education and Sciences, vol. 3, no. 1, pp. 85–95, 2007.

8.   R. C. Lacy, "Considering threats to the viability of small populations using individual-based models,"Ecological Bulletin, vol. 48, pp. 39–51, 2000.

9.   K. Leus, "Captive breeding and conservation," Zoology in the Middle East, vol. 54, supplement 3, pp. 151–158, 2011.

10.   K. A. Wilson, J. Carwardine, and H. P. Possingham, "Setting conservation priorities," Annals of the New York Academy of Sciences, vol. 1162, pp. 237–264, 2009. ·

11.   R. Frankham, J. D. Ballou, and D. A. Briscoe, Introduction to Conservation Genetics, Cambridge University Press, Cambridge, UK, 2002.

12.   T. Getachew, "Biodiversity hotspots: pitfalls and prospects," IBC News Letter, vol. 1, pp. 14–16, 2012.

13.   H. Huang, "Plant diversity and conservation in China: planning a strategic bioresource for a sustainable future," Botanical Journal of the Linnean Society, vol. 166, no. 3, pp. 282–300, 2011.

14.   C. Brutting, I. Hensen, and K. Wesche, "Ex situ cultivation affects genetic structure and diversity in arable plants," Plant Biology, vol. 15, pp. 505–513, 2013.

15.   Institute of Biodiversity Conservation, Ethiopia: Second Country Report on the State of PGRFA to FAO, FAO, Rome, Italy, 2007.

16. E. D. Kjaer, L. Graudal, and I. Nathan, Ex Situ Conservation of Commercial Tropical Trees: Strategies, Options and Constraints, Danida Forest Seed Centre, Humlebaek, Denmark, 2001.

17. J. Hakansson, Genetic Aspects of Ex Situ Conservation: Introductory Paper, Department of Biology, Linköping University, 2004.

18. G. M. Reid, T. C. Macbeath, and K. Csatadi, "Global challenges in freshwater-fish conservation related to public aquariums and the aquarium industry," International Zoo Yearbook, vol. 47, pp. 6–45, 2013.

19. V. A. Melfi, "Ex situ gibbon conservation: status, management and birth sex ratios," International Zoo Yearbook, vol. 46, no. 1, pp. 241–251, 2012.

20. C. Ratledge, "Towards conceptual framework for wildlife Tourism," Tourism Management, vol. 22, pp. 31–40, 2001.

21. J. Balcombe, "Animal pleasure and its moral significance," Applied Animal Behaviour Science, vol. 118, no. 3-4, pp. 208–216, 2009.

22. S. F. Carrizo, K. G. Smith, and W. R. T. Darwall, "Progress towards a global assessment of the status of freshwater fishes (Pisces) for the IUCN Red List: application to conservation programmes in zoos and aquariums," International Zoo Yearbook, vol. 47, pp. 46 64, 2013.

23. R. C. Lacy, "Achieving true sustainability of zoo populations," Zoo Biology, vol. 32, pp. 19–26, 2013.

24. R. L. Eaton, "An overview of zoo goals and exhibition principles," International Journal for the Study of Animal Problems, vol. 2, pp. 295–299, 1981.

25. E. J. Fernandez, M. A. Tamborski, S. R. Pickens, and W. Timberlake, "Animal-visitor interactions in the modern zoo: conflicts and interventions," Applied Animal Behaviour Science, vol. 120, no. 1-2, pp. 1–8, 2009.

26. L. A. Dickie, "The sustainable zoo: an introduction," International Zoo Yearbook, vol. 43, no. 1, pp. 1–5, 2009.

27. M. Claxton, "The potential of the human-animal relationship as an environmental enrichment for the welfare of zoo-housed animals," Applied Animal Behaviour Science, vol. 133, no. 1-2, pp. 1–10, 2011. ·

28. S. E. Williams and E. A. Hoffman, "Minimizing genetic adaptation in captive breeding programs: a review," Biological Conservation, vol. 142, no. 11, pp. 2388–2400, 2009.

29. R. J. Whittington and R. Chong, "Global trade in ornamental fish from an Australian perspective: the case for revised import risk analysis and

management strategies," Preventive Veterinary Medicine, vol. 81, no. 1–3, pp. 92–116, 2007.

30. D. Dudgeon, A. H. Arthington, M. O. Gessner et al., "Freshwater biodiversity: importance, threats, status and conservation challenges," Biological Reviews of the Cambridge Philosophical Society, vol. 81, no. 2, pp. 163–182, 2006.

31. S. Blackmore, M. Gibby, and D. Rae, "Strengthening the scientific contribution of botanic gardens to the second phase of the Global Strategy for Plant Conservation," Botanical Journal of the Linnean Society, vol. 166, no. 3, pp. 267–281, 2011.

32. G. Clarke, "The Frozen Ark Project: the role of zoos and aquariums in preserving the genetic material of threatened animals," International Zoo Yearbook, vol. 43, no. 1, pp. 222–230, 2009.

33. O. A. Ryder, A. McLaren, S. Brenner, Y.-P. Zhang, and K. Benirschke, "DNA banks for endangered animal species," Science, vol. 288, no. 5464, pp. 275–277, 2000.

34. C. W. Vertuccj, "Predicting the optimum storage conditions for seeds using thermodynamic principles," Journal of Seed Technology, vol. 17, pp. 41–52, 1993.

35. R. H. Ellis and E. H. Roberts, "Improved equations for the prediction of seed longevity," Annals of Botany, vol. 45, no. 1, pp. 13–30, 1980.

36. J. B. Dickie, R. H. Ellis, H. L. Kraak, K. Ryder, and P. B. Tompsett, "Temperature and seed storage longevity," Annals of Botany, vol. 65, no. 2, pp. 197–204, 1990.

37. M. L. Moir, P. A. Vesk, K. E. C. Brennan et al., "Considering extinction of dependent species during translocation, ex situ conservation, and assisted migration of threatened hosts," Conservation Biology, vol. 26, no. 2, pp. 199–207, 2012.

38. S. Christie, "Why keep tigers in zoos?" in Tigers of the World: The Science, Politics and Conservation of Panthera Tigris, R. Tilson and P. Nyhus, Eds., pp. 205–214, Elsevier Inc., Amsterdam, The Netherlands, 1998.

39. U. S. Anderson, A. S. Kelling, R. Pressley-Keough, M. A. Bloomsmith, and T. L. Maple, "Enhancing the zoo visitor›s experience by public animal training and oral interpretation at an otter exhibit,"Environment and Behavior, vol. 35, no. 6, pp. 826–841, 2003.

40. H. R. Jensen, A. Dreiseitl, M. Sadiki, and D. J. Schoen, "The Red Queen and the seed bank: pathogen resistance of ex situ and in situ conserved

barley," Evolutionary Applications, vol. 5, no. 4, pp. 353–367, 2012.

41. G. R. Hosey, "How does the zoo environment affect the behaviour of captive primates?" Applied Animal Behaviour Science, vol. 90, no. 2, pp. 107–129, 2005.·

42. K. N. Morgan and C. T. Tromborg, "Sources of stress in captivity," Applied Animal Behaviour Science, vol. 102, no. 3-4, pp. 262–302, 2007.

43. B. O. Braastad, "Effects of prenatal stress on behaviour of offspring of laboratory and farmed mammals," Applied Animal Behaviour Science, vol. 61, no. 2, pp. 159–180, 1998.·

44. S. M. Wisely, R. M. Santymire, T. M. Livieri et al., "Environment influences morphology and development for in situ and ex situ populations of the black-footed ferret (Mustela nigripes)," Animal Conservation, vol. 8, no. 3, pp. 321–328, 2005.

45. K. Helenurm and L. S. Parsons, "Genetic variation and the reintroduction of Cordylanthus maritimus ssp. maritimus to Sweetwater Marsh, California," Restoration Ecology, vol. 5, no. 3, pp. 236–244, 1997. ·

46. D. Lauterbach, M. Burkart, and B. Gemeinholzer, "Rapid genetic differentiation between ex situ and their in situ source populations: an example of the endangered Silene otites (Caryophyllaceae),"Botanical Journal of the Linnean Society, vol. 168, no. 1, pp. 64–75, 2012.

47. Balmford, G. M. Mace, and N. Leader-Williams, "Designing the ark: setting priorities for captive breeding," Conservation Biology, vol. 10, no. 3, pp. 719–727, 1996.

48. R. A. Griffiths and L. Pavajeau, "Captive breeding, reintroduction, and the conservation of amphibians," Conservation Biology, vol. 22, no. 4, pp. 852–861, 2008.

49. R. J. Probert, M. I. Daws, and F. R. Hay, "Ecological correlates of ex situ seed longevity: a comparative study on 195 species," Annals of Botany, vol. 104, no. 1, pp. 57–69, 2009.

50. L. M. Clayton, E. J. Milner-Gulland, D. W. Sinaga, and A. H. Mustari, "Effects of a proposed ex situconservation program on in situ conservation of the babirusa, an endangered suid," Conservation Biology, vol. 14, no. 2, pp. 382–385, 2000.

51. W. G. Conway, "Buying time for wild animals with zoos," Zoo Biology, vol. 30, no. 1, pp. 1–8, 2011. ·

52. N. C. Ellstrand and D. R. Elam, "Population genetic conseqences of small population size: implications for plant conservation," Annual Review of Ecology and Systematics, vol. 24, pp. 217–242, 1993.

53. A. Cunningham, "Disease risks of wildlife translocations," Conservation Biology, vol. 10, no. 2, pp. 349–353, 1996.

54. R. P. Reading, T. W. Clark, and B. Griffith, "The influence of valuational and organizational considerations on the success of rare species translocations," Biological Conservation, vol. 79, no. 2-3, pp. 217–225, 1997.

55. L. I. Wright, T. Tregenza, and D. J. Hosken, "Inbreeding, inbreeding depression and extinction,"Conservation Genetics, vol. 9, no. 4, pp. 833–843, 2008. ·

56. B. Miller, K. Ralls, R. P. Reading, J. M. Scott, and J. Estes, "Biological and technical considerations of carnivore translocation: a review," Animal Conservation, vol. 2, no. 1, pp. 59–68, 1999.

57. L. Laikre, L. C. Larsson, A. Palmé, J. Charlier, M. Josefsson, and N. Ryman, "Potentials for monitoring gene level biodiversity: using Sweden as an example," Biodiversity and Conservation, vol. 17, no. 4, pp. 893–910, 2008.

58. N. I. Vavilov, "The origin, variation, immunity and breeding of cultivated plants," Chronica Botanica, vol. 13, pp. 1–366, 1951.

59. S. Demissew, "How has government policy post-global strategy for plant conservation impacted on science? The Ethiopian perspective," Botanical Journal of the Linnean Society, vol. 166, no. 3, pp. 310–325, 2011.

60. B. Girma, T. Pearce, and D. Abebe, "Biological diversity and current ex situ conservation practices in Ethiopia," in Seed Conservation Turning Science into Practice, R. D. Smith, J. B. Dickie, S. H. Linington, H. W. Pritchard, and R. J. Probert, Eds., pp. 849–856, Kew Publishing, Kew, UK, 2003.

Chapter 5

CITIZENS' PREFERENCES FOR THE CONSERVATION OF AGRICULTURAL GENETIC RESOURCES

Eija Pouta, Annika Tienhaara and Heini Ahtiainen

MTT Agrifood Research Finland, Helsinki, Finland

ABSTRACT

Evaluation of conservation policies for agricultural genetic resources (AgGR) requires information on the use and non-use values of plant varieties and animal breeds, as well as on the preferences for in situ and ex situ conservation. We conducted a choice experiment to estimate citizens' willingness to pay (WTP) for AgGR conservation programmes in Finland, and used a latent class model to identify heterogeneity in preferences among respondent groups. The findings indicate that citizens have a high interest in the conservation of native breeds and varieties, but also reveal the presence of preference heterogeneity. Five respondent groups could be identified based on latent class modeling: one implying lexicographic preferences, two with reasoned choices, one indicating uncertain support and one with a preference for the current status of conservation. The results emphasize the importance of in situ conservation of native cattle breeds and plant varieties in developing conservation policies.

INTRODUCTION

The intensification of agriculture has led to marked changes in the utilization of agricultural genetic resources (AgGR), and many previously common cultivated plant varieties as well as native animal breeds that are of interest in terms of food and agricultural production have become rare or even endangered (Drucker et al., 2001; FAO, 2007, 2010). In Finland, several native breeds, such as the Eastern and Northern Finncattle, the Kainuu Gray Sheep and the Åland Sheep, are endangered according to the FAO classification (FAO, 2007), and the majority of old Finnish crop varieties as well as the Finnish landrace pig are already extinct.

Decisions on the focus and extent of genetic resource conservation should consider both the costs and benefits of conservation. The full benefits of conserving AgGR are not revealed by markets, as the resources are either not traded in the markets or the price of agricultural products does not completely capture their value (Oldfield, 1989; Brown, 1990; Drucker et al., 2001). These market failures result in an inefficient allocation of resources, i.e., the level of conservation is too low as the full benefits are not considered. Although the importance of economic analyses has been recognized, the literature on the monetary value of genetic resources in agriculture is still relatively limited (e.g., Evenson et al., 1998; Rege and Gibson, 2003; Ahtiainen and Pouta, 2011).

Conservation policies for AgGR in Finland, as in many other European countries, are currently based on international agreements such as the Convention on Biological Diversity (1992) and the Global Plan of Action for Animal Genetic Resources (FAO, 2007). National genetic resource programmes were initiated for plants in the year 2003 and for farm animals in 2005 to strengthen the conservation of genetic resources in Finland. Although there has been some progress in putting the programmes into action, they have not been fully implemented. This may reflect, for example, the lack of political interest in the conservation.

To evaluate conservation policies, there is a need for monetary benefit estimates that encompass both use and non-use values associated with genetic resources. Use values refer to the benefits obtained from current and future use of genetic resources in production and breeding, while non-use values are generated from the knowledge that genetic resources, e.g., certain breeds, exist and are saved for future generations. Stated preference methods, such as the discrete choice experiment (CE) method, are capable of estimating both use and non-use values in monetary terms. A choice experiment is a survey-based method whereby respondents are asked to choose between two or more discrete alternatives that are described with attributes. By varying attribute levels and including a price variable as one of the attributes, respondents› willingness to pay (WTP) for a policy alternative or attribute level is indirectly revealed based on the choices they make (e.g.,Hanley et al., 2001). The CE method has been found suitable for valuing genetic resources due to its flexibility and ability to value the different traits that breeds or varieties may have. The CE method can also be used to evaluate the means of conservation in situ (live animals and plants) and ex situ(as seeds, cryopreserved embryos and other genetic material), and both plant genetic resources (PGR) and animal genetic resources (AnGR). Previous choice experiments have focused on valuing breeds or varieties and their attributes, especially related to their use in agriculture (Birol et al., 2006; Ouma et al., 2007), and applications focusing on consumer or citizen values for AgGR are rare. The valuation studies on biodiversity have found heterogeneity in consumer preferences, and even identified lexicographic preferences toward conservation (Hanley et al., 1995; Sælensminde, 2006). Lexicographic preferences imply that people are unwilling to accept any trade-offs for changes in environmental goods, such as biodiversity, and may arise when an individual believes that the environment should be protected without regard to the costs. In the context of AgGR, preference heterogeneity has mainly been studied among farmers (e.g., Ouma et al., 2007; Omondi et al., 2008; Roessler et al., 2008), and there have been only few empirical studies of heterogeneity of citizen preferences (Zander et al., 2013) or lexicographic preferences.

In this paper, we present the results of a choice experiment conducted to estimate the benefits of genetic resource conservation programmes in Finland. We tested the effect of in situ and ex situconservation on citizens› choices between programmes. We also analyzed whether plant varieties and animal breeds are perceived as equally valuable by citizens. As heterogeneity in the preferences for the conservation of AgGR is likely, we tested for the existence of citizen segments that place different values on the conservation of genetic resources.

We expected that AgGR would be rather unfamiliar to some of the respondents of the valuation survey. However, in valuation surveys, respondents are assumed to make "informed" choices when responding to value elicitation questions (e.g., Blomquist and Whitehead, 1998). To obtain informed choices that produce valid estimates of WTP, surveys need to provide a sufficient amount of neutral information on the environmental good while avoiding information overload. Providing more information on the quality (characteristics and services) of an environmental good can increase the stated WTP, have no effect, or in some cases reduce WTP (Blomquist and Whitehead, 1998).

There is a substantial body of literature on the effects of information and respondent effort in contingent valuation studies (e.g., Cameron and Englin, 1997; Blomquist and Whitehead, 1998;Berrens et al., 2004), and some choice experiment studies have also examined the issue, mainly focusing on respondent effort (Hu et al., 2009; Vista et al., 2009). Hu et al. (2009) used data from a choice experiment concerning genetically modified food to simultaneously model voluntary information access and product choices. They found that information was accessed rather infrequently, and that those who held critical views on GM food accessed information more often. There were interlinkages between information access and choices, but they were complex and varied between individuals. Vista et al. (2009) examined the effect of time spent on attribute information, choice questions or completing the survey, finding no significant effects on parameter estimates. Here, we were particularly interested in examining how the use of information differs between respondent segments. In the survey, respondents had the opportunity to obtain additional information on genetic resources by accessing a hyperlink to a web page. The Internet survey allowed us to measure whether the respondents accessed the additional information and how much time they used to read it. Offering the opportunity for voluntary access to information instead of using different information treatments for split samples has the advantage of not assuming that respondents read all the information that is provided (Hu et al., 2009). Furthermore, we tested the effects of response certainty and self-perceived carefulness in filling the survey as sources of preference heterogeneity. The rest of the paper is organized as follows. Section Materials and Methods introduces the data and statistical models used in the analysis. Results are presented in section Results, and section Discussion and Conclusions provides discussion and conclusions.

MATERIALS AND METHODS

Data Collection

The survey data were collected using an Internet survey during the summer of 2011. The sample was drawn from the Internet panel of a private survey company, Taloustutkimus, which comprises 30,000 respondents who have been recruited to the panel using random sampling to represent the population (Taloustutkimus, 2013). After a pilot survey of 138 people, a random sample of 6200 respondents was selected, of which 2426 completed parts of the survey and 1495 completed the survey entirely. These numbers correspond to response rates of 39 and 24%, respectively. Based on the socio-demographic variables, the data represented the population rather well (Table 1).

Table 1: Descriptive statistics (n = **1608**)

	In the data	In the population[a]
Proportion of females, %	48	51
Mean age, years	52	47
Proportion of people with a higher educational level, %	24	23
Proportion of people living in households with a gross income under €40,000, %	43	53
Proportion of people with children (<18 years) in the family, %	35	40
Proportion of people living in South Finland, %	40	41

[a] *Statistics Finland 2010, www.stat.fi.*

Survey Design

In the first section, the survey introduced the most common Finnish native animal breeds and plant varieties by explaining what landraces are and giving examples. After asking the respondents about their familiarity with PGR and AnGR, all respondents were offered a short piece of information on the conservation of these breeds and varieties. Next, the respondents were given the opportunity to obtain further information by clicking on two hyperlinks, one for PGR and the other for AnGR. Providing voluntary access to additional information made it possible to identify those respondents who accessed the information, and the time spent on the information page was also recorded (Hu et al., 2009). The additional information provided in our survey included motives for conservation, descriptions of the in situ and ex situ conservation methods and facts about the sustainable use of genetic resources. After several questions concerning perceptions of genetic resources, the survey proceeded to the choice experiment.

The choice experiment was framed by telling respondents that the conservation of native plant varieties and animal breeds is not yet comprehensive in Finland. The survey presented a programme that would conserve the majority of the varieties and breeds on farms and in gene banks. The operation of gene banks would be extended to missing plants and varieties, and conservation on farms would be enhanced by developing the support provided to farmers for conservation activities. Furthermore, those who are using native varieties in gardens were stated to be supported monetarily and by providing guidance.

The survey explained that the conservation programme would be financed with an increase in income tax between the years 2012 and 2021, and that depending on the extent of the programme, the cost to taxpayers would vary, but all taxpayers would participate in financing the programme. The conservation measures (attributes) of the alternative programmes were illustrated to the respondents using a table.

Table 2 presents the attributes together with their descriptions and levels. The first attribute level is always the level specified in the status quo alternative (current state). The attributes included conservation measures of both plant varieties and animal breeds in gene banks and farms. Instead of having a separate attribute for each native breed, only one attribute for breeds in gene banks and one on farms was included to have the same number of attributes for varieties and breeds, and to ease the cognitive burden of the respondents. The native breeds in gene banks attribute had eight levels and native breeds on farms nine levels, including the status quo attribute level.

Table 2: Attributes of conservation programmes and their levels

Attribute	Description	Current state	Levels (unit)
Native food plant varieties in gene banks	Native food plants are stored in a gene bank, either as seeds or plant parts.	The gene bank contains seeds from about 300 landrace varieties. Plants that are added vegetatively (e.g., berry and apple varieties) are missing.	300, 400, 500 (number of plants)
Farms growing native food plants	Farmers and hobby gardeners cultivate native food plants on farms or in gardens.	Seven farms grow seeds of native food plants with agri-environmental support. Other activities than growing seeds are not supported.	7, 500, 1000 (number of farms)
Native ornamental plant varieties mapped and in gene banks	Scientists identify and register native ornamental plants. Varieties are preserved in a gene bank, either as seeds or plant parts.	Only a small proportion of the native ornamental plants are known. Storage in the official gene bank is not provided.	small proportion, about half, majority (proportion of plants)
Native breeds in gene banks	Landrace breeds are kept in a gene bank as gametes and embryos.	The gene bank contains Western, Eastern and Northern Finncattle, as well as Finn-, Åland and Kainuu sheep. Native chicken, goat and horse breeds are missing from the gene bank.	3 cattle breeds and 3 sheep breeds (status quo level), + all combinations of goat, horse and chicken breeds
Native breeds on farms	Native breeds are kept on farms in their natural environment. A breed is considered to be endangered if the number of females is less than 1000.	Farms secure goat, horse and chicken breeds, Finnish sheep and Western Finncattle. Eastern and Northern Finncattle, as well as Åland and Kainuu sheep, are endangered.	1 cattle breed, 1 sheep breed, goat, horse and chicken (status quo level), + all combinations of additional 1-2 cattle and sheep breeds
Cost	Cost for taxpayers, €/year during 2012–2021.	No additional costs.	0, 5, 20, 40, 80, 100, 150, 300 (€)

After introducing the attributes, the respondents were presented with six choice tasks. Each choice task included three alternatives: the status quo alternative, described as maintaining the current situation, and two policy alternatives describing an improved level of conservation compared to the current level. Each alternative was described with five conservation attributes, their levels and the cost attribute. The status quo alternative was uniform across choice tasks. An example of a choice task is shown in Table 3.

Table 3: Example of a choice set

		Current state	Conservation programme A	Conservation programme B
Native food plant varieties in gene banks		Approximately 300	400	400
Farms growing native food plants		7 farms	2000 farms	1000 farms
Native ornamental plant varieties mapped and in gene banks		Some	Majority	About half
Native breeds in gene banks		3 cattle breeds, 3 sheep breeds	3 cattle breeds, 3 sheep breeds, chicken, goat, horse	3 cattle breeds, 3 sheep breeds, goat
Native breeds on farms		Goat, horse, chicken, 1 cattle breed, 1 sheep breed	Goat, horse, chicken, 3 cattle breeds, 1 sheep breed	Goat, horse, chicken, 2 cattle breeds, 3 sheep breeds
Cost for taxpayers, €/year during 2012–2021	€	€0/year	€80/year	€200/year
I support the alternative		( )	( )	( )

We employed an efficient experimental design to allocate the attribute levels to the choice tasks in the choice experiment survey. Efficient designs aim to generate parameter estimates with standard errors that are as low as possible, and thus produce the maximum information from each choice situation (e.g., Rose and Bliemer, 2009). The generation of efficient designs requires the specification of priors for the parameter estimates. In the pilot survey, we employed zero priors in the design, and used the parameter estimates obtained in the pilot study to construct the final experimental design. In the final study, we employed a Bayesian D-efficient design using Ngene (v. 1.0.2), taking 500 Halton draws for the prior parameter distributions. Bayesian designs take into account the uncertainty related to the parameter priors. Instead of fixed priors, they make use of random priors by specifying a mean and standard deviation for the prior.

In the design phase, animal breeds in gene banks and on farms were treated as separate attributes, but were later combined to the "Native breeds in gene banks" and the "Native breeds on farms" attributes in the choice tasks presented to the respondents. Bayesian priors were employed for the chicken attribute and the number of cattle breeds on the farm attribute, and fixed priors for all other attributes. We generated 180 choice tasks, blocking them into 30 subsets, which resulted in six choice situations presented for each respondent. The final design had a D-error of 0.002.

Statistical Models

The choices between environmental programmes were originally modeled with a conditional logit model (also called a multinomial logit model) (McFadden, 1974). The conditional logit, however, assumes a similar preference structure for all respondents, which implies that they have similar tastes for the attributes of conservation. In this study, we were particularly interested in defining heterogeneous citizen segments, which have a similar preference structure within each segment. One approach that allows this heterogeneity is the latent class model (Boxall and Adamowicz, 2002), which has frequently also been applied in choice experiment models of environmental conservation programmes (e.g., Garrod et al., 2012; Grammatikopoulou et al., 2012). In the latent class model, preferences are assumed to be homogeneous in each segment, but to vary between the segments.

In the modeling, price was treated as a continuous variable and the other attributes were effects-coded, implying that the parameters will sum to zero over the categories of the nominal variable concerned. The status quo attribute levels were thus included in the model, and could obtain either negative or positive coefficients depending on their effect on respondent's utility. Alternative-specific constants (ASC) were included for all alternatives in order to allow systematic choice tendencies not explained with the parameters describing the attributes.

Heterogeneity was statistically included in the latent class model by simultaneously dividing individuals into behavioral groups or latent segments, and estimating a choice model for each of these classes. The estimation was carried out by assuming first one class, then two classes, three classes and so forth. In each step, the explanatory power of the model was assessed to decide on the optimal number of classes. For this purpose, we used the Bayesian information criterion (BIC) and Akaike information criterion (AIC), which are log-likelihood scores with correction factors for the number of observations and the number of parameters. The latent class model also enables the calculation of the WTP for the attributes for each citizen segment.

The relationship between the individual characteristics and the latent classes was examined a posteriori of the actual estimation of the latent class model in order to describe the heterogeneous citizen segments. Thus, the segments were formed solely based on the conservation program choices. The membership in the most probable segment was regressed using a logistic regression to characterize each class compared to the rest of the respondents. The explanatory variables for the class memberships included respondents' socioeconomic characteristics, perceived values and responsibilities, use of provided information, response certainty and self-reported perception of the carefulness of completing the survey. The independent variables in the logistic regression models and their descriptive statistics are presented in Table 4.

Table 4: Variables in the logistic regression models

Characteristic	Description	Mean	Standard deviation	Min	Max
Female	1 if the respondent is female, if male	0.49	0.50	0	1
Year of birth	Respondents year of birth, continuous	1960	15	1931	1992
High income	1 if household income is over €50,000 per year, 0 otherwise	0.45	0.49	0	1
High education	1 if respondents education level is university education, 0 otherwise	0.24	0.46	0	1
Eastern Finnish	1 if respondents lives in Eastern Finland, 0 otherwise	0.11	0.32	0	1
Childhood in city	1 if respondent spent his/her childhood in a city, 0 otherwise	0.41	0.49	0	1
Certainty	Mean of respondent's certainty in the conservation programme choices, on a scale of 10 completely certain—1 not at all certain.	6.85	2.23	1	10
Agri-environmental attitude	Importance of environmental issues in agriculture, mean of nine measures on scales from 1 to 4	3.26	0.44	1	4
Relative importance of preserving AgGR	The importance of preserving native breeds and varieties relative to other environmental protection measures, 1 if both equally important, >1 if preserving native breeds and varieties more important, <1 if other environmental protection measures more important	0.94	0.16	0.36	1.66
Existence value	Factor score based on 8 measures of the importance of existence values, continuous[*]	0.00	1.00	−4.38	2.39
Use values	Factor score based on 8 measures of the importance of use values, continuous[*]	0.00	1.00	−3.78	2.62
Citizen responsibility	Factor score based on 9 measures of stakeholder responsibilities in conservation[*]	0.00	1.00	−3.38	2.30
Consumer responsibility	Factor score based on 9 measures of stakeholder responsibilities in conservation[*]	0.00	1.00	−5.27	2.01
Farmer responsibility	Factor score based on 9 measures of stakeholder responsibilities in conservation[*]	0.00	1.00	−3.12	2.88
Familiarity of products	Familiarity of AgGR products, mean of 10 measures on scales from 1 to 3	2.03	0.42	1	3
Info use (animals) > 0.5 min	1 if respondent used more than 30 s for additional information about breeds, 0 otherwise	0.33	0.47	0	1
Info use (plants) > 0.5 min	1 if respondent used more than 30 s for additional information about varieties, 0 otherwise	0.35	0.48	0	1
Hasty response	1 if respondent evaluated his/her response as hasty, 0 if careful	0.05	0.22	0	1

[*]Detailed description of these variables can be found in Tienhaara et al. (2014).

Results

In 24% of the choice sets, the respondents chose the status quo option, i.e., the current state without any additional program to conserve AgGR. The probability of choosing one of the two alternative conservation programs varied between 46% for the lowest cost level of €5 and 28% for the highest cost level of €300.

Table 5 presents the conditional logit model results for the choice of the conservation programme. As expected, an increase in the programme cost negatively affected the probability of choosing it. Turning to consider the genetic resource attributes, the number of food plants in the gene bank was not statistically significant. All other attributes were significant in determining respondents› choices. A higher number of farms growing native plant varieties increased the choice probability. The larger the number of ornamental plants to be mapped and conserved in gene banks, the more probable it was that the respondent would choose the programme. Conserving native breeds of Finnish goats, horses and chickens in the gene bank all increased the support for the programme. The effect was highest for horse, followed by chicken and goat. The guaranteed existence of cattle breeds on farms had a positive and significant effect on choice. As expected, the effect was greater if the number of conserved cattle breeds was three instead of two. This was also the case with sheep breeds, although the conservation of two breeds did not have a positive effect on choice compared with the status quo of one conserved breed.

Table 5: Conditional logit (CL) model results

Variable	Coefficient	Wald p-value
ASC1 (SQ)	−0.263***	0.000
ASC2	0.291***	
ASC3	−0.028	
Cost	−0.005***	0.000
300 plants in bank (SQ)	0.002	1.000
400 plants in bank	−0.002	
500 plants in bank	0.000	
7 plants on farms (SQ)	−0.199***	0.000
500 plants on farms	0.075***	
1000 plants on farms	0.124***	
Ornamental plants in bank (SQ)	−0.057**	0.008
Ornamental plants in bank L2	−0.004	
Ornamental plants in bank L3	0.061***	
Goats (SQ)	−0.039***	0.005
Goats in bank	0.039***	
Horses (SQ)	−0.075***	0.000
Horses in bank	0.075***	
Chickens (SQ)	−0.047***	0.001
Chickens in bank	0.047***	
1 cattle breed on farms (SQ)	−0.114***	0.000
2 cattle breeds on farms	0.025	
3 cattle breeds on farms	0.089***	
1 sheep breed on farms (SQ)	0.020	0.027
2 sheep breeds on farms	−0.052***	
3 sheep breeds on farms	0.032	
No. of respondents	1608	
No. of observations	9484	
Correct predictions %	48	
R^2	0.04	

z-test: *** 99% significance level; ** 95% significance level.
SQ, attribute level in the status quo alternative.

The alternative specific constants (ASC) capture the tendency to choose one of the alternatives which is not explained by the attributes. The negative ASC1

(SQ) coefficient showed the reluctance to choose the status quo alternative regardless of the attribute levels in the policy alternatives. Furthermore, the ASC2 and ASC3 coefficients differed unexpectedly in sign and significance. The positive coefficient for ASC2 and negative for ASC3 indicated that the conservation programme that was presented first received more support. This was surprising, as the programmes were not presented in a specific order in the survey. The model predicted 48% of the choices right, clearly exceeding the probability of correct random choices of 33%, leading to a relatively weak goodness of fit.

The homogeneity of preferences was tested in the estimation of the latent class models. Based on the AIC and BIC, the estimation process showed that a model of five citizen clusters provided the best fit of the data. Table 6 presents the latent class model results with the cluster names, and the logit model for the membership of each cluster is presented in Table 7.

Table 6: Latent class models for conservation programme choice

	Class 1	Class 2	Class 3	Class 4	Class 5	Overall	
Pseudo R²	0.131	0.288	0.019	0.015	0.472	0.559	
Class size	0.27	0.26	0.17	0.17	0.13		
Class names	Conserva-tionists	Bid-sensitive animal conservers	Uncertain supporters	Status quo preferers	Bid sensitives	Wald p-value	Wald (=) p-value
Attributes				Coefficients and significance levels			
ASC 1 (SQ)	−0.990***	−2.937***	−0.841***	1.668***	−0.554**	0.000	0.000
ASC 2	0.332***	1.499***	1.757***	−0.414**	0.478***		
ASC 3	0.658***	1.438***	−0.916***	−1.254***	0.076		
Cost	0.000	−0.018***	−0.003*	−0.001	−0.041***	0.000	0.000
300 plants in bank (SQ)	−0.162***	0.138**	0.018	0.412**	−0.322***	0.003	0.001
400 plants in bank	0.025	−0.007	0.078	−0.166	0.225*		
500 plants in bank	0.137**	−0.131*	−0.096	−0.245	0.097		
7 plants on farms (SQ)	−0.621***	−0.120*	−0.261***	−0.006	−0.169	0.000	0.000
500 plants on farms	0.125**	0.208***	0.237*	0.003	0.104		
1000 plants on farms	0.496***	−0.088	0.024	0.003	0.066		
Ornamental plants in bank (SQ)	−0.462***	0.015	0.116	−0.004	−0.332**	0.000	0.000
Ornamental plants in bank L2	0.158***	0.002	0.023	−0.053	0.16		
Ornamental plants in bank L3	0.304***	−0.017	−0.139	0.057	0.172		
Goats (SQ)	−0.063***	−0.063***	−0.063***	−0.063***	−0.063***	0.001	C.i.
Goats in bank	0.063***	0.063***	0.063***	0.063***	0.063***		
Horses (SQ)	−0.152***	−0.128***	−0.075	0.447***	−0.256***	0.000	0.000
Horses in bank	0.152***	0.128***	0.075	−0.447***	0.256***		
Chickens (SQ)	−0.062***	−0.062***	−0.062***	−0.062***	−0.062***	0.001	C.i.
Chickens in bank	0.062***	0.062***	0.062***	0.062***	0.062***		
1 cattle breed on farms (SQ)	−0.144***	−0.144***	−0.144***	−0.144***	−0.144***	0.000	C.i.
2 cattle breeds on farms	0.034	0.034	0.034	0.034	0.034		
3 cattle breeds on farms	0.110***	0.110***	0.110***	0.110***	0.110***		
1 Sheep breed on farms (SQ)	−0.213***	0.046	−0.036	0.581***	−0.245**	0.000	0.001
2 Sheep breeds on farms	0.056	−0.04	−0.156	−0.282	0.116		
3 Sheep breeds on farms	0.157***	−0.007	0.192	−0.300	0.128		
No. of respondents	1608						
No. of observations	9484						
Correct predictions %	85						

z-test: *** 99% significance level; ** 95% significance level; * 90% significance level.

SQ, attribute level in the status quo alternative.

C.i., class independent.

Table 7: Logistic regression models profiling consumer classes

Class	Class 1	Class 2	Class 3	Class 4	Class 5
Variable	Coefficients and significance levels				
Constant	−2.76***	−43.31***	48.77***	39.90**	−29.46**
Female	−0.46***				
Year of birth		0.02***	−0.02***	−0.02**	0.02*
High income			−0.39**		
High education				−0.72***	
Eastern Finnish			0.40*		
Childhood in city				−0.68**	
Certainty	0.12***		−0.09**		−0.08**
Agri-environmental attitude	0.37*	0.43**			
Relative importance of AgGR		−1.482***	1.412**	−1.82**	
Existence values	0.32***			−0.50***	
Use values	0.38***			−0.39***	
Citizen responsibility		0.29***	0.21**	−1.06***	−0.43***
Consumer responsibility		0.17**		−0.31**	−0.38***
Farmer responsibility	−0.16**			0.27**	
Familiarity of products					−0.48**
Info use (animals) > 0.5 min				−0.39*	
Info use (plants) > 0.5 min		0.54***	−0.47***		
Hasty response			0.70*		−1.08**
N	1088	1201	1098	1077	1199
Nagelkerke R^2	0.103	0.083	0.071	0.397	0.104
Chi-squared	81.99	71.44	46.48	252.37	68.25
p-Value	0.000	0.000	0.000	0.000	0.000
Correctly classified (cut 0.5)	69.6	71.6	83.9	90.4	86.8

Variables are significant at the *** 99% level, ** 95% level, * 90% level.

The latent class model showed that although preferences for some attributes, such as conserving goat and chicken breeds in gene banks and cattle breeds on farms, did not differ significantly between clusters, there was significant heterogeneity in preferences for most of the attributes. The first class, named as "conservationists," comprised 27% of the respondents. They did not take the personal cost of the conservation programme into account in their decision process, as the coefficient of the cost variable was not significant. Instead, almost all the conservation attributes had significant and positive signs. Contrary to other clusters, most plant-related attributes were significant for conservationists. They also valued the conservation of ornamental plants. Table 5also shows that this cluster perceived higher use and existence values from genetic resource conservation than respondents in other segments, and also higher than average certainty in their responses to the choice tasks. This class contained more men than women and considered the conservation not to be a responsibility of farmers. For this cluster we also tested the effect gardening as a hobby, but it did not turn out to be significant. Thus, it seems that these respondents did not support the program because of the possible private good aspect of measures to support native varieties in gardens.

The second cluster, covering 26% of the respondents, was named as "bid-sensitive animal conservers." This group had a higher tendency to choose the improvement programmes compared to the status quo. The coefficient of the bid was significant and the second smallest of all clusters. In this cluster, the emphasis of preferences was on the conservation of animal breeds. The conservation of plant varieties in gene banks was even valued negatively. These respondents perceived more often than average that citizens and consumers should be responsible for the conservation of genetic resources. They also had positive agri-environmental attitudes. Furthermore, the respondents in this cluster used more than the average time to familiarize themselves with the information available in the survey concerning PGR, and they were slightly younger than the average respondent.

A confusing aspect in the third cluster was the large difference between the ASC for the two conservation programmes. This group, comprising 17% of the respondents, had a considerably greater tendency to choose conservation programme A rather than B or the status quo, although this could not be explained by the experimental design and attribute levels. The bid variable followed expectations, but for the other attributes, only plants on farms and the class-independent variables were significant. The logistic regression revealed that members of this cluster were older and had a lower income, and they emphasized the responsibility of citizens in conservation. Geographically, this cluster had more members who lived in Eastern Finland. The respondents in this group were relatively uncertain of their preferences, used the additional information less, and responded, according to their self-evaluation, less carefully than other respondents. As there were random tendencies in their support for a programme (ASC), but they still preferred an increase in several conservation attributes, they were named as "uncertain supporters."

The fourth class, with 17% of respondents, clearly preferred the status quo option, as the ASC for the programme options were negative. The coefficient of the bid variable was not significant. Among these "status quo preferers," the choice was consistent with their negative attitudes, as the relative importance of AgGR was low, as well as the perceived existence and use values. Citizens and consumers were less frequently seen as those responsible for conservation; instead, it was perceived as a responsibility of the farmers. This class was characterized by an older age, lower educational level and growing up on a farm. The fifth class of respondents (13%), named as "bid sensitives," were the most sensitive to the cost of the programme of all groups. Nevertheless, the ASC revealed that they were interested in conservation, and almost all conservation attributes had significant coefficients. Among these respondents, particularly the ex situ conservation of Finnhorse positively affected their choices. In this class, the conservation of genetic resources was not seen as a responsibility of citizens or consumers. The logit model for this group showed that they evaluated themselves as careful respondents but felt somewhat uncertain of their choices. They were younger than average and less familiar with products from traditional breeds and varieties.

Table 8: Annual willingness to pay (in 2009 €) for attributes

	Conditional logit model	Latent class model, Class 2	Latent class model, Class 3	Latent class model, Class 5
Plants in bank (400)	–	–	–	13
Plants in bank (500)	–	-15	–	–
Plants on farms (500)	60	19	7	–
Plants on farms (1000)	70	–	–	–
Ornamental plants (majority) inventoried and in bank	14	–	–	–
Goats in bank	17	7	105	3
Horses in bank	33	15	–	12
Chickens in bank	20	7	104	3
3 cattle breeds on farms	44	14	211	6
2 sheep breeds on farms	−15	–	–	–

–, Indicates that the estimate is missing due to the non-significance of the cost coefficient.

WTP for different attributes was calculated based on the conditional logit model and the latent class model for those classes for which the cost coefficient was significant (Table 8). WTPs based on the conditional logit model indicated that plants on farms, cattle breeds and horses were most highly valued. In general, there was substantial variation in WTPs between the classes. In class 3, WTPs were higher due to the low importance of the cost attribute.

Discussion and Conclusions

The results of a choice experiment concerning agricultural genetic resource policies showed that citizens are interested in the conservation of native breeds and varieties in agriculture. However, there was considerable variation in preferences between citizen segments. Of the five identified groups, two groups covering over half of the respondents had a high interest in the conservation of native breeds and varieties. Respondents in one of the segments clearly preferred the current state of conservation to additional conservation efforts, while one group had a favorable attitude toward conservation if the expenses were on a low level, and respondents in one segment were supportive but wavering in their preferences. The respondent groups were identified based on their preferences for conservation, and they also differed with respect to the use of additional information, their response carefulness and the certainty of the stated WTP. Similar to previous studies of consumer preferences on biodiversity (e.g., Hanley et al., 1995), we also found lexicographic preferences for conserving AgGR. Those were expressed by the largest group of respondents (27%), as their interest in conservation was high regardless of the costs. Lexicographic choices can occur as a result of simplification if the respondent finds the choice task too difficult to handle or as a result of actual lexicographic preferences (Sælensminde, 2006). In our case, it is difficult to determine whether respondents exhibited lexicographic preferences because they wanted to simplify the choice tasks or because the differences in the attribute levels were large. Respondents in the group which exhibited lexicographic preferences were more certain about their preferences, which supports the phenomenon of actual lexicographic preferences as the reason for their choices. In addition, their positive perceptions concerning the existence and use values of genetic resources support the observation of actual lexicographic preferences.

Due to the preference structures, WTP estimates were only obtained for three respondent groups and some of the attributes. In those groups where the cost variable was significant and meaningful WTP estimates could thus be estimated, the marginal WTPs were considerably lower than the WTPs of the whole sample based on the conditional logit model. This implies that in the whole sample, the results were influenced by the groups that were insensitive to the costs of conservation. Our results can be compared with those obtained by Zander et al. (2013), who assessed the economic value of conservation programs for two Italian cattle breeds using a choice experiment directed to citizens. Zander et al. (2013) also found preference heterogeneity for most of the attributes of the conservation programs, as well as differences in the sensitivity to the cost attribute. According to their findings, 85% of the respondents supported increased conservation, and the mean WTP was 90€ for conserving each breed. The present results can also be linked to previous results of heterogeneity among farmers using native breeds and varieties. Soini et al. (2012)identified a segment of production-oriented farmers among European cattle breeders that would benefit from increased subsidies for keeping native breeds on farms. If the subsidies were increased to correspond to citizens› WTP, it would help particularly this subsidy-dependent group of farmers.

As the survey was Internet-based, we were able to obtain information on the time used for obtaining additional information about plant and AnGR. These variables, combined with certainty, could partly explain the membership in the latent classes. However, similarly to Hu et al. (2009)and Vista et al. (2009), there were no clear tendencies for the use of information to be associated with a lower or higher WTP. Further research is, however, needed to clarify the associations between preferences, uncertainty and information acquisition in the case of genetic resources.

The results provide implications concerning how to direct the conservation policies for AgGR in Finland. The WTP estimates for the attributes of the conservation programmes indicated that the participants valued particularly in situ conservation in the case of PGR, which would also imply the existence of native plant varieties in the landscape. However, a moderate level of this in situconservation would be sufficient, as the highest level increased the WTP only slightly. For the conservation of animal breeds, the results emphasize the importance of in situ conservation of cattle breeds. The weak support for the conservation of sheep breeds compared to cattle breeds was understandable, as Finnsheep breeds are less familiar to the public. However, the low, even negative, WTP for the conservation of sheep breeds is in contradiction with the importance of Finnsheep in breeding (e.g., Thomas, 2010). Ex situ conservation of those animal breeds that are at present insufficiently protected in gene banks was perceived as important, particularly the conservation of the genetic material of the Finnhorse.

Although the cost-effectiveness of AgGR conservation is case-dependent, some previous studies have recommended ex situ conservation in gene banks as a less expensive, less vulnerable and less policy-sensitive method of conservation (Dulloo et al., 2010; Silversides et al., 2012). These cost-effectiveness considerations do not, however, take into account the additional benefits that may be associated with in situ conservation, such as the visibility of local breeds and varieties in the landscape or the opportunity to use local breed products. Thus, taking into account citizens' preferences for in situ and ex situ conservation and using cost-benefit analysis in policy evaluation may shift the priorities of agricultural genetic resource conservation policies.

In this study, the conservation policies were based on equal participation of all citizens, as the policy was financed with taxes. An alternative approach would be to apply market-based incentives, e.g., payments for environmental services (PES) for the conservation of genetic resources (McNeely, 2006; Wunder, 2007; Narloch et al., 2011). PES would imply that actors who are major users of the resources are involved in making and adapting rules for conservation markets. For future experiments of PES, our results of the citizen groups that are most interested provide information for identifying the interested parties for the markets of AgGR.

Conflict of Interest Statement

The Guest Associate Juha Kantanen declares that, despite being affiliated to the same institution as the authors, the review process was handled objectively and no conflict of interest exists. The authors declare that the research was conducted in the absence of any commercial or financial relationships that could be construed as a potential conflict of interest.

REFERENCES

1. Ahtiainen, H., and Pouta, E. (2011). The value of genetic resources in agriculture: a meta-analysis assessing existing knowledge and future research needs. Int. J. Biodivers. Sci. Ecosyst. Serv. Manage. 7, 27–38. doi: 10.1080/21513732.2011.593557

2. Berrens, R. P., Bohara, A. K., Jenkins-Smith, H. C., Silva, C. L., and Weimer, D. L. (2004). Information and effort in contingent valuation surveys: application to global climate change using national internet samples. J. Environ. Econ. Manage. 47, 331–363. doi: 10.1016/S0095-0696(03)00094-9

3. Birol, E., Smale, M., and Gyovai, Á. (2006). Using a choice experiment to estimate farmers' valuation of agrobiodiversity on hungarian small farms. Environ. Resour. Econ. 34, 439–469. doi: 10.1007/s10640-006-0009-9

4. Blomquist, G., and Whitehead, J. (1998). Resource quality information and validity of willingness to pay in contingent valuation.Resour. Energy

Econ. 20, 179–196. doi: 10.1016/S0928-7655(97)00035-3

5. Boxall, C. P., and Adamowicz, L. W. (2002). Understanding heterogeneous preferences in random utility models: a latent class approach. Environ. Resour. Econ. 23, 421–446. doi: 10.1023/A:1021351721619

6. Brown, G. M. (1990). "Valuation of genetic resources," in The Preservation and Valuation of Biological Resources, eds G. H. Orians, G. M. Jr. Brown, W. E. Kunin, and J. E. Swierzbinski (Seattle, WA: University of Washington Press), 203–228.

7. Cameron, T. A., and Englin, J. (1997). Respondent experience and contingent valuation of environmental goods. J. Environ. Econ. Manage. 33, 296–313. doi: 10.1006/jeem.1997.0995

8. Drucker, A. G., Gomez, V., and Anderson, S. (2001). The economic valuation of farm animal genetic resources: a survey of available methods. Ecol. Econ. 36, 1–18. doi: 10.1016/S0921-8009(00)00242-1

9. Dulloo, M. E., Hunter, D., and Borelli, T. (2010). Ex situ and in situ conservation of agricultural biodiversity: major advances and research needs. Notulae Botanicae Horti Agrobotanici Cluj-Napoca 38, 123–135.

10. Evenson, R. E., Gollin, D., and Santaniello, V. (1998). "Introduction and overview: agricultural values of plant genetic resources," in Agricultural Values of Plant Genetic Resources, eds R. E. Evenson, D. Gollin, and V. Santaniello (Wallingford: CABI Publishing), 1–25.

11. FAO, (2007). The State of the World›s Animal Genetic Resources for Food and Agriculture. Rome: Commission on Genetic Resources for Food and Agriculture, FAO.

12. FAO, (2010). The Second Report on the State of the World›s Plant Genetic Resources for Food and Agriculture. Available online at: http://www.fao.org/docrep/013/i1500e/i1500e.pdf [Accessed 11 October 2011].

13. Garrod, G., Ruto, E., Willis, K., and Powe, N. (2012). Heterogeneity of preferences for the benefits of environmental stewardship: a latent-class approach. Ecol. Econ. 76, 104–111. doi: 10.1016/j.ecolecon.2012.02.011

14. Grammatikopoulou, I., Pouta, E., Salmiovirta, M., and Soini, K. (2012). Heterogeneous preferences for agricultural landscape improvements in southern Finland. Landsc. Urban Plan. 107, 181–191. doi: 10.1016/j.landurbplan.2012.06.001

15. Hanley, N., Mourato, S., and Wright, R. E. (2001). Choice modelling approaches: a superior alternative for environmental valuation? J. Econ. Surv. 15, 435–462. doi: 10.1111/1467-6419.00145

16. Hanley, N., Spash, C., and Walker, L. (1995). Problems in valuing the benefits of biodiversity protection. Environ. Res. Econ. 5, 249–272. doi: 10.1007/BF00691519

17. Hu, W., Adamovicz, W. L., and Veerman, M. M. (2009). Consumers› preferences for GM food and voluntary information access: a simultaneous choice analysis. Can. J. Agri. Econ. 57, 241–267. doi: 10.1111/j.1744-7976.2009.01150.x

18. McFadden, D. (1974). "Conditional logit analysis of qualitative choice behaviour," in Frontiers in Econometrics, ed P. E. Zarembka (New York, NY: Academic Press), 105–142.

19. McNeely, J. A. (2006). Using economic instruments to overcome obstacles to in situ conservation of biodiversity. Integr. Zool. 1, 25–31. doi: 10.1111/j.1749-4877.2006.00009.x

20. Narloch, U., Drucker, A., and Pascual, U. (2011). Payments for agrobiodiversity conservation services for sustained on-farm utilization of plant and animal genetic resources. Ecol. Econ. 70, 1837–1845. doi: 10.1016/j.ecolecon.2011.05.018

21. Oldfield, M. L. (1989). The Value of Conserving Genetic Resources. Sunderland, MA: Sinauer Associates Inc.

22. Omondi, I., Baltenweck, I., Drucker, A. G., Obare, G., and Zander, K. K. (2008). Valuing goat genetic resources: a pro-poor growth strategy in the Kenyan semi-arid tropics. Trop. Anim. Health Prod. 40, 583–589. doi: 10.1007/s11250-008-9137-2

23. Ouma, E., Abdulai, A., and Drucker, A. (2007). Measuring heterogeneous preferences for cattle traits among cattle-keeping households in East Africa. Am. J. Agric. Econ. 89, 1005–1019. doi: 10.1111/j.1467-8276.2007.01022.x

24. Rege, J. E. O., and Gibson, J. P. (2003). Animal genetic resources and economic development: issues in relation to economic valuation. Ecol. Econ. 45, 319–330. doi: 10.1016/S0921-8009(03)00087-9

25. Roessler, R., Drucker, A. G., Scarpa, R., Markemann, A., Lemke, U., Thuy, L. T., et al. (2008). Using choice experiments to assess smallholder farmers› preferences for pig breeding traits in different production systems in North-West Vietnam. Ecol. Econ. 66, 184–192. doi: 10.1016/j.ecolecon.2007.08.023

26. Rose, J. M., and Bliemer, M. C. J. (2009). Constructing efficient stated choice experimental designs. Trans. Rev. 29, 597–617. doi: 10.1080/01441640902827623

27. Sælensminde, K. (2006). Causes and consequences of lexicographic choices in stated choice studies. Ecol. Econ. 59, 331–340. doi: 10.1016/j.ecolecon.2005.11.001

28. Silversides, F. G., Purdy, P. H., and Blackburn, H. D. (2012). Comparative costs of programmes to conserve chicken genetic variation based on maintaining living populations or storinf cryopreserved material. Br. Poult. Sci. 53, 599–607. doi: 10.1080/00071668.2012.727383

29. Soini, K., Diaz, C., Gandini, G., de Haas, Y., Lilja, T., Martin-Collado, D., et al. (2012). Developing a typology for local cattle breed farmers in Europe. J. Anim. Breed. Genet. 129, 436–447. doi: 10.1111/j.1439-0388.2012.01009.x

30. Taloustutkimus. (2013). Internet Panel. Available online at: http://www.taloustutkimus.fi/in-english/products_services/internet_panel/ [Accessed 28 August 2013].

31. Thomas, D. L. (2010). Performance and utilization of Northern European short-tailed breeds of sheep and their crosses in North America: a review. Animal 4, 1283–1296. doi: 10.1017/S1751731110000856

32. Tienhaara, A., Ahtiainen, H, and Pouta, E. (2014). "The consumer and citizen roles and motives in valuation of agricultural genetic resources in Finland," in Paper Presented in the 20th Annual Conference of the European Association of Environmental and Resource Economists (Toulouse). Available online at: http://www.webmeets.com/EAERE/2013/m/viewpaper.asp?pid=870

33. Vista, A. B., Rosenberger, R. S., and Collins, A. R. (2009). If you provide, will they read it? Response time effects in a choice experiment. Can. J. Agri. Econ. 57, 365–377. doi: 10.1111/j.1744-7976.2009.01156.x

34. Wunder, S. (2007). The efficiency of payments for environmental services in tropical conservation. Conserv. Biol. 21, 48–58. doi: 10.1111/j.1523-1739.2006.00559.x

35. Zander, K., Signorello, G., De Salvo, M., Gandini, G., and Drucker, A. (2013). Assessing the total economic value of threatened livestock breeds in Italy: Implications for conservation policy. Ecol. Econ. 93, 219–229. doi: 10.1016/j.ecolecon.2013.06.002

Chapter 6

AGRICULTURAL BIOTECHNOLOGY DEVELOPMENT AND POLICY IN CHINA

Jikun Huang and Qinfang Wang

Chinese Academy of Sciences

ABSTRACT

This article provides an overview of China's agricultural biotechnology development policies. Research goals, strategies, priorities, commercialization, and China's organizational framework for agricultural biotechnology development are presented. Included is a description of the evolution of China's biosafety regulations as well as China's research capacity building and public investment—one of the largest public research efforts on agricultural biotechnology in the world.

INTRODUCTION

In the past three years, the growth rate has slowed for acres planted with genetically modified (GM) crops globally, in contrast to its rapid increase in the late 1990s (James, 2002). This slowdown may be due to worldwide conflicting views on biotechnology that not only affect global investment in the biotechnology industry, but also impact farmers' adoption of this technology. Some, but not all, of these issues are present in China; however, China's development of its biotechnology industry has been unique, catalyzed by the active involvement of the public sector.

A survey of China's plant biotechnologists by the authors and their collaborators in 2000 shows that China is developing the largest plant biotechnology capacity outside of North America (Huang, Rozelle, Pray, & Wang, 2002). In 1997, when the National Genetically Modified Organisms (GMO) Biosafety Committee was established, this committee immediately approved 46 cases for field trials, environmental release, and commercialization, which covered 12 GM crops. Among them three cases of cotton, tomato, and petunia were approved for commercialization in certain locations (Huang, Wang, & Keeley, 2001).[1] A number of earlier studies concluded that China adopted a promotional policy to embrace the benefits of biotechnology (Chen, 2000; Huang, Wang, Zhang, & Zepeda, 2001; Paarlberg, 2000). China became one of the world›s leading countries in biotechnology development. China also received criticism from biotechnology opponents for not paying enough attention to biosafety, the environment, consumer and food safety, and the potential impacts of biotechnology on China›s future agricultural trade position.

However, the above perceptions regarding China›s position on agricultural biotechnology lasted for only a few years. In May 2001, China›s State Council decreed a new rule—Regulation on Safety Administration of Agricultural GMOs. And in early 2002, the Ministry of Agriculture (MOA) issued three detailed regulations on the biosafety management, trade and labeling of GM farm products.[2] After these events, China received more criticism than support from both proponents and critics of biotechnology. For example, biotech scientists and biotech industry representatives criticized China's new regulations as too restrictive to provide a favorable environment for the development of biotechnology. They called the period following 1999 as the "winter of biotechnology." Alternatively, Greenpeace and environmental agencies continuously warned China of the potential risks associated with GMOs.

International trade impacts occurred for both imports and exports. New regulations required importers of GM agricultural products to apply for official safety verification approval from China›s Ministry of Agriculture. This led the US government to accuse Beijing of using these new rules to hinder imports and protect Chinese soybean farmers.[3]Pressure was also raised on the export side. China was frequently asked to certify that its agricultural exports to Japan and EU markets were free of GMOs. In addition, there has been growing criticism of China's financial and institutional ability to label its GM farm products.

Additionally, the media has claimed that China had reversed its former enthusiastic embrace of biotechnology by imposing extra restrictions on both domestic and imported varieties of genetically modified crops. These claims stated that China made a decisive shift away from its intentions to become the developing world's leader in biotechnology.[4]After 15 years of nationwide promotion of agricultural biotechnology in China, the current policy debate appears confusing to many observers. The industry wonders whether China will continue to advance its biotechnology, and some scientists question how to proceed in the near future. Given the above background, the objectives of this article are to review the status of China's agricultural biotechnology research and commercialization, and to gain a better understanding of China's policies governing both agricultural biotechnology research and its applications (or commercialization).[5] In order to achieve these objectives, this article is organized as follows. The next section provides an overview of China's agricultural biotechnology development policies. We argue that despite the slight adjustment of GM strategies for commercialization policy in the short run, the overall goal of China's biotechnology development has not been altered. The growth of China's public investment in agricultural biotechnology has not slowed, but instead accelerated. These arguments are further discussed and supported by information provided in the third section of this article, focusing on agricultural biotechnology research capacity building and public investment. The fourth section examines specific cases in China's agricultural biotechnology development, research priorities, and commercialization. The final section provides concluding remarks.

An Overview of China›s Agricultural Biotechnology Development Strategies and Policies

China's leaders have paid great attention to agricultural technology. Among various agricultural technologies, agricultural biotechnology is one of the priority areas that have received the greatest attention. For example, in response to *Science* Editor Ellis Rubenstein's question about concerns in the West regarding GMOs and criticisms of biotechnology, China's President Jiang Zemin stated, «We are also very much concerned about these.... I think it is important to uphold the principle of freedom of science. But advances in science must serve, not harm humankind. The Chinese government is now mulling over new rules and regulations to guide, promote, regulate, and guarantee a healthy development of science. I believe biotechnology—especially gene research—will bring good to humanity...» (Rubenstein, 2000).[6] This statement reflects China's position on biotechnology development: promoting the technology but showing appropriate precaution for biosafety, the environment, food safety, and the commercialization of biotechnology.

Goals and Strategies

Beginning in the early 1980s when China prepared to initiate its national biotechnology program, its biotechnology developmental goals were multifaceted. The government defined its goals of in terms improving the nation's food security, promoting sustainable agricultural development, increasing farmers' income, improving the environment and human health, and raising its competitive position in international agricultural markets along with other public agricultural development programs. And from the point of view of the technology itself, the most frequently stated goal was to create a modern, market-responsive, and internationally competitive biotechnology research and development system in China (Ministry of Science and Technology [MOST], 1990, 2000; State Science and Technology Commission [SSTC], 1990).

To meet these goals, the government's plan to modernize its agricultural biotechnology system was composed of several key measures. These included measures to establish a comprehensive public financed research system, investment to enhance the innovative capacity (both human and physical capacity) of the national biotechnology research program, and creation of institutions and regulations to ensure healthy development of the technology that contributes to human welfare (MOST, 2000).

National Agricultural Biotechnology Research Institutions

The earliest plan to promote biotechnology research was initiated in the beginning of the "Seventh Five-year Plan" (1986-1990) when the first comprehensive National Biotechnology Development Policy Outline was issued (SSTC, 1990). This outline was prepared by more than 200 scientists and officials under the leadership of the Ministry of Science and Technology (MOST), the State Development and Planning Commission (SDPC), and the State Economic Commission in 1985 and further revised in 1986 (Table 1). The outline defined research priorities (see later part of this section), the development plan (e.g., the "863 Plan"), and measures to achieve targets or goals.

Under this outline, a number of high-profile technology programs were launched after the middle 1980s. Some of the most significant programs included the "863 High-tech Plan," the "973 Plan," Natural Science Foundation of China, the Initiative of National Key Laboratories on Biotechnology, the Special Foundation for Transgenic Plants Research and Commercialization, the Key Science Engineering Program, the Special Foundation for High-tech Industrialization (or Commercialization), the Bridge Plan, and others (Table 1).

Table 1: Major policy measures related to biotechnology in China since the early 1980s

Key Breakthrough Science & Technology Projects	Started in 1982 by SDPC. Updated every five years. One of major components of these projects is biotechnology R&D.

Patent system	Patent law promulgated 1985. A total of 1,599 applications on genetic engineering for invention patents were filed between 1985 and 1999.
National Biotechnology Development Policy Outline	Prepared by scientists and officials led by MOST, SDPC, and others in 1985. Formally issued by the State Council in 1988. The Outline defined the research priorities, development plan and measures to achieve targets.
National Key Laboratories (NKLs) on Biotechnology	Started in 1985 under MOST. Thirty National Key Laboratories in biotechnology (15 on agriculture or agriculture related) have been established. NKLs are open laboratories, inviting both domestic and international visiting fellows.
The Climbing Program	A National Program for Key Basic Research Projects, including biotechnology program, initiated in the early 1980s.
High Technology Research and Development Plan (863 Plan)	Approved in March 1986 with 10 billion RMB for 15 years to promote high-technology R&D in China. Biotechnology is one of seven supporting areas, with a total budget of about 1.5 billion RMB from 1986-2000.
Natural Science Foundation of China	Established in 1986 to support basic science research. Life science and agronomy are two support areas related to agrobiotechnology.
Biosafety regulations	MOST issued the Biosafety Regulations on Genetic Engineering in July of 1993, which include the biosafety grading and safety assessment, application and approval procedure, safety control measures, and legal regulations.
Agricultural biosafety regulations	MOA issued the Safety Administration, Implementation, and Regulations on Agricultural Biological Genetic Engineering in July 1996.

973 Plan	Initiated in March 1997 to support basic science and technology research. Life science is one of the key supporting areas.
Agricultural GMO Biosafety Committee	Ministry-level Agri GMO Biosafety Committee was set up in MOA in 1997. The Committee was updated in 2002 to national level with its office in MOA.
Special Foundation for Transgenic Plant Research and Commercialization	A five-year program launched in 1999 by MOST to promote the research and commercialization of transgenic plants in China. The total budget of this program in the first five years is 500 million RMB.
Key Science Engineering Program	Started in the late 1990s under MOST and SDPC to promote basic research, including biotechnology program. The first project on biotech (crop germplasm and quality improvement) was funded in 2000 with 120 million RMB.
Foundation for high-tech commercialization	A special program supported by the SDPC to promote the application and commercialization of technologies, started from 1998.
Seed Regulation and Law	Regulation on the Protection of New Varieties of Plants was issued in 1999. The first Seed Law was issued in 2000.
Updated and amended agricultural biosafety regulations	1996 MOA's biosafety regulation was amended and issued by the State Council in May 2001. Three regulations on the biosafety management, trade, and labeling of GM farm products were issued by MOA to take effect after March 20, 2002.

Foreign investment in GMOs	In April 2002, the SDPC, State Economic and Trade Commission, and MOTEC jointly issued a Guideline List of Foreign Investment, which puts GMO as a prohibited area for foreign investment

The 863 Plan, also called National High-Tech Research and Development Plan, was approved in March 1986. The 863 Plan supports a large number of applied as well as basic research projects with a 10 billion RMB yuan budget (equivalent to US$ 3 billion, based on the official exchange rate of 3.4 in 1985, or US$ 1.2 billion, based on the official exchange rate of 8.27 in 2000) over 15 years to promote high technology research and development (R&D) in China. Biotechnology is one of seven supporting areas, with a budget of 1.3 billion RMB yuan in 1986-2000, with 50% of this budget focused on agricultural biotechnology.

The National Basic Sciences Initiative, also called the 973 Plan, with a total budget of 2.5 billion yuan (US$ 302 million, converted at the1997-2002 average exchange rate) in the period of 1997-2002, was another high-tech research plan initiated in March 1997. This plan is complementary to the 863 and many other national initiatives on high-tech development, as it exclusively supports basic research. Life science, with biotechnology as a priority, constitutes one of the key programs under this plan.

In contrast to the perception that China's biotechnology development is shifting towards a "go slow" approach, our review of recent biotechnology research programs indicates that China instead has accelerated its biotechnology development since the late 1990s. The view suggesting that progress in biotechnology research has slowed is unfounded. For example, a new program aimed at strengthening the national research and industrialization of China's agricultural biotechnology, the Special Foundation of Transgenic Plants Research and Commercialization (SFTPRC), was initiated in 1999 by the Ministry of Science and Technology. This new program is a unique foundation to promote both research and commercialization of transgenic plants. Only those projects that are jointly submitted by research institutes and companies are eligible to receive funding from about half of the programs under SFTPRC. The foundation also requires a significant financial commitment from companies to commercialize technology generated by a project, a reflection of China's aim to accelerate the diffusion of biotechnology. The total budget of SFTPRC during its first five years (1999-2003) was 500 million RMB yuan (about US$ 60 million). Concurrently, the Ministry of Science and Technology and the State Development and Planning Commission jointly sponsored the Key Science

Engineering Program (KSEP), a national program to promote the fundamental construction for research in the late 1990s. As an example, one extremely large biotechnology project on crop germplasm and quality improvement through biotechnology received 140 million RMB yuan (US$ 17 million) from KSEP in 2000. Moreover, the State Council passed a new Agricultural Science and Technology (S&T) Development Compendium in 2001. The compendium reemphasizes the importance of agricultural biotechnology in improving the nation's agricultural productivity, food security, and farmers' income, and has led to a new decision to further increase the research budget for the development of biotechnology. The proposed biotechnology development budget for the Tenth Five-year Plan (2001-2005) is far more than all prior budgets over the past 15 years (see the next section for more detail).

With the above efforts, by 2001 there were about 150 laboratories at national and local levels located in more than 50 research institutes and universities across China working on agricultural (plant and animal) biotechnology. Over the last two decades, China established 30 National Key Laboratories (NKL). Among these NKLs, 12 are exclusively working on, and three have major activities in agricultural biotechnology (Huang, Wang, Zhang, & Zepeda, 2001). Besides NKLs, there are numerous Key Biotechnology Laboratories and programs within ministries and local provinces.

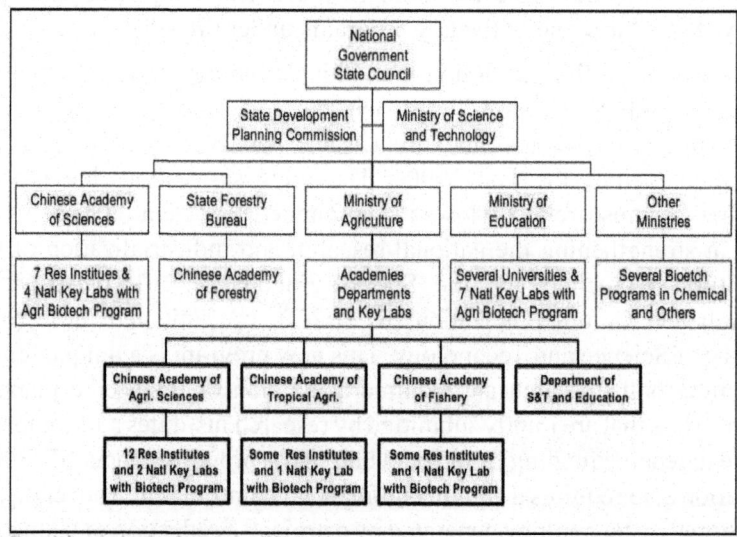

Figure 1: Organization chart for agricultural biotechnology research at national level.

At the national level, the Ministry of Agriculture (MOA), the Chinese Academy of Sciences (CAS), the State Forestry Bureau (SFB), and the Ministry of Education (MOE) are the major authorities responsible for agricultural biotechnology research (Figure 1). Under the Ministry of Agriculture, there are three large academies—the Chinese Academy of Agricultural Sciences (CAAS, which employs about 8,000 research and support staff), the Chinese Academy of Tropical Agriculture (CATA), and the Chinese Academy of Fisheries (CAFi). Among the 37 institutes in CAAS, there are 12 institutes, two National Key Laboratories and five Key Ministerial Laboratories conducting biotechnology research programs. The CAFi and the CATA also have several biotechnology laboratories or programs, and each has one NKL for biotechnology.

Agricultural biotechnology research is also conducted by national institutes external to the Ministry of Agriculture's research system. For example, under the Chinese Academy of Sciences there are at least seven research institutes and four NKLs that focus on agricultural biotechnology. Research institutes within the Chinese Academy of Forestry (CAFo) under the State Forest Bureau and numerous universities (i.e., Beijing University, Fudan University, Nanjing University, Central China Agricultural University, and China's Agricultural University) under the Ministry of Education are examples of other institutions conducting agricultural biotechnology research. There are seven NKLs located in seven leading universities conducting agricultural biotechnology or agriculturally related basic biotechnology research. Other public biotechnology research efforts on agriculturally related topics include agrochemical (e.g., fertilizer) research by institutes in the State Petro-Chemical Industrial Bureau.

Although the programs at the national level presented in Table 1 and Figure 1 and discussed above constitute China's mainstream agricultural biotechnology research, research at the provincial level also contributes to the development of China's agricultural biotechnology. They follow a similar institutional framework to that at the national level (Figure 1). Each province has its own provincial academy of agricultural sciences and at least one agricultural university. Each academy or university at the provincial level normally has one or two institutes or laboratories focused on agricultural biotechnology. Provincial biotechnology research is funded by both local governments (core funding and research projects) and the central government (research projects only).

Finally, it is worth noting that the numbers of both national and provincial biotechnology programs and institutes continue to increase. China is even considering establishing a new national agricultural biotechnology research center—a megaresearch center over the current 150 agricultural biotech laboratories. Based on these developments, if there were shifts in China›s biotechnology developmental plan, it is towards probiotechnology research.

Biosafety Management Institutions and Regulations

Institutional Setting. Although the Ministry of Science and Technology is mainly responsible for biotechnology research, the Ministry of Agriculture is the primary institution in charge of the formulation and implementation of biosafety regulations on agricultural GMOs and their commercialization, particularly after 2000. In order to incorporate representation of stakeholders from different ministries, the State Council established an Allied Ministerial Meeting comprised of leaders from the MOA, the SDPC, the MOST, the Ministry of Public Health, the Ministry of Foreign Economy and Trade (MOFET), the Inspection and Quarantine Agency, and the State Environmental Protection Authority (SEPA). This Allied Ministerial Meeting coordinates key issues related to biosafety of agricultural GMOs, examines and approves the applications for GMO commercialization, determines the list of GMOs for labeling, and establishes import or export policies for agricultural GMOs and their products.

However, routine work and daily operations are handled by the Office of Agricultural Genetic Engineering Biosafety Administration (OGEBA). The National Agricultural GMO Biosafety Committee (BC) is the major player in the process of biosafety management.[7]Currently, the Committee comprises of 56 members.[8] The committee meets twice each year to evaluate all biosafety assessment applications related to experimental research, field trials, environmental release, and commercialization of agricultural GMOs. It provides approval or disapproval of recommendations to OGEBA based on the results of its biosafety assessments. OGEBA is responsible for the final approval of decisions.

The Ministry of Public Health (MPH) is responsible for food safety management of biotechnology products. The Appraisal Committee, consisting of food health, nutrition, and toxicology experts nominated by MPH, is responsible for reviewing and assessing GM foods as they have been designated a Noval Food. The State Environmental Protection Authority participates in GMO biosafety management through the Allied Ministerial Meeting and through their members on the National Agricultural GMO Biosafety Committee. Although SEPA has taken the responsibility of international biosafety protocol, its focus on biotechnology in China is limited to biodiversity.

Concerning the institutional setting of agricultural GMO biosafety management, China has several unique elements compared to the US and the EU. The Ministry of Agriculture in China appears to have more power than its counterparts in the US and the EU. The leaders in the State Council of the previous government believed that the MOA is more familiar with, and has more expertise in agriculture and agricultural GMOs than any other ministry. Moreover, because the MOA is also in charge of pesticide use and its environmental assessment in agricultural production, national leaders consider the MOA a major player in China›s agricultural biosafety management.

Critics (i.e., SEPA) of this system argue that this institutional setting might result in less attention paid to environmental risks of GMOs and may have a potential conflict of interest, as the MOA is primarily responsible for agricultural production, and many biotechnologies are developed under the MOA›s own research system. The debate on whether SEPA or MOA is a more appropriate institution to take a lead role on biosafety has continued since the biosafety management system was set up in 1997. However, under the current national administrative system, it is unlikely for SEPA to take a significant role in biotechnology, unless there is significant reform of the government›s structure by China›s new leaders, who have been in office since early 2003.

The other unique aspect is that China›s National Agricultural GMO Biosafety Committee plays a critical role in the biosafety decision-making process. As most of its 56 current members (29 for GM plants, nine for recombined microorganisms for plant, 12 for transgenic animals and recombined microorganisms for animals, and six for GM aquatic organisms) are experts from various research institutes within the public sector, its GMO biosafety assessment provides key information for decision makers on whether OGEBA should approve or disapprove GMO application cases. However, the weakness of this approach is the time constraint from BC members who often are leading scientists in various disciplines. Other concerns include the heavy service burden of a few key individual scientists and too many biotechnologists on the Biosafety Committee.

Biosafety Regulations. Before 2002, the principle governing China›s agricultural GMO biosafety was to adopt a product-based GMO management system. However, China has attempted to impose labeling regulations on GMOs and GM products since March 2002. By imposing a compulsory labeling policy on GMOs, China›s biosafety management partially shifts towards a process-based GMO management system. This adjustment has led to wide debate within China and between China and many other countries, as we described above in the introduction. Before we discuss this new labeling policy, it is worth reviewing briefly the evolution of China›s agricultural GMO biosafety regulations and policies in the past.

Evolution of China›s Biosafety Regulations. In response to the emerging progress in China›s agricultural biotechnology, the first biosafety regulation, «Safety Administration and Regulation on Genetic Engineering,» was issued by the Ministry of Science and Technology in 1993. This regulation consisted of general principles, safety categories, risk evaluation, application and approval, safety control measures, and legal responsibilities. After the above regulation was decreed, MOST required relevant ministries to draft and issue corresponding biosafety regulations on biological engineering (i.e., the Ministry of Agriculture for agriculture and the Ministry of Public Health for food safety). Following MOST›s guidelines, the MOA issued the Implementation Regulations on Agricultural Biological Engineering in 1996. This regulation is similar in many aspects to the US GMO biosafety regulations. Labeling was not part of this regulation, nor was any restriction imposed on imports or exports of GMO products. The regulation also did not regulate processed food products that use GMOs as inputs.

Under the 1996 GMO biosafety regulation policy, OGEBA received 433 applications for field trials, environmental release, or commercialization in 1997-2000 (Table 2). Among them, 322 cases had been approved, covering more than 60 crops and several animals, as well as numerous microorganisms. It is interesting to note that both the number of cases applied or submitted and cases approved increased persistently over time. Imposing more GMO restrictions did not reduce the number of applications. However, if we decompose this data into different stages of GMO development and by crop, they do show that the numbers of cases approved for commercialization declined in 2000 and that no new GMO crops have been approved since 1999, excluding cotton, tomatoes, sweet peppers, and petunias.

China›s Stance on Biotechnology Development—For or Against? Using solely the above approval case numbers may lead to erroneous conclusions regarding China's stance on GMO development. Indeed, the small number of approvals for commercialization of GM crops in 2000 (seven cases, Table 2) was due to many factors. First, in the prior year, 1999, a large number of cases were approved for commercialization (27 cases, Table 2), almost all for Bt cotton. As expected, there were fewer applications for Bt cotton commercialization in 2000, as most of the Bt cotton varieties (nearly 20 varieties from both CAAS and Monsanto) had been earlier approved for commercialization. Second, as argued by OGEBA and the Biosafety Committee, the existing food-related GM crops were not ready for commercialization due to unclear issues over their food safety. For example, food safety testing managed by Ministry of Public Health has not come to a conclusion on whether the current GM rice is not substantially different from non-GM rice. Research on GM rice›s food safety is still ongoing. Third, the testing on environmental safety and biodiversity has been limited to a very small scale and in few locations. Lastly, Bt cotton had been tested and adopted widely in other countries before China approved its commercialization, and given that cotton is a nonfood crop, the context surrounding GM rice differs than that for cotton. Rice is the most important food crop in both China and the rest of Asia. Additionally, GM rice has never been commercialized anywhere in the world.

Table 2: The number of cases in agricultural (plant, microorganism, and animal) biotechnology submitted and approved for field trials, environmental release, and commercialization in 1997-2000

	1997	1998	1999	2000	Total
Submitted					
Field trial	14	41	28		
Environmental release	37	18	63		
Commercialization	6	9	35		
Total	57	68	126	182	433
Approved					
Field trial	12	40	22	115	189
Environmental release	30	10	34	19	93
Commercialization	4	2	27	7	40
Total	46	52	83	141	322

More recently, our communication with OGEBA's officials and members of the Biosafety Committee reveal that China is badly in need of institutional and capacity building for GMO biosafety management. During the 7th International Symposium on the Biosafety of Genetically Modified organisms held in Beijing in October 2002, an official from OGEBA concluded his speech with five major challenges that OGEBA currently faces: "an appropriate regulatory approach *(to improve current practices)*,[9] a science-based safety assessment, capacity building, transparency, communication and information exchange (Cheng & Peng, 2002).

Given the above discussion, it is no surprise that OGEBA declined three applications for GM rice commercialization in 1999-2000. We believe that China›s current adjustment in biosafety management is just one effort to establish a more comprehensive GMO biosafety management system that provides a firm base for future sustainability. Current adjustment is also partly in response to the growing worldwide debate on GMOs and their potential risks, as well as China›s agricultural trade.[10]

Chinese policymakers are concerned about environmental and food safety in response to the debate on the potential risks of GMOs recently raised by the Chinese media. The debate in China has involved scientists, government officials, and newspaper reporters; responses and reactions vary among stakeholders and change over time as more information becomes available on biotechnology (Huang, Wang, & Keeley, 2001). A consensus seems to be growing in China that the most important task a scientist or biotechnologist can do is to reduce the potential negative effects and demonstrate the safety of GMOs.

As a consequence of this consensus, research budgets allocated to biosafety management and the study of biosafety have increased. Since 1999-2000, nearly all biotechnology research programs have expanded their scope into biosafety issues, particularly for the following programs: 863, 973, and the Special Foundation for Transgenic Plants Research and Commercialization. A number of national institutes under the Ministry of Agriculture, the Ministry of Public Health and the State Environmental Protection Agency have launched various biosafety programs, including capacity building for biosafety management and risk assessment, research studies on environmental safety and food safety, detection technology for GMOs and GMO products, and monitoring of international practices.

However, arguing for a more comprehensive and science-based safety assessment as reasons for the recent adjustment of China's GMO commercialization does not imply that there is no concern over the impacts of GMO development on agricultural trade. Issues such as labeling of GM products and possible trade barriers resulting from biotechnology concerns in countries that follow precautionary and preventive policies do have impacts on the current (short run) pace of GMO commercialization in China. Agricultural trade had been an important contributor to the aggregate Chinese economy and trade.

It appears that international trade concerns may have been one of the important factors, but not the dominant factor, in recent agricultural biotechnology policy processes. The critical event here appears to have been the EU's decision to ban Chinese soy sauce imports produced with GM soybeans imported from the United States. Additionally, the recent decision by Thailand (the world's leading rice exporter) to halt further development of GM rice may also have been significant. It is unclear whether public attitudes toward GMOs in Europe are now softening or whether policies may soon change; hence, a short-run "wait and see" tactic is probable in China.

New Biosafety Regulations. In response to the above concerns, in May 2001 the State Council decreed a new and general rule of Regulation on Safety Administration of Agricultural GMOs to replace an early regulation issued by the Ministry of Sciences and Technologies in 1993 (Safety Administration Regulation on Genetic Engineering, Table 1). The Ministry of Agriculture then announced three new implementation regulations on biosafety management, trade, and labeling of GM farm products that were planned to take effect after March 20, 2002.[11] There were several important changes to existing procedures included in these guidelines, as well as details of regulatory responsibilities after commercialization. These included the addition of an extra preproduction trial stage prior to commercial approval, new processing regulations for GM products, labeling requirements for marketing, new export and import regulations for GMOs and GMO products, and local- and provincial-level GMO monitoring guidelines. In the meantime, the Ministry of Public Health also promulgated its first regulation on GMO food hygiene in April 2002, to take effect after July 2002.

By late 2002, the system of biosafety regulation in China had clearly become progressively more elaborate and sophisticated. Many provinces have established provincial biosafety management offices under provincial agricultural bureaus. These biosafety management offices collect local statistics on and monitor the performance of research and commercialization of agricultural biotechnology in their provinces and assess and approve (or disapprove) all applications of GM related research, field trials, and commercialization in their provinces. Only those cases that are approved by provincial biosafety management offices are submitted to the National Biosafety Committee for further assessment. However, China still has a long way to go before all decreed regulations could be fully implemented. Our three years of Bt cotton farm surveys across five provinces during 1999-2001 found that about half of Bt cotton varieties had been adopted by farmers, but they did not apply to the National Biosafety Committee for commercialization. Seeds are distributed to farmers mainly by local seed companies, the extension system, research institutes, and small traders. The institutions, human capacity, and financial support for implementation of GMO regulations are far away from the necessary requirements. In addition to this, collaboration and coordination between ministries on research, commercialization, and biosafety management needs to be further strengthened.

Agricultural Biotechnology Capacity Building and Public Investment

Creation of a modern and internationally competitive biotechnology research and development system requires substantial investments in human and financial capacities. Since the early 1980s, China's public investment in, and the number of research staff working on biotechnology has increased significantly, in contrast to stagnating trends for general agricultural research expenditures in the late 1980s and early 1990s (Huang, Hu, & Rozelle, 2002). For example, based on our 2000 survey of 29 research institutes in plant biotechnology[12] and on extensive interviews with ministries and research institutes in 2002, we estimate that the number of plant biotechnology researchers tripled in the past 15 years (Table 3). More than 2,100 researchers are now working on plant biotechnology alone. If we include biotechnology from the animal sector, the number of agricultural biotechnology researchers may reach 3,000, and may be one of the largest biotechnology research efforts in the world.

Table 3: Estimated number of research staff and expenditures on plant biotechnology research in China, 1986-2000

| Year | Number of staff | Research expenditure | | |
		Million RMB at current price	Million RMB at 2000 price	Million US$
1986	740	14	38	4.2
1990	1067	40	68	8.3
1995	1447	88	87	10.5
2000	2128	322	322	38.9

Note. Expenditures include both project grants and costs related to equipment and buildings. Both staff and research expenditures are estimated by the authors based on our earlier studies (Huang, Wang, Zhang, & Zepeda, 2001) and recent interviews in China. The results from our recent interviews show that the data in Table 2 are higher than our earlier estimates. Official exchange rate in the corresponding year is used to convert the domestic currency to US dollars.

Similar to other agricultural research programs in China, agricultural biotechnology research is primarily built upon research institutes. Among the 29 institutes surveyed, the number of agricultural biotechnology researchers in universities accounted for only 10% of total research staff.[13] Among total researchers, nearly 60% are professionals, and the share of the professional staff has been increasing over time (Huang, Wang, Zhang, & Zepeda, 2001), again indicating growing human capacity in biotechnology research.

The quality of human capacity to conduct biotechnology research has improved over time. Among professional staff, the share of researchers with Ph.D. degrees increased from only 2% in 1986 to more than 20% in 2000. This share is expected to continue to increase in the future. Although the share of researchers with biotechnology Ph.D. degrees is still low by international standards, it is interesting to note that this share is much higher than those in the general agricultural research system. In China›s national agricultural research system, Ph.D. researchers accounted for only 1.1% of the total professional staff in 1999 (Huang, Hu, & Rozelle, 2001).

Even more dramatic growth has occurred in China›s biotechnology research investment (Table 3). China›s biotechnology research investment was trivial in the early 1980s (MOST, 1990). Although there are no statistics available from official sources, our estimates show that biotechnology investment has grown substantially. For example, the estimated investment in plant biotechnology research was only US$ 4.2 million in 1986 when China formally started its 863 Plan (Table 3). By 1990, China›s investment grew to US$ 8.3 million. During this period, the research project budget nearly tripled, and equipment expenses nearly doubled (Huang, Wang, Zhang, & Zepeda, 2001). Although the growth rate of biotechnology research investment slowed between 1990 and 1995 (this is expected as the large investment in biotechnology equipment was nearly complete in the early 1990s), the annual growth rate in the research project budget in real terms remained as high as 10% during this period.

China›s biotechnology research investment increased considerably from US$ 10.5 million in 1995 to US$ 38.9 million in 2000, representing an annual growth rate of about 30%. This investment in China›s biotechnology is mainly due to government sources. According to our survey of 29 biotech research institutes, public investment accounted for 94% of the total plant biotechnology budget in 1999, and this share has been increasing over our study period, from 1986 to 1999 (Huang, Wang, Zhang, & Zepeda, 2001). Budgets from competitive grants for research projects accounted for two thirds of the total budget and this share also has shown an increase over time, reflecting China›s biotechnology development moving from a capacity-building stage to a research stage. Our recent interviews with officials and research administrators from the Ministry of Science and Technology confirm that the Ministry is accelerating its investment in national biotechnology program: the Tenth Five-year Plan (2001-2005) for biotechnology development. Under this plan, the total investment in agricultural biotechnology is targeted to be four times as much as the total amount spent on agricultural biotechnology in the past 15 years (1985-2000). If this goal is realized, China will account for more than one fourth of the world›s current public spending on agricultural biotechnology.

Agricultural Biotechnology Development, Research Priorities and Commercialization

An Overview

The focus of China's biotechnology development in its early stages (in the early 1970s) was on cell engineering, tissue culture, and cell fusion and emphasized crops such as rice, wheat, maize, cotton, and vegetables (Key Laboratory of Crop Molecular and Cell Biology, 1996). However, the most significant progress in agricultural biotechnology was made following the development of transgenic techniques after 1983. The pace of biotechnology research accelerated significantly after China initiated the 863 Plan in 1986 (Table 1). Bt cotton is a most successful story of agricultural biotechnology in China. In response to rising pesticide use and the emergence of a pesticide resistant bollworm population in the late 1980s, China's scientists began research on GM cotton. Starting with a synthesized gene originally from the bacterium *Bacillus thuringiensis* (Bt), China›s scientists transferred this modified Bt gene into major cotton cultivars by the so-called pollen tube pathway transformation. Greenhouse testing began in the early 1990s. The first commercial use of GM cotton was approved in 1997. During the same year, Bt cotton varieties from publicly funded research institutes and from a joint venture with Monsanto became available to farmers. The release of Bt cotton began China›s first large-scale commercial experience with a product of the nation›s biotechnology research program.

In addition, other transgenic plants with resistance to insects, disease or herbicides, stress tolerance, or plants with improved quality have been approved for field release, and some are nearly ready for commercialization. These include transgenic cotton lines resistant to fungal disease, rice resistant to rice stem borer or bacteria blight, diseases, herbicide, and salt tolerance, wheat resistant to barley yellow dwarf virus (Cheng, He, & Chen, 1997), maize resistant to insects and with improved quality (Zhang, Liu, & Zhao, 1999), poplar trees resistant to gypsy moths, soybeans resistant to herbicides, transgenic potato resistant to bacterial disease or Colorado beetles, among others (Ministry of Agriculture [MOA], 1999; National Center of Biological Engineering Development [NCBED], 2000; Li, 2000).

Progress in plant biotechnology has also been made in recombinant microorganisms such as soybean nodule bacteria (nitrogen-fixing bacteria for rice and corn) and phytase from recombinant yeasts for feed additives (Huang, 2002). Genetically modified nitrogen-fixing bacteria and phytase have been commercialized since 1999. In animals, transgenic pigs and carp have been produced since 1997 (NCBED, 2000). Recently, Chinese researchers also announced the successful sequencing of the rice genome (Yu et al., 2002). They have produced a draft sequence of the rice genome for the most widely cultivated subspecies in China, *Oryza sativa L.* ssp. *indica*, by whole-genome shotgun sequencing.

According to a nationwide survey conducted by the MOA in 1996, Chinese scientists have tried to use more than 190 genes transferring to more than

100 organisms (103 genes used in 47 plants, 32 genes used in 22 animals, 56 genes used in 31 species of microorganisms). These figures have been further expanded after 1996 (Cheng & Peng, 2002). By 2001, there were more than 60 plants under research and 121 genes used for transformation (Peng, 2002). The list of GM crops in trials is also impressive and differs from those being worked on in other countries.

Research Priorities and Products in the Research Continuum

Huang, Wang, Zhang, and Zepeda (2001) summarized research priorities for plant biotechnology identified in various Biotechnology Development Outlines over the past 15 years in China (Table 4). Since the mid-1980s, cotton, rice, wheat, maize, soybean, potato, and rapeseed have been consistently listed as priority crops for biotechnology research funding. Functional genomics for major plants and animals, crop genetic breeding through the application of gene transformation, chromosome hybridization and marker assisted selection, cultivation of crops with improved resistance or quality and genetic breeding of animals, use of animal and plant cells as bioreactors in producing secondary products, special proteins and vaccines, recombined microorganisms to produce biofertilizers and biopesticides, and others have been identified as priority technologies for public funding.

Among crops, cotton is listed as a priority crop not only because of its importance by sown area and its contributions to the textile industry and trade, but also because of the serious problems with the associated rapid increase in pesticide applications to control insects (i.e., bollworm and aphids). Per-hectare pesticide expenditures for cotton production in China increased considerably over recent decades, reaching 834 RMB yuan (approximately US$ 100) in 1995. This amount is much higher than comparable expenditures for grain crop production but lower than horticultural production. Cotton production alone consumed about US$ 500 million annually in pesticides in recent years.

Rice, wheat, and maize are the three most important crops in China. Each accounts for about 20% of the total area planted. Production and market stability of these three crops are a primary concern of the Chinese government, as they are central to China's food security. National food security, particularly related to grains, has been a central goal of China's agricultural and food policy and has been incorporated into biotechnology research priority setting.

Among all traits, pest resistance traits have top priority (Table 4). Recently, quality improvement traits have been included as priority traits in response to increased market demand for quality foods. Quality improvements have been targeted particularly for rice and wheat, as consumer income rises in China. In addition, stress tolerance traits—particularly resistance to drought—are gaining attention, particularly with the growing concern over water shortages in Northern China. In addition, Northern China is a major wheat and soybean production region with significant implications for China's future food security and trade.

Table 4: Research focus of plant biotechnology programs in China

	Prioritized areas
Crops	Cotton, rice, wheat, maize, soybean, potato, rapeseed, cabbage, tomato
Traits	
Insect resistance	Cotton bollworm, boll weevil, and aphids Rice stem borer Wheat aphids Maize stem borer Soybean moth Potato beetle Poplar gypsy moth
Disease resistance	Rice bacteria blight and blast Cotton fungal disease Cotton yellow dwarf Wheat yellow dwarf and rust Soybean cyst nematode Potato bacteria wilt Rapeseed sclerosis CMV and TMV
Stress tolerance	Drought, salinity, cold
Quality improvement	Cotton fiber quality Rice cooking quality Wheat quality Maize quality Corn with phytase or high lysine
Herbicide resistance	Rice, soybean
Functional genomics	Rice, rapeseed, and arabidopsis

Newer research focuses on the isolation and cloning of new disease and insect resistance genes, including the new genes conferring resistance to cotton bollworm (Bt, CpTI and others), rice stem borer (Bt), rice bacterial blight (Xa22 and Xa24), rice plant hopper, wheat powdery mildew (Pm20), wheat yellow mosaic virus, and potato bacterial wilt (cecropin B) (MOA, 1999; NCBED, 2000). These genes have been applied in plant genetic engineering since the late 1990s. Significant progress has also been made in the functional genomics of arabidopsis and in plant bioreactors, especially in utilizing transgenic plants to produce oral vaccines (Biotechnology Research Institute, 2000).

By the end of 2001, GM plants from 13 plant species and more than 50 genes were approved for field trial, environmental release, and commercialization. Thirty-six recombined microorganism species and 51 strains have been involved in research with 89 genes for insect and disease resistance or nitrogen fixation.

Commercialization of Agricultural Biotechnology

By 2002, 18 transgenic cotton varieties generated by Chinese institutions and five varieties from Monsanto with resistance to bollworm have been approved for commercialization in China. Although several GM varieties of tomato, sweet pepper, chili pepper, and petunia have also been approved for commercialization since 1997, the area planted with these four crops remains small. Personal communications with several member of the agricultural Biosafety Committee show that the economic benefits of adopting the current three GM crops are minimal or nonexistent; no private companies have been attracted to invest in their commercialization.

Table 5 presents our most updated estimates of Bt cotton areas sown in China in 1997-2001. After the Bt cotton variety was approved for commercialization in 1997, the total area planted using Bt cotton increased to 0.65 million hectares in 1999. In 2001, the area reached more than 2 million hectares and accounted for 45% of China›s cotton area. China›s GM crop area follows that of the US, Argentina, and Canada. Although less than 4% of the total global area of GM crops was grown in China in 2001, we estimate that nearly 5 million Chinese farmers planted Bt cotton, as the average farm size is only about 0.5 hectares and includes several crops.

Table 5: Bt cotton adoption in China, 1997-2001

Year	Cotton area (000 hectare)		Bt cotton share (%)
	Total	Bt cotton	
1997	4491	34	1
1998	4459	261	6
1999	3726	654	18
2000	4041	1216	30
2001	4810	2174	45

Concluding Remarks

Chinese policymakers consider agricultural biotechnology as a strategically significant tool for improving national food security, raising agricultural productivity, and creating a competitive position in international agricultural markets. Consistent with these aims, China also intends to be one of world leaders in biotechnology research and major domestic supplier of biotechnologies. This objective is closely linked to the perception by Chinese policymakers that there are risks associated with reliance on imported technologies to guarantee national food security. Despite the growing debate worldwide on GM crops, China has developed agricultural biotechnology decisively since the mid-1980s. By 2001, China had the fourth largest sown area of GM crops in the world. Research and development has continued apace, and China now has several genetically modified plants that are in the pipeline for commercialization.

The institutional framework for supporting agricultural biotechnology research program is complex both at national and local levels. The growth of government investment in agricultural biotechnology research has been remarkable. However, coordination among institutions and consolidation of agricultural biotechnology programs will be essential for China to create an even stronger and more effective biotechnology research program in the future.

Examination of the research foci of agricultural biotechnology research reveals that food security objectives and farmers' current demands for specific traits and crops have been incorporated into priority setting. Moreover, the current priority setting for investments in agricultural biotechnology research has been directed at commodities for which China does not have a relative comparative advantage in international markets (such as grain, cotton, and oil crops). This implies that China is targeting its GMO products at the domestic market. The emphasis on developing drought-resistant and other stress-tolerant GM crops also suggests that biotechnological products are not only being geared to high-potential areas, as critics argue, but also at the needs of poorer farmers.

Many competing factors are exerting pressure on Chinese policymakers to continue with research and commercialization of transgenic crops. The demand of producers (for productivity-enhancing technology) and consumers (for cost savings), the current size and rate of increase of research investments, and past success in developing technologies suggest that products from China's plant biotechnology industry are likely to become widespread in China in the near future. Although China is still struggling with issues of environmental and consumer safety, and the system of biosafety regulation has become progressively more elaborate and sophisticated, the system might not work well and might eventually hurt its national biotechnology application in the future if biosafety management capacity is not improved as much as research capacity. Investment in China's biotechnology R&D is essential for the nation to promote its biotechnology industry; investment in biosafety management capacity and policy implementation are also critical factors for health and sustainable development of this industry.

ENDNOTES

[1] In 1998, GM sweet peppers were approved for commercialization.

[2] Also see Marchant, Fang, and Song (2002), in this issue, for information on the evolution of China›s agricultural biotechnology policies.

[3] In 2001, China imported about 14 million metric tons of soybeans from the US, Argentina, and Brazil. Most of these imports were Roundup Ready soybeans. After two months of intensive negotiations between China and the US, an interim agreement was reached in early 2002. China in effect temporarily waived its import and export regulations of GMOs until December 2002, and this was further postponed to September 2003. Concurrently, China has agreed to recognize US assurances that its soybeans are safe for human consumption.

[4] See the recent report in the *Washington Post* (Goodman, 2002), the *New York Times*(Kahn, 2002), and a front-page article in *China Daily* (Zhigang, 2002).

[5] Issues related to impacts of biotechnology are not discussed in this paper. They can be found in a series of papers written by the authors with their collaborators, including Pray, Ma, Huang, and Qiao (2001); Huang, Rozelle, Pray, and Wang (2002); Huang, Hu, Rozelle, Qiao, and Pray (2002); Huang, Hu, Pray, Qiao, and Rozelle (in press); and Pray, Huang, and Rozelle (2002).

[6] In his opening speech at the International Rice Conference held in Beijing on September 15, 2002, President Jiang Zemin restated the importance of agricultural biotechnology in boosting agricultural productivity growth and food security.

[7] The Biosafety Committee was established in 1997 under the Ministry of Agriculture; it was a ministry-level institution. Since June 2002, the Committee was upgraded to a national-level institution.

[8] Biosafety Committee members work part-time for the BC and are scientists from different disciplines including agronomy, biotechnology, plant protection, animal science, microbiology, environmental protection, and toxicology. A few members are also agricultural administrations. All BC members are nominated by the Ministry of Agriculture.

[9] Because biosafety management is a new activity for OGEBA, it is

understandable that they are seeking a more appropriate approach even after years of commercialization of nonfood crops (Bt cotton).

[10] So far, Chinese consumers have not created many problems for GMO development in China.

[11] These three new regulations replaced the Safety Administration, Implementation, and Regulation on Agricultural Biological Genetic Engineering issued by the Ministry of Agriculture in July 1996.

[12] The survey was conduced by the Center for Chinese Agricultural Policy and the International Service for National Agricultural Research; detailed results are reported in Huang, Wang, Zhang, and Zepeda (2001).

[13] In terms of the overall agricultural research system in China, researchers in universities account for about 8% of the nation's total agricultural researchers.

REFERENCES

1.   Biotechnology Research Institute. (2000). *Research achievements of biotechnology*(working paper). Beijing: Chinese Academy of Agricultural Sciences.

2.   Chen, Z. (2000, July-August). *Review of R&D on plant genetic engineering in China*. Paper presented at the China-Asean workshop on transgenic plants, Beijing, China.

3.   Cheng, Z, He, X., & Chen, C. (1997). Transgenic wheat plants resistant to barley yellow dwarf virus obtained by pollen tube pathway-mediated transformation. In *Chinese Agricultural Science for the Compliments to the 40th Anniversary of the Chinese Academy of Agricultural Science* (pp. 98-108). Beijing: China Agricultural SciTech Press.

4.   Cheng, J., & Peng, Y. (2002, October). *Biosafety regulation in China*. Paper presented at the 7th International Symposium on the Biosafety of Genetically Modified Organisms, Beijing, China.

5.   Goodman, P. (2002, September 25). China's new economy begins on the farm: Growers bear burden of being first as trade brings opportunity, risk. *Washington Post*, p. A01.

6.   Huang, D. (2002, October). *Research and development of recombinant microbial agents and biosafety consideration in China*. Paper presented at the 7th International Symposium on the Biosafety of Genetically Modified Organisms, Beijing, China.

7.   Huang, J., Hu, R., & Rozelle, S. (2002). *China's agricultural research investment: Challenges and prospects*. Beijing: China Financial and Economic Press.

8. Huang, J., Hu, R., Pray, C., Qiao, F., & Rozelle, S. (in press). Biotechnology as an alternative to chemical pesticides: A case study of Bt cotton in China. *Agricultural Economics*.

9. Huang, J., Hu, R., Rozelle, S., Qiao, F., & Pray, C.E. (2002). Transgenic varieties and productivity of smallholder cotton farmers in China. *Australian Journal of Agricultural and Resource Economics*, 46(3), 367-388.

10. Huang, J., Rozelle, S., Pray, C., & Wang, Q. (2002). Plant biotechnology in China. *Science*, *295*, 674-677.

11. Huang, J., Wang, Q., & Keeley, J. (2001). *Agricultural biotechnology policy processes in China* (working paper). Beijing: Chinese Academy of Sciences Center for Chinese Agricultural Policy.

12. Huang, J., Wang, Q., Zhang, Y., & Zepeda, J. (2001). *Agricultural biotechnology development and research capacity in China* (working paper). Beijing: Chinese Academy of Sciences Center for Chinese Agricultural Policy.

13. James, C. (2002). *Global review of commercialized transgenic crops: 2001*. International Service for the Acquisition of Agro-Biotech Applications.

14. Kahn, J. (2002, October 22). The science and politics of super rice. *The New York Times*.

15. Key Laboratory of Crop Molecular and Cell Biology. (1996). *The research and prospects of crop genetic engineering*. Beijing: Chinese Agricultural Science and Technology Press.

16. Li, N. (2000, July-August). *Review on safety administration implementation regulation on agricultural biological genetic engineering in China*. Paper Presented at the China-ASEAN Workshop on Transgenic Plants, Beijing, China.

17. Marchant, M.A., Fang, C., & Song, B. (2002). Issues on adoption, import regulations, and policies for biotech commodities in China with a focus on soybeans. *AgBioForum*, 5(4), 167-174.

18. Ministry of Agriculture [China]. (1990). *The guideline for the development of science and technology in middle and long terms: 1990-2000*. Beijing: MOA.

19. Ministry of Agriculture [China]. (1999). The application and approval on agricultural biological genetic modified organisms and its products safety. *Administrative Office on Agricultural Biological Genetic Engineering, 4*, 35-37.

20. Ministry of Science and Technology [China]. (1990). *Biotechnology development policy*. Beijing: China S&T Press.

21. Ministry of Science and Technology [China]. (2000). *Biotechnology development outline*. Beijing: MOST.

22. National Center of Biological Engineering Development. (2000). The research progress in biotechnology. *Biological Engineering Progress, 20.*

23. Paarlberg, R.L. (2000). *Governing the GM crop revolution: Policy choices for developing countries* (2020 Discussion Paper #33). Washington DC: International Food and Policy Research Institute.

24. Peng, Y. (2002, October). *Strategic approaches to biosafety studies in China*. Paper presented at the 7th International Symposium on the Biosafety of Genetically Modified Organisms, Beijing, China.

25. Pray, C., Ma, D., Huang, J., & Qiao, F. (2001). Impact of Bt cotton in China. *World Development, 29*(5), 813-825.

26. Pray, C., Huang, J., & Rozelle, S. (2002). Five years of Bt cotton production in China: The benefits continue. *The Plant Journal, 31*(4), 423-430.

27. Rubenstein, E. (2000). China's leader commits to basic research, global science. *Science,288*, 1950-1953.

28. State Development and Planning Commission [China]. (2003). *Development of biotechnology industry in China—2002*. Beijing: Chemical Industry Press.

29. State Science and Technology Commission [China]. (1990). *Development policy of biotechnology*. Beijing: The Press of Science and Technology.

30. Yu, J., Hu, S., Wang, J., Wong, G.K.S., Li, S., Liu, B., et al. (2002). A draft sequence of the rice genome (Oryza sativa L. ssp. Indica). *Science, 296*, 79-108.

31. Zhang, X., Liu, J., & Zhao, Q. (1999). Transfer of high lysine-rich gene into maize by microprojectile bombardment and detection of transgenic plants. *Journal of Agricultural Biotechnology, 7*(4), 363-367.

Chapter 7

CRYOPRESERVATION OF SPICES GENETIC RESOURCES

K. Nirmal Babu[1], G. Yamuna[1], K. Praveen[1], D. Minoo[2], P.N. Ravindran[1] and K.V. Peter[1]

[1] Indian Institute of Spices Research, Kerala
[2] Providence Women's College, Kerala India

INTRODUCTION

Plant genetic resources - constituting genotypes or populations of cultivars (landraces, advance/improved cultivars), genetic stocks, wild and weedy species, which are maintained in the form of plants, seeds, tissues, etc. - hold key to food security and sustainable agricultural development (Iwananga, 1994). They are non-renewable and are among the most essential of the world's natural resources. Due to deforestation, spread of superior varieties and selection pressure, genetic variability is gradually getting eroded. This demands priority action to conserve germplasm be it at species, genepool or ecosystem level, for posterity (Frankel, 1975). Whilst ecologists focused on in situ conservation might argue that ex situ conserved germplasm cannot offer the advantages afforded by selection and adaptation as a result of environmental pressures, there is no denying that if species are under threat—or worse, near extinction—then ex situ conservation of even limited germplasm is preferable to extinction. The opportunities offered by conservation biotechnology should not be missed or restricted by lack of interconnectivity between traditional and contemporary conservation practitioners.

SPICES AND GERMPLASM CONSERVATION

Spices and herbs are aromatic plants–fresh or dried plant parts like foliage, young shoots, roots, bark, buds, seeds, berries and other fruits of which are mainly used to flavour our culinary preparations, confectionary. They are also major ingredients in indigenous medicine and perfumery. Spices and herbs are grown throughout the world–different plant species in different regions.

Peninsular India is a rich repository of spices and over 100 species of spices and herbs are grown. The other major spice growing countries are Brazil, China, Guatemala, Indonesia, Madagascar, Nigeria, West Indies, Malaysia, Sri Lanka, Spain, Turkey, Mediterranean region and the Central America. Black pepper, cardamom, ginger, turmeric, vanilla, capsicum, cinnamon, clove, nutmeg, tamarind, coriander, cumin, fennel, fenugreek, dill, caraway, anise and herbs like saffron, lavender, thyme, oregano, celery, anise, sage and basil are important as spices. India being the native home of many spices, their conservation and characterization are one of the priority programmes. Deforestation, habitat degradation and overexploitation caused considerable loss of diversity in spices.

In many spices, conventional seed storage can satisfy most of the conservation requirements. But in crops with recalcitrant seeds and those having conservation needs cannot be satisfied by seed storage, have to be stored in vitro. Most field gene banks are prone to high labour cost, vulnerable to hazards like natural disasters, pests and pathogens attack (especially viruses and systemic pathogens), to which they are continuously exposed and require large areas of space. This supports in vitro and cryo conservation. In addition, other resources like continuous supply of standard stock cultures for experiments to examine physiological and biochemical processes, cell and callus lines developed for in vitro synthesis of valuable secondary products, flavours and other important compounds will benefit strongly from in vitro cultures. Most of the spice crops are either vegetatively propagated or have recalcitrant seeds. The spices germplasm is mostly conserved in field gene banks. Most of the spices are plagued by destructive and epidemic diseases caused by viruses, bacteria and fungi. This makes germplasm conservation in field gene bank risky. Thus in vitro and cryo storage system becomes important in the overall strategy of conserving genepool. Each technology should be chosen on the basis of utility, security and complementarily to other components of the strategy. A balance needs to be struck between seed, field gene bank, in vitro and cryo conservation of propagules, tissues, pollen, cell lines and DNA storage for overall objective of conserving gene pool.

METHODOLOGIES

Micropropagation

Plant regeneration and successful cloning of genetically stable plantlets in tissue culture is an important pre-requisite in any conservation effort of recalcitrant species. These techniques form the base for establishing tissue cultures and developing in vitro and cryo conservation technology for conservation. Simultaneously these tissue-cultured plants should be evaluated for their morphological and genetic stability in culture. The in vitro storage experiments, as much as possible, use growth regulators free media to reduce the rate of multiplication which in turn will reduce the extent of variation. Micropropagation (culture initiation, multiplication, plant regeneration and in vitro rooting) form the cycle of events that form the backbone of cryopreservation studies. For initial culture establishment earlier protocols developed by Nirmal Babu et al., 1997 can be used. Murashige and Skoog (1962), Woody Plant (McCown and Amos, 1979) and Schenk and Hildebrandt (1972) media can be used depending upon the crop for micropropagation Table 1. The miniaturized in vitro grown shoots can be used for cryopreservation. Micropropagation protocols for stable cloning of elite genotypes of spice crops were standerdised. Protocols were available for black pepper and its related species cardamom, ginger, turmeric and related genera, large cardamom, kasturi turmeric, mango ginger, Kaempferia galanga, K. rotunda, Alpinia spp, large. Cardamom, vanilla and related species, cinnamon, camphor, cassia seed and herbal spices like lavender, celery, thyme, mint, anise, savory, spearmint and oregano (Nirmal Babu et al., 1997, 2005, Minoo 2002). These techniques form the base for establishing tissue cultures and developing in vitro technology for conservation.

Table 1: Composition of MS*, WPM* and SH* basal media

Composition	Molecular formula	Concentration (mgl-1) MS	Concentration (mgl-1) WPM	Concentration (mgl-1) SH
Macronutrients				
Ammonium nitrate	NH_4NO_3	1650.00	400.00	-
Ammonium phosphate	$NH_4H_2PO_4$	-	-	300.00
Potassium nitrate	KNO_3	1900.00	-	2500.00
Calcium chloride	$CaCl_2.2H_2O$	440.00	-	-
Calcium chloride	$CaCl_2$	-	72.50	151.00
Calcium nitrate	$Ca(NO_3)_2.4H_2O$	-	386.00	-
Potassium di hydrogen orthophosphate	KH_2PO_4	170.00	170.00	-
Potassium sulfate	K_2SO_4	-	990.00	-
Magnesium sulphate	$MgSO_4.7H_2O$	370.00	180.70	195.40
Micronutrients				
Sodium EDTA	Na_2EDTA	37.30	37.30	20.00
Ferrous sulphate	$FeSO_4.7H_2O$	27.80	27.800	15.00
Boric acid	H_3BO_3	6.20	6.20	5.00
Manganese sulphate	$MnSO_4.4H_2O$	22.30	22.30	10.00
Potassium iodide	KI	0.83	-	1.00
Zinc sulphate	$ZnSO_4.7H_2O$	8.60	8.60	1.00
Sodium molybdate	$Na_2MoO_4.2H_2O$	0.25	0.25	0.10
Copper sulphate	$CuSO_4.5H_2O$	0.025	0.25	0.20
Cobalt chloride	$CoCl_2.6H_2O$	0.025	-	0.10
Vitamins				
Myo-inositol	$C_6H_{12}O_6$	100.00	100.00	1000
Thiamine HCl	$C_{12}H_{17}CIN_4OS.HCl$	0.10	0.50	100
Nicotinic acid	$C_6H_5NO_2$	0.50	0.025	1.00
Pyridoxine HCl	$C_6H_{11}NO_3.HCl$	0.50	0.025	1.00
Amino acid				
Glycine	$C_2H_5NO_2$	2.00	1	-

*Murashige and Skoog, 1962, McCown and Amos, 1979, Schenk and Hildebrandt 1972

The basal media used are MS (Murshige and Skoog, 1962) for crops like cardamom, ginger, turmeric, kasturi turmeric, mango ginger, large cardamom, Kaempferia, Vanilla spp. seed and herbal spices and WPM-Woody Plant Medium (Mc Cown and Amos, 1979) for black pepper and its related species, cinnamon, camphor and cassia. Simultaneously these tissue-cultured plants are being evaluated for their morphological and genetic stability in culture (Luckose et al, 1993, Chandrappa et al, 1997, Nirmal babu et al 2003, Madhusoodanan et al 2005). Though micropropagation protocols were standardized using growth regulators, all the in vitro storage experiments were carried out using growth regulators free media to reduce the rate of multiplication which in turn will reduce the extent of variation.

Protocols are available for micropropagation and multiplication of many endangerd species like Piper hapnium, P. silent vallyensis, P.schmidtii, P. wightii, P. barberi , Vaniilla aphylla, V. pilifera, V. walkyrie, V. wightiana, K. rotunda and Alpinia galanga are available (Peter et al 2002, Minoo 2002, Nirmal Babu et al 1999, 2005). Bertaccini et al (2004), Du et al (2004) reported micropropagation and establishment of mitebrone virus-free garlic.

Callus and Cell Culture Systems

Quatrano (1968) and Nag and Street (1973) reported the first successful experiments on cryopreservation of plant cells. Since then a large number of cell suspension and calli cultures have been successfully cryopreserved (Engelmann et al 1994). In general, callus cultures are more difficult to cryopreserve than cell suspensions, because of the relative volume of the callus, its slow growth rate and the cellular heterogeneity (Withers 1987). One successful cryopreservation procedure that is applicable to all different cell suspensions or calli cultures has not been developed yet. Research focuses on optimizing the factors on which successful cryopreservation of plant organs cells suspensions and calli depends, such as: (i) starting material, (ii) pretreatment, (iii) cryopreservation procedure, and (iv) postthaw treatment. Plant cells cultured in vitro produce wide range of primary and secondary metabolites of economic value. Production of phytochemicals from plant cell cultures has been presently used for pharmaceutical products. Production of flavour components and secondary metabolites in vitro using immobilised cells is an ideal system for spices crops. Production of saffron and capsaicin was reported using such system (Ravishankar et al., 1988; 1993, Johnson et al., 1996; Venkataraman and Ravishankar 1997). Johnson et al (1996) reported biotransformation of ferulic acid vanillamine to capsacin and vanillin in immobilised cell cultures of Capsicum frutescens. Reports on the in vitro synthesis of crocin, picrocrocin and safranel from saffron stigma (Himeno and

Sano, 1995) and colour components from cells derived from pistils (Hori et al, 1988) are available for further scaling up. Callus and cell cultures were established in nutmeg, clove, camphor, ginger, lavender, mint, thyme, celery etc. Cell immobilization techniques have been standardized in ginger, sage, anise and lavender (Ilahi and Jabeen, 1992; Ravindran et al, 1996; Sajina et al, 1997). Studies on conservation of cell lines is yet o become popular in spices. Suspensions of embryogenic cell lines of fennel, conserved at 4 0C for up to 12 weeks produced normal plants upon transfer to normal laboratory conditions (Umetsu et al, 1995).

Somatic Embryogenesis and Plant Regeneration

In black pepper primary embryogenic cultures can be established as per the method described by Nair and Dutta Gupta (2003). Culture the surface sterilized seeds on agar gelled full-strength, PGR-free SH (Schenk and Hildebrandt, 1972) medium containing 3.0% (W/V) sucrose under darkness. Primary somatic embryos (PEs) derived from micropylar tissues of germinating seeds after 90 days could be utilized for inducing secondary somatic embryogenic cultures. Primary somatic embryo clumps having pre-globular to torpedo shaped embryos (5–6 visible embryos per seed) were carefully detached and inoculated on half strength PGR-free SH medium containing 1.5 % sucrose and gelled with 0.8% agar (Bacteriological grade, Himedia). The pH of the medium was adjusted to 5.9 prior to autoclaving. Cultures were maintained at darkness at a temperature of 25±2oC. The culture conditions remained the same for all further experiments unless otherwise specified. While inoculating, the PEs were uniformly spread on the surface of the medium. Secondary embryogenic cultures were further maintained by subculturing on SH medium containing 1.5% sucrose at intervals of 20 d. The proliferating SEs were spread periodically on the surface of the medium, to facilitate proliferation.

Somatic Embryogenesis and Plant Regeneration

In black pepper primary embryogenic cultures can be established as per the method described by Nair and Dutta Gupta (2003). Culture the surface sterilized seeds on agar gelled full-strength, PGR-free SH (Schenk and Hildebrandt, 1972) medium containing 3.0% (W/V) sucrose under darkness. Primary somatic embryos (PEs) derived from micropylar tissues of germinating seeds after 90 days could be utilized for inducing secondary somatic embryogenic cultures. Primary somatic embryo clumps having pre-globular to torpedo shaped embryos (5–6 visible embryos per seed) were carefully detached and inoculated on half strength PGR-free SH medium containing 1.5 % sucrose and gelled with 0.8% agar (Bacteriological grade, Himedia). The pH of the medium was adjusted to 5.9 prior to autoclaving. Cultures were maintained at darkness at a temperature of 25±2oC. The culture conditions remained the same for all further experiments unless otherwise specified. While inoculating, the PEs were uniformly spread on the surface of the medium. Secondary embryogenic cultures were further maintained by subculturing on SH medium containing 1.5% sucrose at intervals of 20 d. The proliferating SEs were spread periodically on the surface of the medium, to facilitate proliferation.

Pollen Storage

Pollen storage can be considerable value supplementing the germplasm conservation strategy by facilitating hybridisation between plants with different time of flowering and to transport pollen across the globe for various crop improvement programmes in addition to developing haploid or homozygous lines. No significant work was done in India, except a few initial reports. The technique of pollen storage is comparable with that of seed storage, since pollen can be dried (less than 5% moisture content on a dry weight basis) and stored below 0oC. There are limited reports on the survival and fertilizing capacity of cryopreserved pollen more than five years old. Pollen might represent an interesting alternative for the long-term conservation of problematic species (IPGRI, 1996). However, pollen has a relatively short life compared with seeds (although this varies significantly among species) and viability testing can be time-consuming and uneconomical. Other disadvantages of pollen storage are the small amount produced by many species, the lack of transmission of organelle genomes via pollen, the loss of sex-linked genes in dioecious species and the general inability to regenerate into plants. Pollen, therefore, has been used to a limited extent in germplasm conservation (Hoekstra ,1995). An advantage is that pests and diseases are rarely transferred by pollen (excepting some virus diseases). This allows safe movement and exchange of germplasm as pollen.

Cryo Preservation

For long-term conservation of the problem species, cryopreservation is the only method currently available. Dramatic progress has been made in recent years in the development of new cryopreservation techniques and cryopreservation protocols have been established for over 100 different plant species. Cryopreservation is an attractive option for long-term storage. Liquid nitrogen (–196°C) is routinely used for cryogenic storage, since it is relatively cheap and safe, requires little maintenance and is widely available. Below –120°C the rate of chemical or biophysical reactions is too slow to cause biological deterioration (Kartha 1985). Only in the long term might there be a small risk of ionising radiation causing genetic changes in materials stored at cryogenic temperatures (Grout 1995). An array of plant material could be considered for cryopreservation as dictated by the actual needs vis-a-vis preservation. These include meristems, cell, callus and protoplast cultures, somatic and zygotic embryos, anthers, pollen or microspores and whole seeds (Withers, 1985; Kartha, 1985). Plant germplasm stored in liquid nitrogen (-196°C) does not undergo cellular divisions. In addition, metabolic and most physical processes are stopped at this temperature. As such, plants can be stored for very long time periods and both the problem of genetic instability and the risk of loosing accessions due to contamination or human error during subculturing are overcome. Most cryopreservation endeavours deal with recalcitrant seeds, in vitro tissues from vegetatively propagated crops, species with a particular gene combination (elite genotypes) and dedifferentiated plant cell cultures. Care must be taken to avoid ice crystallisation during the freezing process, which otherwise would cause physical damage to the tissues. The existing cryogenic strategies rely on air-drying, freeze dehydration, osmotic dehydration, addition of penetrating cryoprotective substances and adaptive metabolism (hardening), encapsulation, vitification or combinations of these processes. Cryopreservation methods have been developed for more than 80 different plant species in various forms like cell suspensions, calluses, apices, somatic and zygotic embryos (Kartha and Engelmann, 1994; Engelmann, 1997, 2000, Engelmann et al 1994, 1995). However, their routine utilisation is still restricted almost exclusively to the conservation of cell lines in research laboratories For small volumes, long-term storage is practicable through storage of cultures in cryopreservation at ultra-low temperature, usually by using liquid nitrogen (-196oC). At this temperature all cellular divisions and metabolic processes are virtually halted and consequently, plant material can be indefinitely stored without alteration or modification. The normal approach of tissue culture is to find a medium and set of conditions that favour the most rapid rate of growth with a subculture interval of 20 – 30 days. For cryopreservation storage biological materials are stored in liquid nitrogen for long term with out subculturing. Cryopreservation, i.e., the storage of biological material at ultra low temperature usually that of liquid nitrogen (-1960C) can be achieved by different techniques like direct freezing, encapsulation- dehydration, encapsulation- vitrification and vitrification.

Encapsulation – Dehydration

A simplified methodology for vitrification is given below (Yamuna 2007). The in vitro plants already established were used as mother plants for source of explants. This in turn facilitates the reduction in size of the plantlets and smaller somatic embryos which made them suitable for cryopreservation.

- Suspend in vitro grown shoots/ somatic embryos in MS basal medium supplemented with 4% (w/v) Na alginate, 2M Glycerol and 0.4 M sucrose.

- Drop the mixturecontining microshoots, with a sterile pipette into 0.1M CaCl2 solution containing 2M Glycerol and 0.4M sucrose and left for 20 min to form beads about 4 mm in diameter, each bead containing at least one shoot.

- Preculture the encapsulated shoots – stepwise - on MS medium enriched with different concentration of 0.3, 0.5, 0.75 and 1.0M for four days with one day on each.

- Place the precultured beads on sterile fitter paper in Petridishes (diameter 90mm) and dehydrated by air drying on a flow bench (at room temperature and humidity) for periods of 0-10 h to determine the optimal dehydration time.

- Measure the water content of the beads was by weighing them prior and after drying in an oven at 800C for 48h.

- Transfer the dehydrated beads into a 2 ml cryovial (ten beads per tube) and directly immerse in liquid nitrogen for 24h.

Vitrification

A simplified methodology for vitrification is given below (Yamuna 2007).

- Shoots (1-2mm)/ somatic embryos were excised and cultivated on MS medium supplemented with 0.3 M sucrose for 24h at 250C.

- The treated explants were then cultured on MS medium supplemented with sucrose at 0.75 M for 1 day in the same conditions.

- After pretreatments explants were transferred to a cryovial with 1.8 ml of loading solution (2 M Glycerol + 0.4 M sucrose) and kept for 15 min.

- Different incubation periods in PVS2 (40-100 minutes) were tested for osmoprotected explants

- Cryovials containing 8-10 explants were directly immersed in liquid nitrogen and kept for 24 h.

Encapsulation – Vitrification

A simplified methodology for encapsulation - vitrification is given below (Yamuna 2007).

- Suspend pre-cultured shoots (1-2mm)/ somatic embryos with 2-3 apical domes on 0.3M sucrose for 16h in MS basal medium supplemented with 4% sodium alginate and 0.3 M sucrose.

• Dispense the mixture including shoots, were with a sterile pipette into MS medium supplemented with 0.1M CaCl2 and 0.4 to 1.0M sucrose, with or without 2M Glycerol gently shaken (20 rpm) on a rotary shaker for 1h at 250C.

• The encapsulated and osmo-protected shoots were dehydrated with 20 ml PVS2 in a 100 ml Erlenmeyer flask at 250C and plunged into LN and held for at least 24 h at - 1960C.

Thawing and Recovery of Conserved Materials

After LN storage, cryovials warm rapidly in a 40 0C water bath for 2-3 minutes. The solution was drained from the cryovials and replace twice at 10 min intervals with 1 ml 1.2 M sucrose solution in the case of encapsulation-vitrification and vitrification methods. The composition of recovery medium was MS/WPM/SH basal medium supplemented with 2.22 – 4.44 µM and BA, 2.69- 5.37 µM NAA. In the Encapsulation - dehydration, Encapsulation - vitification and vitrification procedures, surviving shoots can be identified by greening of explants following 2 weeks of post culture. Regrowth can be defined as the shoots that regenerated to shoots in 6 weeks of postculture. Elongated shoots can be used for micropropagation and rooting and subculture was done every 4 weeks. For rooting well grown shoots can be transferred to solid MS medium used for multiplication.

Genetic Stability of Conserved Materials

An important prerequisite for any conservation technique is that the regenerants produced from the conserved material should be true-to-type. There are ample evidences to indicate that under certain culture conditions the materials undergo genetic changes (somaclonal variations) and as a consequence lose their integrity and uniformity. This would be highly undesirable in spices varieties where the purpose is not only to conserve a genotype but also retain its specific quality traits. Thus testing for the genetic stability of in vitro conserved materials is of utmost importance. Besides morphology, cytology and isozyme profiling sophisticated biochemical and DNA-based techniques have enabled more critical analysis of the genetic stability of in vitro materials.

RAPD, ISSR and SSR analysis can be done to evaluate genetic fidelity of the cryopreserved lines of Spices. DNA isolation can be done as per CTAB method (Ausubel et al., 1995 or Sambrook et al. 1989). RAPD and ISSR, SSR profiles were developed as per the method suggested by Williams et al., (1990),

Nirmal babu et al., (2003, 2007) and Ravindran et al., (2004). Morphological characters coupled with RAPD profiles using 24 operon primers have indicated genetic fidelity among randomly selected micropropagated plants of Subhakara and Aimpiriyan, indicating that micropropagation protocol can be used for commercial cloning of black pepper (Nirmal Babu et al., 2003). Genetic uniformity of micropropagated Piper longum using RAPD profiling was reported by Ajith (1997) and Parani et al. (1997) for conservation. Peter et al (2001) and Ravindran et al (2004) reported that the conserved materials of all the species conserved by them showed normal rate of multiplication when transferred to multiplication medium after storage. The normal sized plantlets when transferred to soil established with over 80% success. They developed into normal plants without any deformities and were morphologically similar to mother plants. RAPD profiling of these conserved plants also showed their genetic uniformity. Ravindran et al (2004), Yamuna et al (2007) and Yamuna (2007) reported genetic uniformity was observed in cryo preserved and recovered plants of cardamom, ginger, black pepper and endangered species of Piper, P. barberi based on RAPD and ISSR profiling.

Status of Cryo Conservation in Spices

Reports on cryopreservation of spices are meager and limited. The present status of cryo preservation in major spices is given Table 2. The number of accessions conserved in cryo genebank at the National Bureau of Plant Genetic Resources (NBPGR), New Delhi are given in Table 3.

Table 2: Present status of information on cryo conservation of spices

Application	Technique	Reference
Black pepper (*Piper nigrum*) and related species	Meristem culture	Philip *et al.*, 1992
Disease eradication propagation	Shoot Culture Leaf/root	Broome and Zimmerman, 1978
Cryopreservation	Seeds	Chaudhury and Chandel ,1994
Cryopreservation	synseeds	Ravindran *et al.*, 2004; Nirmal Babu *et al.*, 2007; Yamuna 2007
Cryopreservation	Seed	Decruse and Seeni, 2003
Slow growth storage and cryopreservation	Plantlets and shoot tips	Ravindran *et al.*, 2004; Nirmal Babu *et al.*, 2007; Yamuna 2007
***Allium* Spp**		
Disease eradication	Meristem culture and themotherapy	Conci and Nome, 1991
Cryopreservation	Shoot culture,	Keller 1991

Application	Technique	Reference
	microbullbets	
Cryopreservation	Shoot tips	Niwata 1995
Cardamom (*Elettaria cardamomum* Maton)		
Disease eradication	Meristem culture	Nadagauda *et al.* 1983
Cryopreservation	Seeds	Chaudhury and Chandel, 1995
Slow growth storage and cryopreservation	Plantlets and shoot tips	Ravindran *et al.*, 2004; Nirmal Babu *et al.*, 2007; Yamuna 2007
***Zingiber* spp.**		
Disease eradication	Shoot cultures, shoot buds	Balachandran *et al* .,1990
Propagation	Somatic embryo regeneration	Hosoki and Sagawa ,1977, Nirmal Babu, 1997
Cryopreservation	Synseeds	Sharma *et al.*, 1994
Slow growth storage and cryopreservation	Plantlets and shoot tips	Ravindran *et al.*, 2004; Nirmal Babu *et al.*, 2007; Yamuna *et al* 2007 ; Yamuna 2007
***Curcuma* spp**		
Slow growth storage and cryopreservation	Plantlets and shoot tips	Ravindran *et al.*, 2004 ; Nirmal Babu *et al.*, 2007
***Vanilla* spp.**		

Disease eradication	Apical meristem	Cereveta and Madrigal, 1981
Cryo preservation	Synthetic seeds	Ravindran *et al.*, 2004
Pollen Cryo preservation	Pollen	Minoo, 2002; Minoo *et al* 2011
Slow growth storage and cryopreservation	Plantlets and shoot tips	Ravindran *et al.*, 2004; Nirmal Babu *et al.*, 2007, Minoo and Babu 2009
Herbal spices		
Slow growth storage	*In vitro* plantlets	Nirmal Babu *et al.*1996
Capsicum		
Cryopreservation	Seed	Peter *et al* 2002 ; Ravindran *et al* 2004
Cryopreservation	Pollen	Alexander *et al.*, 1991
Cryopreservation	Pollen	Rajasekharan and Ganeshan, 2003
Fennel (*Foeniculum vulgare*)		
Cold storage	Embryogenic	Umetsu *et al.*, 1995

Application	Technique	Reference
	suspension cells	
Coriander (*Coriandrum sativum*)		
Cryopreservation	somatic embryos	Elena *et al.,* (2010)
Mint (*Mentha* spp.)		
Cryopreservation	Somatic embryos	Leigh and Remi 2003
***Ocimum* spp**		
Slow growth	Encapsulated beads	Mandal *et al* (2000)
Syzygium francissi		
	Shoot tips	Shatnawi *et al* (2004).
Armoracia rusticana		
Cryopreservation	Hairy root cultures	Phunchindawan *et al*
***Crocus* spp.**		
	Encapsulated calluses	Chand *et al* (2000); Baghdadi et al., (2010)

*Ashmore, 1997, 2002 and Nirmal Babu *et al*, 1999, 2007, Yamuna 2007; Yamuna et al 2007

Table 3: Present status of Spices in in vitro and Cryo genebank at NBPGR

Species	No.of accessions
Maintained as in vitro cultures	
Spices and industrial crops	380 accessions (7 genera, 27 species)
Medicinal and Aromatic plants	169 accessions (21 genera, 28 species)
Maintained in cryo bank	
Spices and Condiments	148 accessions
Medicinal and Aromatic plants	5 accessions
Total	702

Source: Annual Report NBPGR 2010-11

Black Pepper and Related Species

Cryopreservation of black pepper (Piper nigrum L.)seeds in liquid nitrogen (LN2) was reported by Choudhary and Chandel, (1994), and Choudhury and Malik (2004). Pepper seeds are recalcitrant and the seed viability decreases with reduction in moisture content. Seeds desiccated to 12% & 6%moisture contents were successfully cryopreserved in liquid nitrogen at −1960C, with a survival rate of 45% & 10.5% respectively (Chaudhury and Chandel 1994).

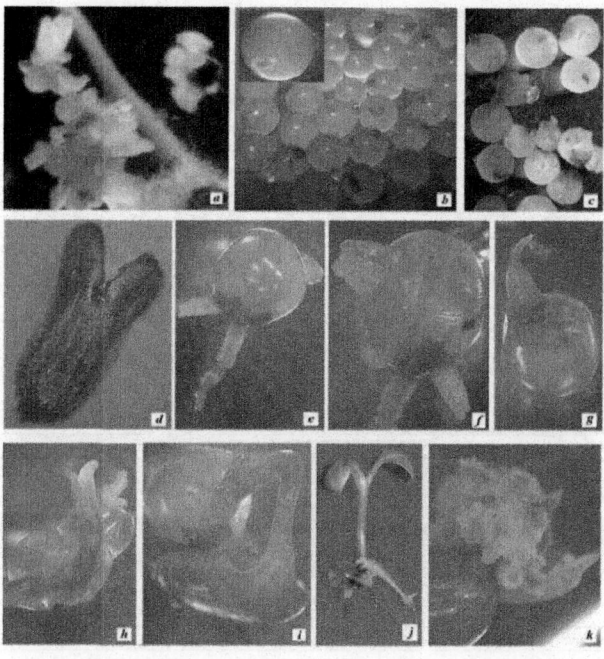

Figure. 1: Cryopreservation of black pepper somatic embryos by encapsulation de-

hydration. a) Somatic embryos used for cryopreservation, b) Somatic embryos en-capsulated in Naalginate, c) Encapsulated and dehydrated somatic embryos, d) Viable somatic embryo stained in red colour after cryopreservation, e), f), g), h) & i) Various stages of development of somatic embryos to plantlet after cryopreservation, j) Fully developed plantlet from a somatic embryo cryopreserved by encapsulation dehydra-tion, k) A cluster of somatic embryos at different stages of development, originated from an embryogenic line after cryopreservation

laminar air flow for 6 h which resulted in 21 % moisture content. In the vitrification procedure, the somatic embryos were precultured for 3 days on SH basal medium containing 0.3 M sucrose and subjected to vitrification treatment for 60 minutes at 250C resulted in 71 % survival after cryopreservation. The study concluded that the embryogenic lines of Piper nigrum cultivar karimunda can be successfully cyopreserved following an encapsulation dehydration/ desiccation procedure (62 % success). This success rate can be enhanced to 71 % using a vitrification/one step freezing in liquid nitrogen (Fig. 1).This was mainly because of the nature of somatic embryos which is more suitable to cryopreservation compared to shoot buds. The genetic stability of the conserved somatic embryos was proved by RAPD and ISSR profiling. Cryopreservation of encapsulated shoot buds of endangered Piper barberi was reported by Peter et al (2001) and Ravindran et al (2004). Encapsulated shoot tips of Piper barberi were cryopreserved with 60% success using vitrification technique. In encapsulation vitrification the encapsulated shoot tips were precultured on MS medium, supplemented with 0.3 M, 0.5 M and 0.7 M sucrose (pH 5.8) for three days followed by dehydration with PVS2 solution (100%) at 00 C for 3 hours. After dehydration the beads (10 encapsulated shoot tips in 0.8 ml PVS2 solution per 1.5 ml cryotube) were frozen rapidly by direct immersion in to liquid nitrogen (- 196 0C) and kept for one hour (Peter et al 2001 and Ravindran et al 2004). Yamuna 2007 also reported that studies on cryopreservation of endangered P.barberi shoot tips revealed that, the encapsulation- vitrification procedure produced higher survival (70 %) of cryopreserved shoot tips (Fig. 2) compared to encapsulation - dehydration which gave 40 % survival. Genetic fidelity studies showed that the regenerated plants were similar to the controls. Thus encapsulation - vitrification as a simple and efficient method for long term preservation of P.barberi propagules.

Cardamom and Related Species

Choudhary and Chandel (1995) attempted cryo-conservation of cardamom (Elettaria cardamomum Maton.) seed. They tried to conserve seeds at ultra-low temperature by suspending seeds in cryovials in vapor phase of liquid nitrogen (-150oC) by slow freezing and also by direct immersion in liquid nitrogen

(-196oC) by fast freezing. The result showed that seeds possessing 7.7-14.3% moisture content could be successfully cryo-preserved with 80% germination when tested after one-year storage in vapor phase of liquid nitrogen (at-150oC). Shoot tips(1.0-2.0mm) from in vitro grown plantlets of cardamom were subjected to progressive increase of sucrose concentrations (0.1, 0.3, 0.5, 0.7, 0.9, and 1.0) for two days each under the same cultural conditions as the parent plantlets. These shoot tips were transferred to 1.8ml cryotube containing ice cold PVS2 solution (30%(v/v) glycerol + 15% (v/v) ethylene glycol + 15% (v/v) DMSO in culture medium with 0.4 M sucrose, pH (5.8)) at 00C for 3 hours. After 3 hours equilibration at 00C, the shoot tips were directly immersed into liquid nitrogen for 1 hour. Vials were thawed in 400C water for 1 minute. The cryoprotectant was removed and the shoot tips were washed 2-3 times in 1.2M sucrose solution. About 70%Shoot tips were recovered on MS medium supplemented with BAP and NAA. But the encapsulation vitrification method gave only 60% success (Ravindran et al 2004).

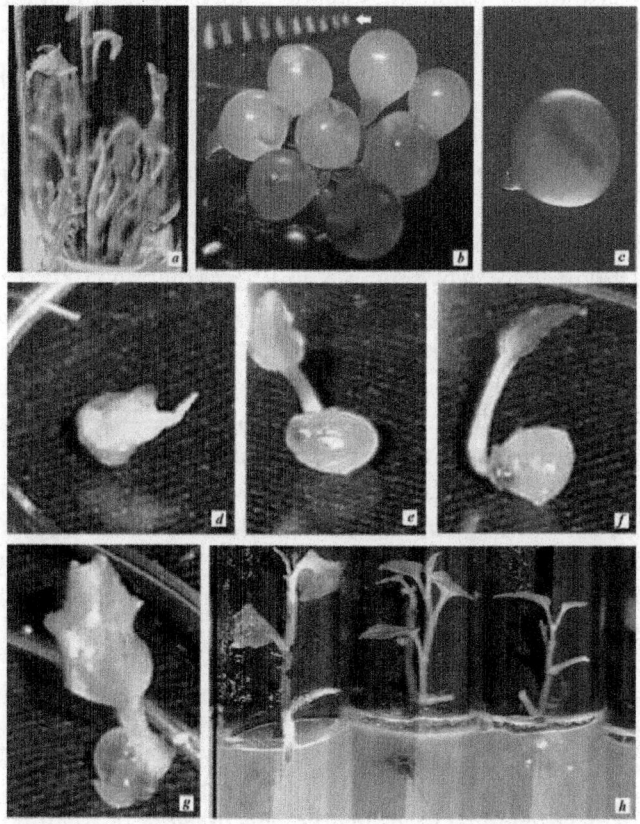

Figure. 2: Cryopreservation of Piper barberi by encapsulation vitrification. a) In vitro

culture of P. barberi, b) & c) Shoot tips encapsulated in Na-alginate, arrow indicates shoot tip used as explants, d), e), f) & g) Various stages of development of cryopreserved shoot tips after post culturing, h) Regenerated plantlets after 3 months of post culturing

Yamuna (2007) tested the effect of encapsulation – dehydration, encapsulation vitrification and vitrification methods on cryopreservation of cardamom. In the vitrification treatment, to enhance tolerance to vitrification solution (PVS2), a two step sucrose preculture with 0.3 M and 0.75 M sucrose for one day each and an osmo protection step with a loading solution (LS) of 2 M glycerol and 0.4 M sucrose were performed prior to PVS2 treatment. The shoots

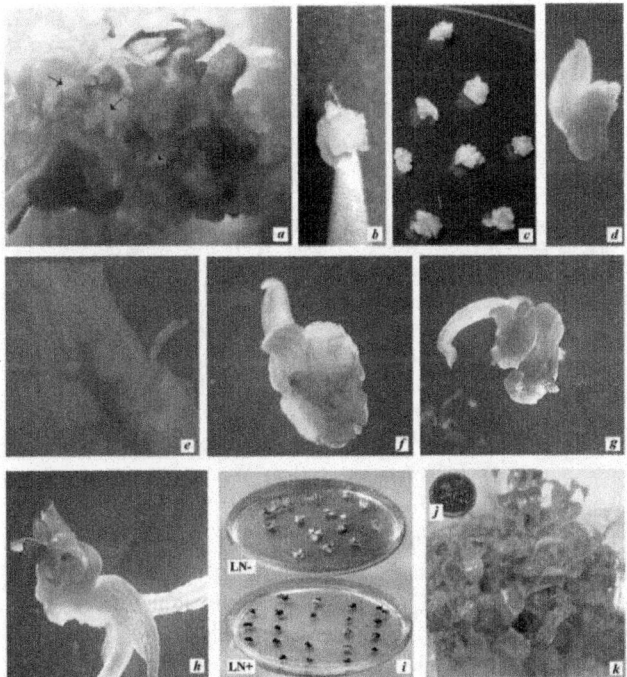

Figure. 3: Plant regeneration from cryopreserved miniature shoots of cardamom by vitrification. a) Cardamom culture with miniature shoots, b) & c) Excised meristematic clumps used for cryopreservation, d) Explant turned brown after cryopreservation, e) Viable tissues stained in TTC after cryopreservation, f), g), h), & i) Shoot development after 10, 14 and 25 days of post culturing , j) regenerating shoot buds in a petridish, k) Development of multiple shoots after 4 months of post culturing

dehydrated with PVS2 for 60 min retained a high level of shoot formation (70 %). The vitrification procedure resulted in higher regrowth (70 %) (Fig.3) when compared to encapsulation vitrification (62 %) and encapsulation dehydration

(60 %). In all the three cryopreservation procedures tested, shoots grew after cryopreservation without intermediary callus formation. The genetic stability of cryopreserved cardamom shoots were confirmed using ISSR and RAPD profiling.

Figure. 4: Plant regeneration from cryopreserved shoot buds of ginger by encapsulation vitrification. a) In vitro culture, b) A typically excised shoot bud used for cryopreservation, c) & d) Shoot buds encapsulated in Na-alginate, e) & f) Shoot buds turned brown after thawing, g) Viable apical dome stained in red colour after liquid nitrogen storage (TTC staining), h) Regenerating shoot bud 20 days after post culturing, i) & j) Elongated shoot with no intermediary callus formation, k) & l) Regenerating shoot buds in petriplates, m) Plantlets regenerating from cryopreserved shoot bud

Ginger, Turmeric and Related Species

Cryopreservation of Ginger (Zingiber officinale Rosc) and turmeric (Curcuma longa L.) shoot tips was successfully done with 80% of recovery using vitrification method. But the rate of recovery was only 40% when encapsulated shoot tips were dehydrated in progressive increase of sucrose concentration together with 4- 8 hrs. of desiccation (Peter et al 2001 and Ravindran et al 2004). Efficient cryopreservation techniques were developed for in vitro grown shoots of ginger based on encapsulation dehydration, encapsulation vitrification and vitrification procedures (Yamuna et al 2007 and Yamuna 2007. The vitrification procedure resulted in higher regrowth (80 %) when compared to encapsulation vitrification (66 %) and encapsulation dehydration (41 %). The genetically stability of shoot apices was confirmed by molecular profiling. The RAPD and ISSR assays performed suggested that no genetic aberrations originated in ginger plants during culture and cryopreservation (Fig. 4).

Vanilla and Related Species

Technology for cryopreservation of vanilla germplasm - using encapsulation and vitrification methods – were available. Encapsulated in vitro grown shoot tips of vanilla could be cryo preserved with 70% success when pretreated with progressive increase of sucrose concentration (0.1M-1.0M) for one day each and dehydrated for 8 hrs (Peter et al 2001; Minoo 2002 and Ravindran et al 2004) (Fig. 5).

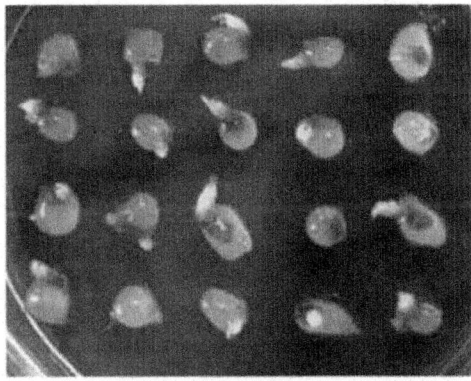

Figure. 5: Germination of cryopreserved encapsulated shoot tips protocorms of vanilla

Ginzalez-Arnao, et al., (2009) attempted to cryo-preserve V. planifolia Andr. using in vitro fragmented explants (IFEs) and the apices derived from them. Cryopreservation of apices from in vitro grown plants was achieved using the droplet vitrification protocol. Maximum survival (30%) and further regeneration (10%) of new shoots were obtained for apices derived from clusters of in vitro plantlets produced from microcuttings through a three-step droplet vitrification protocol: 1-d preculture of apices on solid MS medium with 0.3 M sucrose; loading with a 0.4 M sucrose + 2 M glycerol solution for 20–30 min; and exposure to plant vitrification solution PVS3 for 30 min at room temperature. Minoo (2002) reported cryopreservation of vanilla pollen for conservation (Fig. 6) of haploid genome as well as assisted pollination between species that flower at different seasons and successful fertilisation using cryopreserved pollen (Minoo, 2002, Minoo et al 2011). Pollen from two asynchronously flowering species of Vanilla viz., cultivated V. planifolia and its wild relative V. aphylla, were cryopreserved after desiccation to 12 % moisture content, pretreated with cryoprotectant Dimethyl sulphoxide (5%) and cryopreserved -196°C in Liquid Nitrogen. This cryopreserved pollen was latter thawed and tested for their viability both in vitro and in vivo. A germination percentage of 82.1% and 75.4% in V. planifolia and V.aphylla pollen respectively were observed indicating their viability(Fig.6). This cryopreserved pollen of V. planifolia was used successfully to pollinate V.aphylla flowers resulting in fruit set (Fig.7). The seeds thus obtaines were sussfully cultured to develop hybrid plantlets. This system is of great importance and can be used for conserving the haploid gene pool of Vanilla in cryobanks and their subsequent utility in crop improvement (Fig. 6 and 7)

Figure. 6: Germination of cryopreserved Vanilla pollen

Figure. 7: Fruit set after pollination with cryopreserved pollen

Capsicum

Plants could be successfully regenerated (Fig 8) from cryopreserved seeds of capsicum (Peter et al 2001 and Ravindran et al 2004). Alexander et al (1991) and

Rajasekharan and Ganeshan. (2003) reported freeze preservation of capsicum pollen (Capsicum annuum) in liquid nitrogen (–1960C) for 42 months

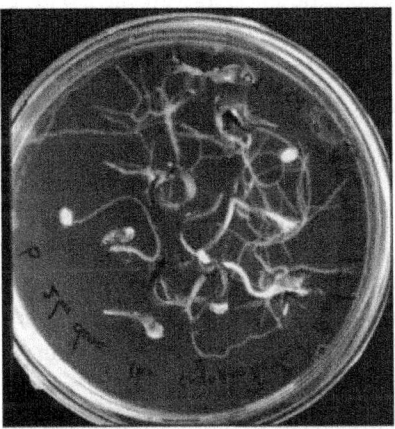

Figure. 8: Successful germination of cryopreserved seeds of capsicum

Seed Herbal and Other Spices

Elena et al., (2010) successfully cryopreserved coriander (Coriandrum sativum L.) somatic embryos using sucrose pre-culture and air desiccation procedure utilized embryo clumps (ECs). The regrowth after cryopreservation and average number of new embryos developed from cryopreserved ECs were retained at the level of the untreated control (98% and 13 embryos per clump, respectively). Both normal and abnormal plants were produced from control and cryopreserved cultures, indicating that appearance of abnormalities was not related to cryopreservation. The regenerants with normal phenotype showed the same peaks of relative DNA content regardless of cryopreservation. The results suggest that simple desiccation method is effective for cryopreservation of coriander somatic embryos with subsequent regeneration. Plants could be regenerated from cryopreserved seeds of Anise.(Peter et al 2001). Successful Cryopreservation of seeds, meristems, somatic or zygotic embryos were reported in Allium Spp (Niwata, 1995, Hyung et al 2003, Haeng et al 2003, 2004.2005, Jung et al 2005, Gayle et al 2004). Preliminary success was reported in cryo preservation of Mint (Leigh and Remi 2003). Most of the reports are confined to a few genotypes and hence the techniques standardized needs to be extended to more genotypes before adopting them for routine conservation. Reports of cryoconservation of spices like Ocimum, Lavendula, Salvia are available from National Bureau of Plant Genetic Resources (NBPGR), New Delhi.

Mandal et al (2000) reported propagation and conservation of four pharmaceutically important herbs, Ocimum americanum L. syn. O. canum Sims. (hoary basil); O basilicum L. (swett basil); O. gratissimum L. (shrubby basil); and O. sanctum L. (sacred basil) using synthetic seed technology. Synthetic seeds were produced by encapsulating axillary vegetative buds harvested from garden-grown plants of these four Ocimum species in calcium alginate gel. The gel contained Murashige and Skoog (MS) nutrients and 1.1-4.4 μM benzyladenine (BA). Shoots emerged from the encapsulated buds on all six planting media tested. However, the highest frequency shoot emergence and maximum number of shoots per bud were recorded on media containing BA. Of the six planting media tested, both shoot and root emergence from the encapsulated buds in a single step was recorded on growth regulator-free MS medium as well as on vermi-compost moistened with halfstrength MS medium. Rooted shoots were retrieved from the encapsulated buds of O. americanum, O. basilicum, and O. sanctum on these two media, whereas shoots of O. gratissimum failed to root. The encapsulated buds could be stored for 60 d at 4°C. Plants retrieved from the encapsulated buds were hardened off and established in soil. An efficient procedure for the in vitro propagation and cryogenic conservation of Syzygium francissi was developed by Shatnawi et al (2004). Shoot tips excised from in vitro-grown plants were successfully cryostoraged at −196°C by the encapsulation-dehydration method. A preculture of formed beads on MS medium containing 0.75 M sucrose for 1 d, followed by 6 h dehydration (20% moisture content) led to the highest survival rate after cryostorage for 1h. This method is a promising technique for in vitro propagation and cryopreservation of shoot tips from in vitro-grown plantlets of S. francissi germplasm. Hairy root cultures of Armoracia rusticana Gaertn. Mey. et Scherb. (horseradish) were successfully cryopreserved by two cryogenic procedures (Phunchindawan et al., 1997). Encapsulated shoot primordia were precultured on solidified Murashige-Skoog medium supplemented with 0.5M sucrose for 1 day and then dehydrated with a highly concentrated vitrification solution (PVS2) for 4 h at 0°C prior to a plunge into liquid nitrogen. The survival rate of encapsulated vitrified primordia amounted to 69%. In a revised encapsulation-dehydration technique, the encapsulated shoot primordia were precultured with a mixture of 0.5M sucrose and 1M or 1.5M glycerol for 1 day to induce dehydration tolerance and then subjected to air-drying prior to a plunge into liquid nitrogen. The survival rate of encapsulated dried primordia was more than 90%, and the revived primordia produced shoots within 2 weeks after plating. A long-term preservation of shoot primordia was also achieved by the technique. Thus, this revised encapsulation-dehydration technique appears promising as a routine method for the cryopreservation of shoot primordia of hairy roots The effect of sucrose concentration and dehydration period on survival and regrowth of encapsulated calluses were also studied in 2 species of Crocus (Chand et al 2000). Highest survival (83.3; 88.9%) and regrowth (77.6; 83.3%) rates were obtained when encapsulated unfrozen calluses of Crocus hyemalis and C. moabiticus precultured with 0.1 M sucrose for two days without further air dehydration. After cryopreservation, the highest survival (55.6; 61.1%) and regrowth (16.7; 27.8%) rates were achieved when calluses of C. hyemalis and C. moabiticus were pretreated with 0.5 M sucrose for two days after two hours of dehydration. Viability of crocus decreased with increased sucrose concentration and dehydration period. Dehydration of encapsulated calluses of C. hyemalis and C. moabiticus with silica gel for one hour prior to freezing resulted in maximum rates of survival (77.8; 83.3%) and re-growth (33.3; 72.1%). However, further studies should be initiated to improve regrowth of surviving embryogenic calluses and to study genetic stability after cryopreservation.

DNA BANK

Concurrent with the advancements in gene cloning and transfer has been the development of technology for the removal and analysis of DNA. DNAs from the nucleus, mitochondrion and chloroplast are now routinely extracted and immobilized onto nitrocellulose sheets where the DNA can be probed with numerous cloned genes. In addition, the rapid development of polymerase chain reaction (PCR) now means that one can routinely amplify specific oligonucleotides or genes from the entire mixture of genomic DNA.These advances, coupled with the prospect of the loss of significant plant genetic resources throughout the world,have led to the establishment of DNA bank for the storage of genomic DNA. The conserved DNA will have numerous uses viz, molecular phylogenetics and systematics of extinct taxa, production of previously characterized secondary compounds in transgenic cell cultures, production of transgenic plants using genes from gene families, in vitro expression and study of enzyme structure and function and genomic probes for research laboratories. The vast resources of dried specimens in the world's herbaria may hold considerable DNA that would be suitable for PCR. It seems likely that the integrity of DNA would decrease with the age of specimens. Because there are many types of herbarium storage environments, preservation and collections, there is a need for systematic investigations of the effect of modes of preparation, collection and storage on the integrity of DNA in the world's major holdings. The advantage of storing DNA is that it is efficient and simple and overcomes many physical limitations and constraints that characterize other forms of storage (Adams 1988, 1990, 1997, Adams and Adams 1991, Adams et al 1994). The disadvantage lies in problems with subsequent gene isolation, cloning and transfer but, most importantly, it does not allow the regeneration of live organisms (Maxted et al., 1997). DNA banking is yet to catch up in spices. DNA samples of over 600 genotypes of spices is stored in the DNA bank of Indian Institute of Spices Research (IISR), Calicut.

FUTURE FOCUS

In contrast to the prevailing attitude among conservation biologists, globally there is considerable interest among cryobiologists in the use of in vitro, cold and ultra-cold technology for germplasm conservation. The procedures for plant material are given indepth coverage by Reed et al. (2004) who stress equally the ecological and plant/germplasm health aspects preceeding and following storage. Panis and Lambardi (2006) discussed the evolution of technologies for plant material, covering cell suspensions and callus cultures of herbaceous species, pollen, shoot meristems, woody species, as well as seed and embryonic axes. The ex situ gene bank at Gatersleben in Germany houses 986 potato accessions are cryopreserved and trials on other species are performed (Börner 2006). The National Bureau of Plant Genetic Resources (NBPGR), New Delhi has over 702 accesion of various spices, medicinal and aromatic crops in its cryo gene bank (Table. 3). Keller et al. (2008) make the point that cryopreservation affords the best of conditions for the long-term maintenance of plant material, particularly for vegetatively propagated species. Cryopreservation is the only viable method available for long-term preservation of the both plant and animal origin species. As an ultimate aim of cryoconservation is the reintroduction of preserved material into the field, it is appropriate at this point to consider the concept of restoration a little more closely. In terms of ultimate ecosystem restoration, the possibilities raised by in vitro conservation, including cryoconservation, do not mean that species selection should merely take random advantage of what germplasm has or can be conserved as there are many genetic, physiological and phenotypic considerations to be taken into account (Kramer and Havens 2009). The establishment and maintenance of biological resource centers (BRCs) or germplasm conservatories requires careful attention to implementation of reliable preservation technologies and appropriate quality control to ensure that recovered cultures and other biological materials perform in the same way as the originally isolated culture or material. There are many types of BRC that vary both in the kinds of material they hold and in the purposes for which the materials are provided. All BRCs are expected to provide materials and information of an appropriate quality for their application and work to standards relevant to those applications. There are important industrial, biomedical, and conservation issues that can only be addressed through effective and efficient operation of BRCs in the long term. This requires a high degree of expertise in the maintenance and management of collections of biological materials at ultra-low temperatures, or as freeze dried material, to secure their long-term integrity and relevance for future research, development, and conservation. The application of cryogenic preservation in biotechnology and medicine has recently been a topic of interest. The use of cryogenic preservation in this area has given new horizon to this field of applications.

REFERENCES

1. Adams, R.P. and Adams, J.E., 1991. Conservation of Plant Genes: DNA Banking and In Vitro Biotechnology. Academic Press, New York.

2. Adams, R.P., 1988. The preservation of genomic DNA: DNA Bank Net. Amer. J. Bot., 75: 156.

3. Adams, R.P., 1990. The preservation of Chihuahuan plant genomes through in vitro biotechnology:DNA Bank-Net, a genetic insurance policy.

4. In: Third Symposium on Resources of the Chihuahuan Desert Region, (Eds.) Powell, A.M., Hollander, R.R., Barlow, J.C., McGillivray, W.B. and Schmidly, D.J. Printech Press, Lubbock, TX, pp.1–9.

5. Adams, R.P., 1997.Conservation of DNA: DNA banking. In: Biotechnology and Plant Genetic Resources: Conservation and Use, (Eds.) Callow, J.A. Ford-Loyd, B.V. and Newbury, H.J. Biotechnology in Agriculture Series, No. 19. CAB International, pp. 163–174.

6. Adams, R.P., Miller, J.S., Golenberg, E.M. and Adams, J.E., 1994. Conservation of Plant Genes 11: Utilization of Ancient and Modern DNA. Missouri Botanical Garden Press. St.Louis, MI, 276 pp. 227.

7. Ajith, A., 1997. Micropropagation and genetic fidelity studies in Piper longum L. In: Biotechnology of Spices, Medicinal and Aromatic Plants, (Eds.) Edison, S., Ramana,K.V., Sasikumar, B., Nirmal Babu K. and Santhosh J. Eapen. Indian Society for Spices, Calicut, India, p. 94–97.

8. Alexander, M.P., Ganeshan, S. and Rajasekharan, P.E., 1991. Freeze preservation of capsicum pollen (Capsicum annuum) in liquid nitrogen (–196°C) for 42 months:

9. Effect on viability and fertility. PlantCell Incompatibility Newsletter, 23: 1–4. NBPGR2011 Annual Report, National Bureau of Plant Genetic Resources, 2010-11

10. Ashmore, S.E., 1997. Status report on the development and application

of in vitro techniques for conservation and use of plant genetic resources. International Plant Genetic Resources Institute, Rome, Italy. 67 p.

11.   Ausubel F M, Brent R, Kingston R E, Moore D D, Seidman, J G, Smith J A & Struhl K 1995 Short Protocols in Molecular Biology, John Wiley & Sons, New York, Ch.2.4.

12.   Balachandran, S.M., Bhat, S.R. and Chandel, K.P.S., 1990. In vitro clonal multiplication of turmeric (Curcuma longa) and ginger (Zingiber officinale Rosc.). Plant Cell Reports, 3: 521–524.

13.   Bertaccini, A., Botti, S., Tabanelli, D., Dradi, G., Fogher, C., Previati, A., da Re, F., Nicola, S., Nowak, J. and Vavrina, C.S., 2004. Micropropagation and establishment of miteborne virus-free garlic Allium sativum), Acta Horticulturae, 631: 201–206.

14.   Börner A. Preservation of plant genetic resources in the biotechnology era. Biotech J 1: 1393– 1404; 2006.

15.   Broome, O.C. and Zimmerman, R.N., 1978. In-vitro propagation of black pepper. Hort. Sci., 43: 151–153.

16.   Chand, P.K., Mandal, J. and Pattnaik, S., 2000. Alginate encapsulation of axillary buds of Ocimum americanum L. (hoary basil), O.basilicum L. (sweet basil), O. gratissimum L.

17.   (shrubby basil), and O. sanctum L. (sacred basil). In vitro Cellular and Developmental Biology (Plant), 36(4): 287–292.

18.   Chandrappa, H.M., Shadakshari, Y.G., Sudharshan, M.R. and Raju, B. ,1997. Preliminary yield trial of tissue cultured cardamom selections. In: Biotechnology of Spices, Medicinal and Aromatic Plants,(Eds.) Edison, S., Ramana, K.V., Sasikumar, B., Nirmal Babu K. and Santhosh J. Eapen. Indian Society for Spices, Calicut, India,p. 102–105.

19.   Chaudhury, R. and Chandel, K.P.S., 1994. Germination studies and cryopreservation of seeds of black pepper (Piper nigrum L.): A recalcitrant species. Cryoletters, 15: 145–150.

20.   Choudhary, R. and Chandel, K.P.S., 1995. Studies on germination and cryopreservation of cardamom (Elletaria cardamomum Maton.) seeds. Seed Science and Biotechnology,23(1): 235–240.

21.   Choudhury, R. and Malik, S.K., 2004. Genetic conservation of plantation crops and spices using cryopreservation. Ind. J. Biotechnology, 3: 348–358.

22.   Conci, V.C. and Nome, S.F., 1991. Virus free garlic (Allium sativum L.) plants obtained by thermotherapy and meristem tip culture. J. Phytopathol., 132: 186–192.

23. Decruse, S W and Seeni, S 2003. Seed cryopreservation is a suitable storage procedure for arange of Piper species. Seed Sci. Technol. 31, 213-217.

24. Du, Y.Q., Zhu, J.Z. and Shen, W.P., 2004. Study on virus elimination technology of garlic in Jiading by shoot tip culture. Acta Agriculturae Shanghai, 20(1): 9–12.

25. Elena Popova, Haeng-Hoon Kim and Kee-Yoeup Paek 2010. Cryopreservation of coriander (Coriandrum sativum L.) somatic embryos using sucrose pre-culture and air desiccation Scientia Horticulturae 124 (4), 522-528.

26. Engelmann F. Use of biotechnologies for conserving plant biodiversity. Acta Hort 812: 63–82; 2009.

27. Engelmann, F., 1997. In vitro conservation methods. In: Biotechnology and Plant Genetic Resources:Conservation and Use, (Eds.) Callow, J.A., Ford-Loyd, B.V. and Newbury, H.J. Biotechnology in Agriculture Series, No. 19. CAB International, UK, pp. 119–161.

28. Engelmann, F., 2000. Importance of cryopreservation for the conservation of plant genetic resources. In: Cryopreservation of Tropical Plant Germplasm, (Eds.) Engelmann, F. and Takagi, H. Current Research Progress and Application. Japan International Research Centre for Agricultural Sciences, Japan/International Plant Genetic Resources Institute, Rome, Italy, pp. 8–20.

29. Engelmann, F., Benson, E.E., Chabrillange, N., Gonzalez-Arnao, M.T., Mari, S., MichauxFerriere, N., Paulet, Glazmann, J.C. and Charrier, A., 1994. Cryopreservation of several tropical plant species using encapsulation/dehydration of apices. In. Proc.VIIIth IAPTC Meeting, Firenze, Italy.

30. Engelmann, F., Dumet, D., Chabrillange, N., Abdelnour-Esquivel, A., Assy-Bah, B., Dereuddre, J. and Duval, Y., 1995. Cryopreservation of zygotic and somatic embryos from recalcitrant and intermediate-seed species. Plant Genet. Resources Newsletter, 103: 27–31.

31. Frankel, O.H., 1975. Genetic resources survey as a basis for exploration. In: Crop Genetic Resources for Today and Tomorrow, (Eds.) O.H. Frankel and J.H. Hawkes. Cambridge University Press, Cambridge. pp. 99.

32. Gayle, M., Volk, Nicholas Maness and Kate Rotindo, 2004. Cryopreservation of Garlic (Allium sativum L) using Plant Vitrification Solution 2. CryoLetters, 25: 219–226.

33. Gonzalez-Arnao M.T, Claudia Esther Lazaro-Vallejo, Florent Engelmann, Roberto GamezPastrana, Yolanda Maria Martinez-Ocampo, Miriam

Cristina Pastelin-Solano and

34. Carlos Diaz-Ramos 2009. Multiplication and cryopreservation of vanilla (Vanilla planifolia 'Andrews') In Vitro Cellular & Developmental Biology - Plant 45 (5): 574-582.

35. Grout B W W 1995 Introduction to the in vitro preservation of plant cells, tissues and organs. In: Grout, B.W.W. (Ed.), Genetic Preservation of Plant Cells in Vitro.Springer–Verlag, Berlin. pp : 1–20.

36. Haeng-Hoon Kim, Eun-Gi Cho, Hyung-Jin Baek, Chang-Yung Kim, ER Joachim Keller and Florent Engelmann, 2004. Cryopreservation of garlic shoot tips by vitrification:

37. Effects of dehydration, Rewarming, unloading and regrowth conditions CryoLetters, 25: 59–70.

38. Haeng-Hoon Kim, Jung-Bong Kim, Hyung-Jin Baek, Eun-Gi Cho, Young-Am Chaand Florent Engelmann, 2004. Evalution of DMSO Concentration in Garlic Shoot tips during a Vitrification Procedure. CryoLetters, 25: 91–100.

39. Haeng-Hoon Kim, Ju-Won Yoon, Jung-Bong Kim, Florent Engelmann and Eun-Gi Cho, 2005. Thermal analysis of Garlic Shoot tips during a Vitrification Procedure.CryoLetters, 26(1): 33–44.

40. Himeno, H. and Sano, K., 1995. Synthesis of crocin, picrocrocin and safranal by saffron stigma like structures proliferated in vitro. Agricultural Biology and Chemistry, 51 (9):2395–2400.

41. Hoekstra, F.A., 1995. Collecting pollen for genetic resources conservation. In: Collecting Plant Genetic Diversity: Technical Guidelines, (Eds.) Guarino, L., Rao, V.R. and Reid, R.CAB International, Wallingford, UK, pp. 527–550.

42. Hori, H., Enomoto, K. and Nakaya, H., 1988. Induction of callus from pistils of Crocus sativus L. and production of colour components in the callus. Plant Tissue Culture Letters, 5:72 –77.

43. Hosoki, T. and Sagawa, Y., 1977. Clonal propagation of ginger (Zingiber officinale Rosc.) through tissue culture. Horticulture Science, 12: 451-452.

44. Hyung-Jin Baek, Haeng-Hoon Kim, Eun-Gi Cho, Young-Am Chae and Florent Engelmann, 2003. Importance of explant size and origin and of preconditioning treatments for cryopreservation of garlic shoot apices by vitrification. CryoLetters,24: 381–388.

45. Ilahi, I. and Jabeen, M., 1992. Tissue culture studies for micropropagation and extraction of essential oils from Zingiber officinale Rosc. Pakistan

Journal of Botany, 24(1): 54–59.

46. IPGRI, 1996 Pragamme activities, germplsm maintenance and use. In: Annual Report, IPGRI. Rome. pp. 56–65.

47. Iwananga, M., 1994. Role of International organisations in global genetic resource management. In:Proc. 27th International Symposium on Tropical Agriculture Research, Japan International Research Centre for Tropical Agricultural Sciences, Ministry of Agriculture, Forestry and Fisheries, Tsukuba, Japan, August, 25–26,1993, p.1–6.

48. Johnson, T.S., Ravishanker, G.A and Venkataraman, L.V., 1996. Biotransformation of ferulic acid vanillamine to capsaicin and vanillin in immobilised cell cultures of Capsicum frutescens PlantCell, Tissue and Organ Culture, 44(2): 117–123

49. Jung-Bong Kim, Haeng-Hoon Kim, Hyung-Jin Baek, Eun-Gi Cho, Yong-Hwan Kim and Florent Engelmann, 2005. Changes in sucrose and glycerol content in garlic shoot tips during freezing using pvs3 solution. CryoLetters, 26(2): 103–112.

50. Kartha K K & Engelmann F 1994 Cryopreservation and germplasm storage. In Plant Cell and Tissue Culture. Vasil, I. K. and Thorpe, T. A. (eds.), Kluwer Academic Publishers, Dordrecht/Boston/London. pp : 195230.

51. Kartha, K.K., 1985. In: Cryopreservation of Plant Cells and Organs, (Ed.) Kartha, K.K. CRC Press, Boca Raton, Florida, pp. 243–267.

52. Keller E. R. J.; Kaczmarczyk A.; Senula A. 2008. Cryopreservation for plant genebanks—a matter between high expectations and cautious reservation. CryoLetts 29: 53–62.

53. Keller, J., 1991. In vitro conservation of haploid and diploid germplasm in Allium cepa L. Acta Hortic.,289: 231–232.

54. Leigh, E. Towill and Remi Bonnart, 2003. Cracking in a vitrification solution during cooling or warming does not effect growth of cryopreserved mint shoot tips. CryoLetters, 24: 341–346.

55. Lukose, R., Saji, K.V., Venugopal, M.N. and Korikanthimath, V.S., 1993. Comparative field performance of micropropagated plants of cardamom (Elettaria cardamomum).Indian Journal of Agricultural Sciences, 63(7): 417–418.

56. Madhusoodanan, K.J., Kuruvilla, K.M., Vadiraj, B.A., Radhakrishnan, V.V. and Thomas, J., 2005. On farm evaluation of tissue culture vanilla plants vis-à-vis vegetative cuttings. Proceedings of ICAR National Symposium on Biotechnological Interventions for Improvement of

Horticultural Crops: Issues and Strategies, Kerala Agricultural University, Trissur, Kerala, India, p. 89–90.

57. Mandal, J., Pattnaik, S.and Chand, P.K. 2000. Alginate encapsulation of axillary buds of Ocimum americanum L. (hoary basil), O. Basilicum L. (sweet basil), O. Gratissimum L.

58. (shrubby basil), and O. Sanctum. L. (sacred basil) In Vitro Cellular & Developmental Biology - Plant 36(4): 287-292.

59. Maxted N, Brian, F and Hawkes, J G. 1997. Plant Genetic Conservation : the in situ approach. Springer. 446 pp

60. Mc Cown, B.H. and Amos, R., 1979. Initial trials of commercial micropropagation with birch. Proc.Inter. Plant Pro. Soc., 29: 387–393.

61. Minoo Divakaran and Nirmal Babu, K. 2009.Micropropagation and In Vitro Conservation of Vanilla (Vanilla planifolia Andrews) pp 129-138. In SM Jain and PK Saxena (eds)

62. Springer Protocols, Methods in Molecular Biology 547, Protocols for In Vitro Cultures and Secondary Metabolite Analysis of Aromatic and Medicinal Plants, The Humana Press, (Springer), USA.

63. Minoo Divakaran, Nirmal Babu K. and Pete,r K. V. 2011. Cryopreservation of pollen and inter specific hybridization in important orchid species V. planifolia and V. aphylla.

64. National Consultation for Production and Utilisation of Orchids 19-21 February, 2011 National Research Center for Orchids, Pakyong- 737106, Sikkim. Abstract.pp.98-99.

65. Minoo Divakaran, Nirmal Babu K and Michel Grisoni 2010. Biotechnological applications In Vanilla, pp. 51-73 In Eric Odoux and Michel Grisoni (eds) Vanilla, CRC Press,Boca Raton, USA.

66. Minoo, D., 2002. Seedling and somaclonal variation and their characterization in Vanilla. Ph.D. Thesis, Calicut University, Kerala, India.

67. Murashige, T. and Skoog, F., 1962. A revised medium for rapid growth and bioassays with tobacco tissue cultures. Physiol Plant, 15: 473–493.

68. Nadagauda, R S., Mascarenhas, A F & Madhusoodanan K J. 1983. Clonal multiplication of Elettaria cardamomum Maton. by tissue culture. J. Plantn Crops 11 : 60-64.

69. Nag, K.K. and Street, H.E., 1975. Freeze preservation of cultured plant cells, 11. The freezing and thawing phases. Physiol. Plantarum, 340: 254–260

70. Nair RR and Dutta Gupta 2003 Somatic embryogenesis and plant

regeneration in black pepper (Piper nigrum L.): Direct somatic embryogenesis from tissues of germinating seeds and ontogeny of somatic embryos. J. Hortic. Sci. Biotechnol, 78 : 416-421

71. Nirmal Babu, K., Geetha, S.P., Minoo D., Yamuna, G., Praveen, K., Ravindran, P.N. and Peter, K.V. 2007. Conservation of Spices Genetic Resources through In Vitro Conservation and Cryo- Preservation. pp. 210-233. In KV Peter and Z Abraham (ed). Biodiversity in Horticultural Crops Vol 1, Daya Publishing house, New Delhi Nirmal Babu, K., Geetha, S.P., Minoo, D., Ravindran, P.N. and Peter, K.V., 1999. In vitro conservation of germplasm. In: Biotechnology and its Application in Horticulture, (Ed.) S.P. Ghosh. Narosa Publishing House, New Delhi, pp. 106–129.

72. Nirmal Babu, K., Minoo, D., Geetha, S.P., Ravindran, P.N. and Peter, K.V., 2005. Advances in Biotechnology of Spices and Herbs. Ind. J. Bot. Res., 1(2): 155–214.

73. Nirmal Babu, K., Ravindran, P.N. and Peter, K.V., 1997. Protocols for Micropropagation of Spices and Aromatic Crops. Indian Institute of Spices Research, Calicut, India, p. 35.

74. Nirmal Babu, K., Ravindran, P.N. and Sasikumar, B., 2003. Field evaluation of tissue cultured plants of spices and assessment of their genetic stability using molecular markers. Final Report submitted to Department of Biotechnology, Government of India, pp. 94.

75. Nirmal Babu, K., Rema, J., Sree Ranjini, D.P., Samsudeen, K. and Ravindran, P.N., 1996. Micropropagation of an endangered species of Piper, P. barberi Gamble and its conservation. Journal of Plant Genetic Resources, 9(1): 179–182.

76. Niwata, E., 1995. Cryopreservation of apical meristems of garlic (Allium sativum L.) and high subsequent plant regeneration. CryoLetters, 16: 102–107.

77. Panis B.; Lambardi M. Status of cryopreservation technologies in plants (crops and forest trees). In: Ruane J.; Sonnino A. (eds) 2006.The role of biotechnology in exploring and protecting agricultural genetic resources. FAO, Rome, pp 61–78;

78. Parani, M., Anand, A. and Parida, A., 1997. Application of RAPD finger printing in selection of micropropagated plants of Piper longum for conservation. Current Science, 73(1): 81–83.

79. Peter, K.V., Ravindran, P.N., Nirmal Babu, K., Sasikumar, B., Minoo, D., Geetha, S.P. and Rajalakshmi, K., 2002. Establishing In vitro Conservatory of Spices Germplasm. ICAR Project report. Indian Institute

of Spices Research, Calicut, Kerala, India, pp. 131.

80. Philip, V.J, Joseph, D., Triggs, G.S. and Dickinson, N.M., 1992. Micropropagation of black pepper (Piper nigrum L.) through shoot tip cultures. Plant Cell Report, 12: 41-44.

81. Phunchindawan, M., Hirata, K., Sakai, A. and Miyamot K. Cryopreservation of encapsulated shoot primordia induced in horseradish (Armoracia rusticana) hairy root cultures Plant Cell Reports 16(7): 469-473. DOI: 10.1007/BF01092768

82. Quatrano, R.S., 1968. Freeze preservation of cultured flax cells using DMSO. Plant, 43: 2057–2061.

83. Rajasekharan, P.E. and Ganeshan, S., 2003. Pollen cryopreservation in Capsicum species: A feasibility study. Capsicum and Eggplant Newsletter, 22: 87–90.

84. Ravindran, P.N., Nirmal Babu, K., Saji, K.V., Geetha, S.P., Praveen, K. and Yamuna, G., 2004. Conservation of Spices genetic resources in in vitro gene banks. ICAR Project report. Indian Institute of Spices Research, Calicut, Kerala, India pp. 81.

85. Ravishankar, G A., Sarma K S., Venkataraman L V. and Kalyan A K. 1988. Effect of nutritional stress on capsaicin production in immobilized cell cultures of Capsicum annuum. Curr. Sci. 57 : 381-383.

86. Ravishankar, G.A., Sudhakar, J.T. and Venkataraman, L.V., 1993. Biotechnological approach of in vitro production of capsaicin. In: Proceedings of the National Seminar on Post Harvest Technology of Spices, Trivandrum, p. 75–82.

87. Reed B. M.; Engelmann F.; Dulloo M. E.; Engels J. M. M. 2004. Technical guidelines for the management of field and in vitro germplasm collections. IPGRI Handbooks for Genebanks No. 7. IPGRI, Rome, Italy.

88. Rekha Chaudhury and Malik, S.K., 2004. Genetic conservation of plantation crops and spices using cryopreservation. Indian Journal of Biotechnology, 3(3): 348–358.

89. Sajina, A., Minoo, D., Geetha, S.P., Samsudeen, K., Rema, J., Nirmal Babu, K., Ravindran, P.N. and Peter, K.V., 1997. Production of synthetic seeds in a few spice crops. In: Biotechnology of Spices, Medicinal and Aromatic Plants, (Eds.) Edison, S., Ramana, K.V., Sasikumar, B., Nirmal Babu, K. and Santhosh, J.E. Indian Society for Spices, Calicut, India, p. 65–69.

90. Sambrook J, Fritsch E F & Maniatis T 1989 Molecular Cloning – a laboratory manual. Vol.3.Cold Spring Harbor Laboratory Press, New

York.

91. Schenk R U & Hildebrandt A C 1972 Medium and techniques for induction and growth of monocotyledonous and dicotyledonous plant cell cultures. Can. J. Botany, 50:199-204.

92. Sharma, T.R., Singh, B.M. and Chauhan, R.S., 1994. Production of encapsulated buds of Zingiber officinale Rosc. Plant Cell Reports, 13: 300–302.

93. Shatnawi, M. A., Johnson K.A. and Torpy F.R. (2004) In vitro propagation and cryostorage of Syzygium francissi (Myrtaceae) by the encapsulation-dehydration method, In Vitro Cellular & Developmental Biology - Plant 40 (4): 403-407.

94. Baghdadi, S.H, Shibli, R.A, Syouf M.Q., Shatnawi MA., Arabiat, A. and Makhadmeh I.M (2010) Cryopreservation by encapsulation-vitrification of embryogenic callus of wild crocus (Crocus hyemalis and C. moabiticus) Jordan J. Agricultural Sciences. 6 (3) 436-442.

95. Umetsu, H., Wake, H., Saitoh, M., Yamaguchi, H. and Shimomura, K., 1995. Characteristics of cold preserved embryogenic suspension cells in fennel Foeniculum vulgare Miller.Journal of Plant Physiology, 146(3): 337–342.

96. Venkataraman, L.V., Ravishanker, G.A., Sarma, K.S. and Rajasekaran, T., 1989. In vitro metabolite production from saffron and capsicum by plant tissue and cell cultures.

97. In: Tissue Culture and Biotechnology of Medicinal and Aromatic plants, (Eds.) Kukreja et al. CIMAP, Lucknow, India, p. 147–151.

98. Williams J G K, Kubelik A R, Livak K J, Rafalski J A & Tingey S V 1990 DNA polymorphisms amplified by arbitrary primers are useful as genetic markers.Nucleic acid research 18:6531-6535

99. Withers, L.A., 1985. Cryopreservation of cultured cells and meristems. In: Cell Culture andSomatic Cell Genetics of Plants, Vol. 2: Cell Growth, Nutrition, Cyto-differentiation and Cryopreservation, (Ed.) I.K. Vasil. Academic Press, Orlando, Florida, pp. 253–316.

100. Withers, L.A., 1987. Long-term preservation of plant cells, tissues and organs. Oxford Surveys of Plant Mol. and Cell Biology, 4: 221–272.

101. Withers, L.A., 1991. Biotechnology and plant genetic resources Conservation. In: Plant Genetic Resources Conservation and Management: Concepts and Approaches, (Eds.) R.S. Paroda and R.K. Arora. IBPGR, New Delhi, pp. 273–297.

102. Yamuna, G. 2007. Studies on Cryopreservation of Spices Genetic

Resources Ph. D Thesis, University of Calicut, Kerala, India.

103. Yamuna, G, Sumath,i V., Geetha, S. P., Praveen, K., Swapna, N. and Nirmal Bab, K. 2007. Cryopreservation of In Vitro grown shoot of Ginger (Zingiber officinale Rosc). CryoLetters.28(4):241-252

Chapter 8

DIVERSITY AND GENETIC EROSION OF ANCIENT CROPS AND WILD RELATIVES OF AGRICULTURAL CULTIVARS FOR FOOD: IMPLICATIONS FOR NATURE CONSERVATION IN GEORGIA (CAUCASUS)

Maia Akhalkatsi, Jana Ekhvaia and Zezva Asanidze

Ilia State University Georgia

INTRODUCTION

The interpretation of a healthy diet is one of the dilemmas for our modern civilization. Advances in agriculture are mainly directed at increasing food production to solve problems of a growing human population. However, food security remains a problem to ensure healthy food and to prevent human disease. These two tendencies often do not coincide. At present, the selective breeding programs of crops are mainly oriented toward the production of high-yielding varieties of genetically enhanced cultivars of cereals that have increased growth rates, increasing the percentage of usable plant parts and resistance against crop diseases. This initiative is linked to what began in the 1960s and was named by William Gaud (of USAID) a "Green Revolution" (Davies, 2003). It was a product of globalization as evidenced in the creation of international agricultural research centres to introduce new crop varieties around the world. This process caused a significant increase in total cereal production and daily calorie supply in developing countries between the 1960s and 1990s (Davies, 2003). However, this process has caused the gradual replacement of traditional crop varieties, and as a result has had a dramatic effect on agrobiodiversity in many countries. Particularly impacted have been the traditional landraces used by local peoples for thousands of years and this has affected the health of these communities. Georgia, located in the South Caucasus, owns one of the oldest agricultural traditions. The name of the country is "Sakartvelo" in the Georgian language but its common name "Georgia" is semantically linked to Greek (γεωργία, transliterated geōrgía) and Latin (georgicus) roots meaning "agriculture" (Javakhishvili, 1987). Many Georgian endemic species and local varieties of wheat, barley, legumes, grapevine and fruits are known (Ketskhoveli, 1957). The traditional use of local cultivars is considered to be a reason for human longevity in the Caucasus region (Fox, 2004). Over five percent of the male Georgian centenarians were reputed to have been over age 120 in 1959 (Garson, 1991). The percentage of males over age 70 was 0.9% in 1959 and 1.07% of women were over 70. However, these values had diminished by 1970 to 0.66% and 0.86%, respectively. At present, no exact data are available, but longevity has obviously diminished (Fox, 2004).

Archaeological data clearly show that the Caucasus region (and Georgia in particular) was settled from prehistoric time and agriculture was developed during the early Neolithic era (Javakhishvili, 1987). The information about the wide chronological intervals in the archaeological materials connected with the history of mankind in the Caucasus starts from the Early Pleistocene. The 1.7-Myr-old specimens of small-brained hominids are found in the Caucasus at Dmanisi, located in Southern Georgia (Fig.1), which is the earliest known hominid site outside of Africa (Gabunia & Vekua, 1995; Finlayson, 2005). This speciman has been classified as Homo erectus senso lato, which is a very early type of H. ergaster and/or a new taxon, H. georgicus (Gabunia et al., 2002). The next chronological interval in the archaeological materials is connected with the period of Late Middle Palaeolithic and Early Upper Palaeolithic periods demonstrating patterns of mobility, land-use, and hunting of Neanderthal and modern human competition within the South Caucasus (Adler & Bar-Oz, 2009). Neanderthals invaded the Caucasus region at an unknown time and modern humans may have occupied the region alongside them from ~40 Ka before the present (BP). According to the archaeological material from different caves in Georgia (Tushabramishvili, 2011) and the northern Caucasus (Ovchinnikov et al., 2000), the final replacement of the Neanderthals by modern humans might be occurred here ~28 Ka BP. The Upper Palaeolithic archaeological findings at Dzudzuana Cave (Fig.1), Imereti region, Georgia, revealed remnants of wool (Capra caucasica) and dyed fibers of wild flax (Linum usitatissimum) dated to ~36–34 Ka BP (Adler & Bar-Oz, 2009). The Dzudzuana Cave flax fibers have clearly been modified, cut, twisted and dyed black, gray, turquoise and pink, most likely with locally available natural plant-derived pigments (Kvavadze et al., 2009). E. Kvavadze and colleagues (2009) surmise that this represents the production of colourful textiles for some purpose, perhaps clothing. In general, it is supposed that the microscopic flax fibres are the remains of linen and thread, which would have been used in clothing for warmth, for shoes, to sew together pieces of leather or to tie together packs. The archaeological findings from

Neolithic and Early Bronze periods are rich with plant fossils and seeds of both wild species and local landraces. The ancient findings from Neolithic period of cereal grains in Georgia were discovered from Trialeti Range, Kvemo Kartli region (Arukhlo excavations, Bolnisi district; Fig.1) and Samegrelo region (Dikha-Gudzuba, Zugdidi district; Fig.1) from 6th up to 2nd millennium BC (Melikishvili, 1970). Seven species of cultivated wheat - Triticum aestivum, T. spelta, T. carthlicum, T. macha, T. monococcum, T. dicoccum, T. compactum and one wild relative Aegilops cylindrica have been discovered in Arukhlo, Kvemo Kartli region. Other cereals: millet - Panicum milleaceaum, barley - Hordeum vulgare, Italian millet - Setaria italica, oats - Avena sativa, wild lentil - Lens ervoides and pea -Pisum sativum have been found in the same site. The wheat fields in Arukhlo were irrigated. Very recent studies on einkorn wheat domestication using amplified fragment length polymorphism (AFLP) show that T. boeoticum was domesticated in southeast Turkey in the Karacadag Mountains close to Diyarbakir (Heun et al., 1997). Old Georgian kingdom Diauehi (Diaokhi) is adjacent region to this place. Therefore, it might be considered to be an area where cultivation of cereals occurred in very early historical time. The earliest archaeological finding of cultivar grapevine pips are found in Shulaveri (Fig.1), located near Dmanisi in southern Georgia and dated to ~8.000 years BP (Ramishvili, 1988). A wide range of carbonised seeds, including wild and domesticated grape (Vitis vinifera, V. vinifera subsp. sylvestris), wheat (Triticum sp.), pea (Pisum sativum), rowan (Sorbus sp.) and walnut (Juglans regia) are found in soil samples in Nokalakevi (Fig.1), Western Georgia, dated to the Hellenistic period (Grant et al., 2009).

Figure. 1: Map of Georgia. The administrative regions: 1. Abkhazia; 2. Samegrelo-Upper Svaneti; 3. Guria; 4. Adjara; 5. Racha-Lechkhumi; 6. Imereti; 7. Meskheti- Javakheti; 8. Shida Kartli; 9. Kvemo Kartli; 10. Mtskheta-Mtianeti; 11. Kakheti. The places of archaeological excavations are indicated: Dikha-Gudzuba, Nokalakevi, Dzudzuana cave, Arukhlo, Dmanisi and Shulaveri.

According to N. I. Vavilov (1992), the origin of ancient crop varieties and landraces in Georgia coincides with the period of their primary domestication. Georgia is often considered part of the Near East where many field crops were domesticated. N. I. Vavilov (1992) determined 8 centres of crop origin and diversity. Among them was the fourth centre, which included the South Caucasus, Asia Minor, Iran and Turkmenistan. The main crops domesticated in this centre (which includes Georgia) are wheat, rye, oats, seed and forage legumes, herbs, fruits, and grapes for winemaking; 83 species all tolled. The problem is that there are no concrete data to assess either the current status of local varieties or information about the domestication process in Georgia. The fundamental work on domestication and origin of wheat and barley in this region was done by the famous Georgian botanist V. Menabde (1938, 1948). The agricultural evidence was reported by several other Georgian authors (Ketskhoveli, 1957; Khomizurashvili, 1973; Akhalkatsi et al., 2010). We have studied domestication of wild grapevine (Vitis vinifera subsp. sylvestris) and wild pear (Pyrus caucasica) using morphometric and systematic molecular methods (Ekhvaia & Akhalkatsi, 2010; Ekhvaia et al., 2010; Asanidze et al., 2011) confirming genetic relationships between wild populations and local cultivars of grape and pear. However, complete evaluation of diversity of Georgian local cultivars and crop wild relatives (CWR) has not yet been complete. There are many threats to these oldest of crops in the modern period. In our opinion, the main threat to agrobiodiversity in Georgia is the loss of landraces and ancient crop varieties. Protection measures in the country are still not being implemented at an appropriate rate. National policies and comprehensive measures are urgently needed to address the problem of conserving the genetic resources of ancient crops in Georgia. Thus, we suggest that it is necessary to establish a general overview of the types of crops that are current landraces and primitive forms occurring in Georgia and to publish lists of indigenous landraces and CWRs of cereals, legumes, vegetables and fruits representing direct ancestors, and endemic, rare or endangered species, in order to evaluate the sustainability of their traditional use in terms of nature conservation.

LANDRACE ASSESSMENT

Agriculture in Georgia is characterized by a great diversity of local landraces, varieties and even endemic species of crops. These varieties reveal a high level of adaptation to local climatic conditions and often have high resistance to diseases. Georgians have used these crops for a very long time and their healthy life, reflected by the longevity of individuals in the population, was considered to be connected to their good food. However, there are many threats to these oldest of crops in the modern period, particularly since the

1950s. The loss of local and ancient crop varieties should be considered to be the main threat to agrobiodiversity in Georgia. The known diversity and distribution of local landraces is based on data obtained from archaeological reports, historical manuscripts, ethnography, and botanical field expeditions in different regions of Georgia since 1920s. The oldest known text about Georgian cultivars is a XVII century by the work of Vakhushti Batonishvili "Geographic Description of Georgian Kingdom" (Batonishvili, 1991). Active investigation of Georgian crops began in the 1920s (Ketskhoveli, 1928, 1957; Menabde, 1938, 1948; Dekaprelevich, 1947). These investigations revealed that ancient cultivars of grapevine, fruits, wheat, barley, rye, oats, common millet, Italian millet, legumes, flax, and a number of herbal and spice plants, were still being cultivated in Georgia. The rapid loss of local cultivars of cereals, legumes and flax began in the 1950s and reached an extreme in the 1990s (Akhalkatsi et al., 2010). At present, almost all of Georgia's ancient crops are maintained in conservatory collections and seed banks, but none are present in peasant house gardens in lowland areas. Only the mountain areas contain depositories of the ancient crops of Georgia, where some number of landraces still exist. The process of genetic erosion of ancient crop varieties, however, has begun even in these regions since the 1990s and this presents great concern about the loss of aboriginal crops adapted to high mountain areas (Pistrick et al., 2009). Monitoring of crop diversity is now conducted by international nature conservation institutions and Georgian scientific and nongovernmental organizations to preserve the genetic resources of local cultivars. One of the problems is the deficit of information about the current state of ancient crops and recommendations for their conservation are inadequate. Therefore, it is necessary to assess research needs and implications for conservation and to formulate recommendations for the conservation and on-farm maintenance of Georgian landraces.

Diversity of Ancient Crop Varieties

Reports of the diversity of local landraces in Georgia has to present been published primarily in Georgian- and Russian-language scientific publications (Ketskhoveli, 1928, 1957; Menabde, 1938, 1948; Dekaprelevich, 1947; Kobakhidze, 1974). Databases and international periodicals lack descriptions of this diversity, taxonomy and discussions of the conservation value of landraces. In our opinion it is important to spread information about diversity and conservation needs of local cultivars of Georgia worldwide to support the evaluation of their roles in healthy life of human. Some crops, such as grapevine, wheat, barley and fruit trees are characterized by the highest diversity of landraces in Georgia. Grapevine - Vitis vinifera L. (Vitaceae) shows greatest genetic and morphological variability. About 500 names of autochthonous grapevine varieties known from Georgia are characterized by a wide range of colour gamma and shapes of berries and pips (Javakhishvili, 1987; Ketskhoveli et al., 1960), which points to an evolutionary centre in this region (Vavilov, 1992). These cultivars showed great ampelometric variability and broad adaptability to different climate and soil conditions (Ketskhoveli et al., 1960). Each province of Georgia possesses its own grapevine cultivars adapted to local climate. The varieties are of three forms: 1) Babilo is an old grapevine with stem more than 20 cm in diameter clambering on trees (Fig.2A). 2) Maghlari represents varieties that climb tree trunks (alder, persimmon, mulberry, cherry, beech, chestnuts, etc.) distributed mainly in peasant orchards in western and southern Georgia (Fig.2B). 2) Dablari is used to create typical vineyards (Fig.2C) found in commercial areas. The total area of vineyards in Georgia was 40.000 ha in the 1980s. It has diminished to ca. 25.000 ha today (Bedoshvili, 2010). Forty-four percent of this territory is located in Kakheti region, 26% in Imereti, 15% in Kartli and 15% in almost all regions of Georgia except in the high mountain regions of Khevi, Khavsureti and Tusheti. Forty-one cultivars of grapevine are used as commercial varieties in Georgia. Among them, 27 are technical varieties used for winemaking and 14 are table grapes (Bedoshvili, 2010). Ninetyseven percent of total annual yield is used for winemaking and only 3% as table grapes. Wine is made from landraces: 'Rkatsiteli' (55%); 'Tsolikauri' (10.2%); 'Chinuri' (7%); 'Saperavi' (4%); 'Kakhuri Mtsvane' (3.3%); and, several local and introduced cultivars (20.5%).

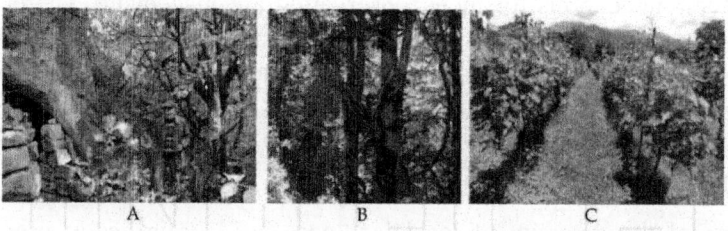

Figure. 2: Different types of vineyeards: A - Babilo, an old grapevine 'Meskhuri Shavi' (diameter 32 cm) belonged to the family of Gogi Natenadze in village Chachkari, Meskheti; B - Maghlari, clambering introduced cultivar Vitis labrusca 'Odessa' on beech tree in village Naghvarevi, Adjara. C - Dablari, typical vineyeard of landraces 'Rkatsiteli' and 'Saperavi' in village Shilda, Kakheti. Photos by Maia Akhalkatsi.

The first known threat to the grapevine in Georgia occurred during the occupation of Georgian territory by the Muslims during the medieval period. They destroyed vineyards and/or led to the destruction of human settlements, where until now are presented local wilding grape cultivars clambering on trees. The vineyards grow on the terraces of Mediterranean type in the historic province of Tao-Klarjeti located now in southern Georgia and in the province of Artvin, Turkey. Since 15th century, the Seljuk Turks occupied this territory and vine terraces disappeared and were covered with trees or grasses. However, we have found peasants in some villages of Meskheti province searching for old cultivars in abandoned settlements. They are replanting naturalized grape cultivars to house gardens. We have found ancient grapevine varieties 'Meskhuri Shavi' and 'Meskhuri Mtsvane' to be frost resistant and growing in high mountain areas in villages Zemo Vardzia (1322 m a.s.l.) and Chachkari (1264 m a.s.l.), Aspindza district; village Zazalo, (1486 m a.s.l.), Adigeni district; and, Karzameti castle near boundary to Turkey, 1450 m a.s.l. One of the oldest Georgian grape cultivar 'Krikina', which is morphologically nearly identical to wild grapevine (V. vinifera subsp. sylvestris [C. C. Gmel.] Hegi), was found in village Rveli, Borjomi district, in house garden of Gaioz Tabatadze who replanted this cultivar from the ruins of the historic village Baniskhevi where his ancestors lived. The grape variety 'Shonuri' ('Lushnu' in Svanetian) adapted to high mountain areas grows in Upper Svaneti province from 1045 to 1400 m a.s.l. Landraces adapted to high elevations are rare and are usually replaced with the introduced grape cultivar V. labrusca 'Odessa'. It is widespread in mountainous villages in all regions of Georgia. Some other rare grapevine cultivars - 'Kachichi', 'Saperavi', 'Sebeli', 'Jvarisa', have been found on lower elevations in village Gvimbrala, Lentekhi district, Lower Svaneti province. Rare grape cultivars - 'Aladasturi', 'Tsulukidzis Tetra' and 'Tskhvedanis Tetra' have been found in village Tabori and 'Usakhelouri' in village Zubi, Tsageri district, Lechkhumi province. Rare landrace 'Chkhaveri' was found in village of Merisi, at 474 m a.s.l. Adjara province. In 1860, the V. vinifera was virtually wiped out in the places of its origin, when an aphid, Phylloxera vastatrix was accidentally introduced into France, and within a few years had ravaged all vineyards in Europe and in Georgia as well. Currently, almost all Georgian grape varieties are grafted on rootstocks of American grapevines - V. riparia, V. rupestris and V. berlandieri and their hybrids are resistant to Phylloxera. This disaster made it necessary to undertake urgent steps for ex situ conservation of old, endangered and autochthonous grapevine varieties by establishing living collections in Georgia; this was begun in the 1930s. The collections of plant genetic resources were established in research institutes, which have been under reforms since 1990s and operating with diminishing funding to maintain the collections. In 2003, 929 varieties were protected in the living collections. Among them, 701 were cultivars obtained from selective breeding and only 248 of the 524 autochthonous Georgian varieties remain. These collections of the State Agricultural University were located in Dighomi (573 cultivars) and Mukhrani (155 cultivars), and, the collections in Telavi (226 cultivars) and Skra (75 cultivars) belonged to the Georgian research Institute of Horticulture, Viticulture and Winemaking (Maghradze, et al., 2010). Recently, these collections have been closed. Nevertheless, some effort has been made to establish new collections in Telavi (573 accessions), Skra (440) and Vachebi (312) in 2008. Three other new collections were set up by Saguramo "Centre for Grapevine and Fruit Tree Planting Material Propagation" (ca 400 accessions), "Kindzmarauli" and "Shumi" wineries (as a total 149 accessions). Two new collections were established in Italy by the University of Milan (Maghradze, et al., 2010). Some Georgian cultivars are in living collections abroad in Russia, Moldova and Germany as well.

Table 1: List of wheat species distributed in Georgia by V. Menabde (1948, 1961). The accepted names are added from web-page: http://www.ars-grin.gov/cgibin/npgs/html/splist.pl?28515. The status of species is based on phylogenetic studies of V. Menabde (1948, 1961): EG - endemic of Georgia; W- wild; PS - primary species; SP - secondary species; IS - introduced species. Ploidy levels are indicated

N	Taxon names by Menabde, 1948, 1961	Taxon accepted name by ARS-GRIN, USDA 2011	Status	Ploidy levels
	T. aegilopoides Balansa ex Körn. (=T. boeticum Boiss.)	T. monococcum subsp. aegilopoides (Link) Thell.	W	2n
	T. monococcum L.	T. monococcum L.	PS	2n
	T. timopheevii Zhuk.	T. timopheevii (Zhuk.) Zhuk. subsp. timopheevii	EG, PS	4n
	T. chaldicum Menabde	T. timopheevii (Zhuk.) Zhuk. subsp. armeniacum (Jakubz.) Slageren	W	4n
	T. dicoccoides Körn.	T. turgidum L. subsp. dicoccoides (Körn. ex Asch. & Graebn.) Thell.	W	4n
	T. palaeo-colchicum Menabde	T. turgidum L. subsp. palaeocolchicum Á. Löve & D. Löve	EG, SP	4n
	T. dicoccum Schuebl.	T. turgidum L. subsp. dicoccon (Schrank) Thell.	SP	4n
	T. durum Desf.	T. turgidum L. subsp. durum (Desf.) Husn.	SP	4n
	T. turgidum L.	T. turgidum L. subsp. turgidum	SP	4n
	T. carthlicum Nevski (=T. ibericum Menabde; T. persicum Vavilov ex Zhuk.)	T. turgidum L. subsp. carthlicum (Nevski) Á. Löve & D. Löve	EG, SP	4n
	T. polonicum L.	T. turgidum L. subsp. polonicum (L.) Thell.	SP	4n
	T. turanicum Jakubz.	T. turgidum L. subsp. turanicum (Jakubz.) Á. Löve & D. Löve	IS	4n
	T. abyssinicum Vavilov	T. dicoccon subsp. abyssinicum Vavilov	IS	4n
	T. vulgare Villars	T. aestivum L.	SP	6n
	T. macha Dekapr. & Menabde	T. aestivum L. subsp. macha (Dekapr. & V.L. Menabde) Mackey	EG, PS	6n
	T. spelta L.	T. aestivum subsp. spelta (L.) Thell.	IS	6n
	T. sphaerococcum Percival	T. aestivum subsp. sphaerococcum (Percival) Mackey	IS	6n
	T. compactum Host	T. aestivum subsp. compactum (Host) Mackey	SP	6n
	T. zhukovskyi Menabde & Ericzjan	T. zhukovskyi V.L. Menabde & Eritzjan	EG, SP	6n

A small living grapevine collection exists in the G. Eliava National Museum in Martvili district, Samegrelo province, founded in 1972 and containing 24 old Colchic grapevine varieties (Eliava, 1992). Seven cultivars of Meskheti region were collected in the research station of Biological Farming Association

Elkana in village Tsnisi, Akhaltsikhe district. Many grape landraces are extinct and do not exist even in living collections. Wheat - Triticum L. (Poaceae) also shows high diversity in Georgia. Nineteen species of wheat from the 26 known species of the genus Triticum have been historically distributed in Georgia (Tab.1). Some of them are endemic species: T. timopheevii, T. zhukovskyi, T. macha, T. carthlicum and T. palaeo-colchicum. Sixteen species, 144 varieties, and 150 forms of wheat were registered in Georgia in the 1940s (Menabde, 1948).

According to V. Menabde (1948), three species from the list are wild – T. boeticum (2n=14), T. dicoccoides (2n=28), T. timopheevii subsp. armeniacum (2n=28); they were mixed with cultivars in the wheat fields and did not exist in natural habitats in Georgia. Sites of T. boeoticum are concentrated in south-eastern Turkey, where this species was probably domesticated (Heun et al., 1997). The current distribution indicates that its weedy races have spread with cultivated cereals far to the west and east. There is evidence that it was found in fields with T. monococcum in Georgia (Menabde, 1948). Since the 1930s their number has diminished and all of these species had disappeared after the 1960s, when non-aboriginal cultivars were introduced in kolkhozis—agricultural farming corporations in Soviet times, changing the species composition in wheat fields. At present, none of these species occur in agricultural fields of Georgia. Three species from the list (Tab.1) are considered by V. Menabde (1948) as primary species (close to the first domesticated species): T. monococcum (2n=14), T. timopheevii (2n=28) and T. macha (2n=42). First two species, T. monococcum - 'Gvatsa Zanduri' and T. timopheevii - 'Cheltha Zanduri', in Georgian, are old autochthonous wheat species distributed mainly in western Georgia - Racha-Lechkhumi, Imereti and Samegrelo. T. timopheevii was growing in a small area in western Georgia together with its hexaploid derivative - T. zhukovskyi, and cultivated einkorn - T. monococcum (Menabde, 1961). These three species represent polyploid series of wheat Zanduri, which was possible to find in peasant farms till 1990s. T. macha is archaeological findings in Dikha-Gudzuba and Shulaveri excavations dated by Neolithic period and was cultivated in Racha-Lechkhumi, Imereti and Samegrelo up to 1950s (Dekaprelevich, 1947). Nine native species of wheat - T. palaeo-colchicum, T. dicoccum, T. durum, T. turgidum, T. carthlicum, T. polonicum, T. aestivum, T. zhukovskyi and T. compactum, are considered by V. Menabde (1948, 1961) as secondary species originated by hybridization with wild and primary species of Triticum, Aegilops spp., Thinopyrum intermedium (Host) Barkworth & D. R. Dewey subsp. intermedium (=Agropyron glaucum [Desf. ex DC.] Roem. & Schult.), and Thinopyrum elongatum (Host) D. R. Dewey (=Agropyron elongatum [Host] P. Beauv.). T. aestivum, T. carthlicum and T. durum have many varieties and cultivars. The four species in the list (Tab.1)

- T. abyssinicum, T. spelta, T. sphaerococcum and T. turanicum represent geographical races introduced from different regions in the historically different times. The traditional wheat fields in all regions of Georgia usually contain several species and varieties (Eritzjan, 1956; Zhizhilashvili & Berishvili, 1980). Bread wheat fields contain: T. aestivum var. erythrospermum 'Tetri dolis puri', T. aestivum var. ferrugineum 'Tsiteli dolis puri', T. aestivum var. lutescens 'Upkho tetri dolis puri', T. aestivum var. milturum 'Upkho tsiteli dolis puri', T. compactum 'Kondara khorbali'. Usually, this combination of wheat taxa is associated with wild weed Makhobeli - Cephalaria syriaca (L.) Schrad. ex Roem. & Schult. (Dipsacaceae) occurring most often in such wheat fields. The seeds of this species are of the same size as wheat and after threshing remain in the harvest. Seeds are ground into a powder and used with wheat to make bread, cakes, etc. It adds a nice flavour but quickly goes rancid. Another combination of varieties was dominated by T. durum 'Shavpkha' composed by T. durum var. apulicum, T. durum var. leucurum, T. durum var. murciense, T. aestivum var. erythrospermum, T. aestivum var. pseudo-barbarossa, T. aestivum var. lutescens, T. compactum var. erinaceum. This population is adapted to dry climate in the lowland areas and in the high elevations up to 1800 m a.s.l. in Javakheti Plateau, where it is sown in early spring. The same character of adaptation to high elevation is typical for the wheat species, T. carthlicum 'Dika', sown on high mountain areas in spring. The combination of varieties dominated by 'Dika' is as follows: T. carthlicum var. rubiginosum, T. carthlicum var. stramineum, T. aestivum var. erythrospermum, T. aestivum var. ferrugineum, T. compactum var. erinaceum.

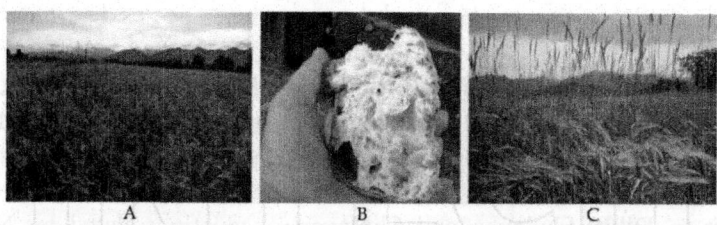

Figure. 3: A- Wheat field of the Georgian endemic Triticum carthlicum 'Dika' in research station of the Biological Farming Association Elkana, village Tsnisi, Meskheti; B - Traditional wheat bread in Meskheti; C- Six row barley field with mixture of wild rye Svila (Secale segetale) in village Shilda, Kakheti. Photos by Maia Akhalkatsi.

Wheat fields were planted throughout Georgia at elevations from 300 m to 2160 m a.s.l. We have found this highest location of soft wheat field in the Eastern Greater Caucasus, village Chero in Tusheti (Akhalkatsi et al., 2010). At present, almost none of these traditional wheat varieties and species occur in the territory of Georgia. Only aboriginal varieties of bread wheat still exist in several high mountain regions like Tusheti, Meskheti, Javakheti and Svaneti (Pistrick et al., 2009). Living collections and gene banks preserve the local varieties. The living collection of the Biological Farming Association Elkana has many landraces in village Tsnisi, Akhaltsikhe district (Fig.3A). In 2010, they sowed a 10 ha wheat field. The harvest from this field contained local cultivar T. aestivum var. ferrugineum 'Akhaltsikhis tsiteli dolis puri' and weed Makhobeli. The flour was baked as bread in Tbilisi and as traditional bread in Meskheti (Fig.3B). Barley – Hordeum vulgare L. (Poaceae) is an ancient agricultural crop in Georgia. It was the second most important cereal in Georgia after wheat and main crop in high mountain regions used for bread, forage and production of beer, as well as an attribute of religious rituals and in the folk medicine (Javakhishvili, 1987). Two different names were used for barley in Georgian language - Krtili and Keri. Krtili denotes six-row winter barley (H. vulgare subsp. hexastichon [L.] Čelak.) that was sowed in autumn; Keri refers to two-row summer barley (H. vulgare subsp. distichon [L.] Körn.) sowed in spring (Menabde, 1938). Six-row barley was sown in lowland areas but was cultivated up to 2130 m a.s.l. in Svaneti. Two-row barley was cultivated mainly in high mountain regions. The cultivars of two-row barley H. vulgare var. nutans ‹Akhaltesli› and H. vulgare var. nigrum Willd. ‹Dzveltesli shavpkha› are distributed up to 2100 m a.s.l. in all high mountain areas. H. vulgare var. nutans is mixed in the field with wheat - T. carthlicum ‹Dika›, and the flour is produced from mixed wheat and barley seeds. H. vulgare var. nudum Spenn. ‹Kershveli› was cultivated in Meskheti and Svaneti. Four-row barley (H. vulgare subsp. tetrastichon [Stokes] Čelak.) is rare and the cultivar - H. vulgare var. pallidum Ser. ‹Tetri Keri› occurs only in the high mountain region of Meskheti, Tusheti and Svaneti up to 2100 m a.s.l. These cultivars persist today only in high mountain regions. However, their distribution has been seriously diminished. At present, introduced varieties of barley are widely cultivated in the lowlands and their names are unknown to the local population. Rye – Secale cereale L. (Poaceae) is only a local cultivar of high mountain regions of Georgia (1800-2200 m a.s.l.). Fields of S. cereale (2n=14) are now found only in Upper and Lower Svaneti and Meskheti. Rye was used for making alcohol and as forage. The wild species, S. segetale (Zhuk.) Roshev. (2n=42), called Svila is widespread in wheat and barley fields and is harvested together with them (Fig.3C). The bread of wheat with Svila is considered to be very nutricious and has good taste. An endemic species of rye is S. vavilovii Grossh. (2n=14). It is also called Caucasian rye. This species was found in wheat field in Georgia (Bockelman et al., 2002). We have monitored the place in village Beghleti, Khashuri district in 2008, where Georgian botanists had noted the presence of this species in the wheat fields, but cultivated plots no longer exist in that area. The village has lost of most of its residents and no agriculture is undertaken there. Introduced cultivars and commercial varieties of rye are not used in Georgia. Oats – Avena sativa L. (Poaceae) is a traditionally cultivated plant distributed from 400 to 1400 m a.s.l. It is used only as forage for horses and poultry. Two varieties of oats have been described for Upper Svaneti (Ketskhoveli, 1957) - A. sativa var. aurea Körn. and A. sativa var. krausei Körn. In lowlands, usually, the origin of the seeds is unknown to local farmers. It is purchased in the market and farmers receive no information about their origin. Millet – Panicum miliaceum L. (Poaceae) is very old agricultural plant cultivated in all regions of Georgia. It was used as a supplementary feed (for animals and poultry) and for making alcoholic drinks. At present, it is cultivated only in high mountain regions (1000-1800 m a.s.l.). Several varieties are described in Upper and Lower Svaneti: P. miliaceum var. aureum V.M. Arnold & Shibaiev. - grain yellow or cream; P. miliaceum var. subaereum Körn. - grain grey; P. miliaceum var. griseum Körn. - grain brown; P. miliaceum var. atrocastaneum Batalin ex V.M. Arnold & Shibaiev. - grain black; P. miliaceum var. badium Körn. - grain white (Zhizhilashvili & Berishvili, 1980). The acreage of millet fields declined after introduction of maize in Georgia in 17th century. Italian millet - Setaria italica (L.) P. Beauv. (Poaceae) was cultivated in Colchis, Samegrelo since ancient times. The cultivar - S. italica subsp. colchica (Dekapr. & Kaspar.) Maisaya & Gorgidze was represented with 32 landraces (Maisaia et al., 2005). It can currently be found in the Samegrelo region of western Georgia. Another subspecies - S. italica subsp. moharia (Alef.) H. Scholz., is called Kvrima in Georgian. It was cultivated for a long time but was replaced by maize.

Legumes - peas, lentils, chickpeas, faba beans, common vetch, bitter vetch, chickling vetch, alfalfa, sainfoin and blue fenugreek are traditional crops in Georgia (Tab.2). Green Pea (Pisum sativum) is originated in the South Caucasus. It is grown in house gardens in small amounts for food today. Two species of pea are cultivated in Georgia - P. sativum with white flowers, round white or yellow seeds, and P. arvense with purple flowers, ridged dark coloured seeds. The third wild species P. elatius Steven ex M. Bieb. with dark purple flowers is often found in locations of old settlements, ruins of monasteries and churches and inside castle walls. The local cultivar of green pea, P. sativum subsp. transcaucasicum Govorov, has 14 varieties (Kobakhidze, 1974). Local varieties of Chickpea (Cicer arietinum) are rarely cultivated today. Three subspecies and 24 varieties were available in western Georgia - Racha-Lechkhumi, Svaneti and Imereti up to 1920s (Dekaprelevich & Menabde, 1929). In the 1970s, the same three subspecies - C. arietinum subsp. mediterraneum G. Pop., C. arietinum subsp. eurasiaticum G. Pop. and C. arietinum subsp. orientalis G. Pop., remained, but included only 6 of 24 varieties - C. arietinum subsp. mediterraneum var. ochroleucum A.

Kob., C. arietinum subsp. mediterraneum var. rozeum G. Pop., C. arietinum subsp. eurasiaticum var. aurantiacum G. Pop., C. arietinum subsp. orientalis var. fulvum G. Pop., C. arietinum subsp. orientalis var. rufescens G. Pop., and C. arietinum subsp. orientalis var. rufescens brunneopunctatus A. Kob. (Kobakhidze, 1974). Chickpeas were traditionally available in Svaneti, but by the 1970s only one farmer was sowing it in Kala community village Khe (Zhizhizlashvili & Berishvili, 1983). The Biological Farming Association Elkana is producing local cultivars of chickpea and selling them in market. Lentil (Lens culinaris) was represented in Georgia by two subspecies - L. culinaris subsp. macrosperma N.F. Mattos and L. culinaris subsp. microsperma N.F. Mattos; and 15 varieties (Kobakhidze, 1974). The last subspecies with small seeds was sown in high mountain areas in Javakheti. It was available in Meskheti till 1970s. Lentils were cultivated in Upper Svaneti from prehistoric times, but, at present, it is nearly extinct. In 1980s, three cultivars were described in Svaneti - 1. L. culinaris var. persica Bar. - reddish-brown seeds; 2. L. culinaris var. ochroleucus nigro-punctulata A. Kob. - light brown seeds with black dots; 3. L. culinaris var. nigro-marmorata A. Kob. - seeds have reddish-yellow background with black marbling (Zhizhizlashvili & Berishvili, 1983). The Biological Farming Association Elkana is producing local cultivars of lentil for the market. Faba bean (Vicia faba) is one of the oldest cultivated plants. Faba bean is ancient agricultural plant in western and southern Georgia. Three varieties and 31 subvarieties were described in Georgia with small (V. faba var. minor Beck.), medium (V. faba var. equina Pers.) and large (V. faba var. major Harz.) seeds (Kobakhidze, 1974). At present, the large seed Faba bean is widely distributed in Upper and Lower Svaneti. Two varieties are found in the Lower Svaneti: 1. V. faba var. minor subvar. straminea A. Kob. - compressed on sides, tip obtuse, colour light cream. 2. V. faba var. equina subvar. ochroleucus A. Kob. - slightly compressed on sides, tip rounded, colour yellowish (Zhizhizlashvili & Berishvili, 1983). Chickling vetch (Lathyrus sativus) is used as human food in a soup to called shechamandi. It is also a green forage, used as silage and fed as seed flour to pigs and poultry. It is now available only at the research station of the Biological Farming Association Elkana. Bitter vetch – Vicia ervilia is distributed in Meskheti and Javakheti. There are cultivated and wild forms of this species. It is used as a forage and for soil enrichment with nitrogen. Common vetch (Vicia sativa) is used as forage and for hay, especially in Upper and Lower Svaneti and Javakheti. It is a valuable forage crop, rich in proteins. More often it appears as a weed in the fields of high mountain areas among grain crops – millet, barley, rye. Sainfoin (Onobrychis spp.), alfalfa (Medicago sativa) and clover (Trifolium spp.) are forage legumes. A local variety of Onobrychis transcaucasica Grossh. ‹Akhalkalakuri›, is widely used. Blue fenugreek (Trigonella caerulea) is traditional spice plant used in almost all of the foods of Georgian cuisine. It is available in all regions of Georgia.

Flax – Linum usitatissimum L. (Linaceae), was one of the oldest and important field crops in Georgia. Since prehistoric times, it was used to produce excellent linens (Kvavadze et al., 2009) and to make oil from its seeds. Big millstones were used to extract the oil from the flax seeds and they remain in many historical ruins. Until recently, flax was cultivated only in Javakheti, where flax seeds were used to produce pharmacologically pure oil for medicines. According to the eighteenth century Georgian scientist and geographer Vakhushti Batonoshvili (1991), several volatile oil-bearing plants were cultivated in Georgia - roses, camphor, lavender and basil. Kenaf - Cannabis sativa L. (Cannabaceae), was used to produce fiber for cord and thread for sacks. The seeds were used to produce oil. A traditional use of kenaf seeds was to mix them with wheat flour and making breads that had antidepressant effects. Traditional vegetables (Tab.2) are represented by sugar beets, spinach, carrots, radishes, turnips, onions, Welsh onion, leeks and garlic. Beet - Beta vulgaris, is an ancient cultivated plant whose tubers and young leaves were used in Georgian cooking. Leaves primarily came from the variety B. vulgaris subsp. cicla (L.) W.D.J. Koch ‹Tsiteli Mkhali› that was grown in lower elevations up to 1400 m a.s.l. Another beet variety - B. vulgaris L. subsp. vulgaris ‹Sasufre Charkhali› is rare. Carrot - Daucus carota, was edible as a wild species in Georgia since prehistoric times. The cultivated carrot is widespread in peasant›s house gardens in lowland areas. Onion - Allium cepa and garlic - A. sativum, are ancient cultivated plants available in all regions of Georgia. Red onions are very popular in Georgian people. A. sativum is called ‹Georgian garlic›. Another variety is ‹Russian garlic› representing A. ampeloprasum L. Leek - A. porrum, is typical in western Georgia. Welsh onion - A. fistulosum is currently grown in several high mountain areas. Until the 1970s, it was widespread in Imereti, but at present, Chinese shallot - A. cepa var. aggregatum G. Don has completely supplanted Welsh onion. Radish - Raphanus sativus, is grown in lower elevations in gardens and is cultivated by farmers for the market.

Table 2: Seed and forage legumes and traditional vegetables of Georgia

N	Latin name	Family	English common name	Georgian common name
	Allium cepa L.	Liliaceae	*Onion*	*Khakhvi*
	Allium fistulosum L.	Liliaceae	*Welsh Onion*	*Chlakvi*
	Allium porrum L.	Liliaceae	*Leek*	*Prasi*
	Allium sativum L.	Liliaceae	*Garlic*	*Niori*
	Beta vulgaris L. subsp. *vulgaris*	Chenopodiaceae	*Beet*	*Charkhali*
	Brassica rapa L. subsp. *rapifera* Metzger	Brassicaceae	*Turnip*	*Talgami*
	Cannabis sativa L.	Cannabaceae	*Kenaf*	*Kanafi*
	Cicer arietinum L.	Fabaceae	*Chickpea*	*Mukhudo*
	Daucus carota L.	Apiaceae	*Carrot*	*Stafilo*
	Lathyrus sativus L.	Fabaceae	*Chickling vetch*	*Tsulispira*
	Lens culinaris Medik.	Fabaceae	*Lentil*	*Ospi*
	Linum usitatissimum L.	Linaceae	*Flax*	*Seli*
	Medicago sativa L.	Fabaceae	*Alfalfa*	*Ionja*
	Onobrychis transcaucasica Grossh.	Fabaceae	*Sainfoin*	*Espartseti*
	Pisum arvense L.	Fabaceae	*Pea*	*Barda*
	Pisum sativum L.	Fabaceae	*Pea*	*Barda*
	Raphanus sativus L.	Brassicaceae	*Radish*	*Boloki*
	Spinacia oleracea L.	Chenopodiaceae	*Spinach*	*Ispanakhi*
	Vicia ervilia (L.) Willd.	Fabaceae	*Bitter vetch*	*Ugrekheli*
	Vicia faba L.	Fabaceae	*Faba bean*	*Tsertsvi*
	Vicia sativa L.	Fabaceae	*Common vetch*	*Tsertsvela*

Herbs are represented by numerous species (Tab.3) - parsley, coriander, tarragon, sweet basil, savory, gardencress pepperweed, dill, fennel, celery, garden lettuce, peppermint. Herbs are cultivated in small sections of house gardens even in urban settlements. Sometimes, people have herbs indoors in pots.

Table 3: List of traditionally cultivated herbs in Georgia

N	Latin name	Family	English common name	Georgian common name
	Anethum graveolens L.	Apiaceae	*Dill*	*Kama*
	Apium graveolens L.	Apiaceae	*Celery*	*Niakhuri*
	Artemisia dracunculus L.	Asteraceae	*Tarragon*	*Tarkhuna*
	Coriandrum sativum L.	Apiaceae	*Coriander*	*Kindzi*
	Foeniculum vulgare Mill.	Apiaceae	*Fennel*	*Didi Kama*
	Lactuca sativa L.	Asteraceae	*Garden lettuce*	*Salati*
	Lepidium sativum L.	Brassicaceae	*Gardencress pepperweed*	*Tsitsmati*
	Mentha piperata L.	Lamiaceae	*Peppermint*	*Pitna*
	Ocimum basilicum L.	Lamiaceae	*Sweet basil*	*Rehani*
	Petroselinum crispum (Mill.) A.W. Hill	Apiaceae	*Parsley*	*Okhrakhushi*
	Satureja hortensis L.	Lamiaceae	*Savory*	*Kondari*
	Trigonella caerulea (L.) Ser.	Fabaceae	*Blue fenugreek*	*Utskho Suneli*

Fruits are valuable cultivars in Georgia. Wild and cultivated fruit crops reveal high species and genetic diversity in Georgia and represent rich material for future breeding activities. Many fruits have wild relatives representing the same species and direct ancestors of local cultivars (Tab.4,5).

Introduced Cultivated Plants

Georgia is located at the crossroads of Europe and Asia. Many cultivated plants have been introduced since ancient times to Georgia from other regions of the world (Javakhishvili 1987). Some introduced crops have become very popular and widespread. They are introduced from different countries. Such crops as cucumber (Cucumis sativus), found in Georgia since medieval times, eggplant (Solanum melongena), marigold (Tagetes patula), used in almost all traditional meals; and black pepper (Piper nigrum) were introduced from India. Watermelon (Citrullus lanatus) from South Africa was cultivated in the Caucasus since medieval times. Maize (Zea mays), sunflower (Helianthus annuus), tomato (Solanum lycopersicum), bean (Phaseolus vulgaris), pepper (Capsicum annuum), and potato (Solanum tuberosum) were introduced to Georgia from the Americas at about the same time as in Europe (Javakhishvili, 1987). Tea (Camellia sinensis) and citrus fruits (Citrus limon, Citrus reticulata, Citrus sinensis) came from China in the 1830s (Bakhtadze, 1947). Nicotiana rustica, (tutuni in Georgian) has been cultivated for a long time and is found in the most regions, including high mountain areas, of Georgia. N. tabacum, was introduced during the Soviet period and was cultivated in kolkhozis for commercial use. Georgia has become a secondary centre of diversity for most of these crops. Landraces of bean, maize, potato, tomato, and cucumber that do not exist in their countries of origin can be found in Georgia. Bean (Phaseolus spp.; Vigna spp.) was introduced via Turkey to Guria and Samegrelo during the second half of the XVI century (Javakhishvili, 1987). At present, 61 varieties and 406 forms of common bean had originated in Georgia due to widespread distribution and hybridization of different species of bean: Phaseolus vulgaris L., P. lunatus L., P. coccineus L. (=P. multiflorus Lam.), P. acutifolius A. Gray, Vigna radiata (L.) R. Wilczek var. radiata (=P. aureus Roxb.), V. angularis (Willd.) Ohwi & H. Ohashi var. angularis (=P. angularis [Willd.] W. Wight) and V. umbellata (Thunb.) Ohwi & H. Ohashi (=P. calcaratus Roxb.) (Kobakhidze, 1965). Beans are cultivated in gardens in large amounts providing sufficient food for families and representing a cash crop for additional income. Diversity of beans remains high. Maize (Zea mays L.) was introduced to Georgia in 1633-1650 (Javakhishvili, 1987). The Georgian name Simindi originated from the old name for flour Samindo as flour was introduced earlier to Georgia than the initial cultivation of maize. Besides landraces such as 'Kazha simindi' from Svaneti there are some cultivars that originated in Georgia: 'Ajametis tetri', 'Abashis kviteli', 'Kartuli kruki', 'Gegutis kviteli', 'Imeruli hibridi' and 'Lomtagora'. Many cultivars were introduced from Russia, Hungary, Yugoslavia, etc., during Soviet time. The last introduction occurred in 2011, when the high-yield US corn hybrid 'Pioneer' was sown in Georgia. Corn had replaced common and Italian millet and is used as an everyday food, especially in western Georgia. Potato (Solanum tuberosum L.) was introduced to Georgia during the second half of the XIX century. Several landraces of high quality are grown in high-mountain regions: Svaneti, Racha-Lechkhumi, Khevsureti, Khevi and Adjara. Breeder's cultivars were introduced into lowland areas during Soviet time. Recently, genetically modified potato cultivars have been introduced in Georgia by international seed-distribution organizations. These modern cultivars have almost completely supplanted local landraces even in high mountain regions. Tea and citrus had high commercial value in Georgia, but in the 1990s these crops were abandoned and tea was not produced in Georgia until recently. Citrus (lemons, oranges and mandarins) were sold only in the local Georgian market. At present, this business is restored. Information about introduced varieties has been published annually during the XX century. The latest official edition of the Catalogue of the Georgian Released Varieties of 1997 (published in 1996) listed 195 varieties of field and vegetable crops and 195 varieties of fruits. These varieties were part of the collections that existed at the end of the 1980s and beginning of the 1990s. At present, only a few of these varieties exist. The data about modern breeder's varieties introduced into Georgia during last decade are usually absent and a number of varieties have been cultivated in Georgia without going through the official procedures for release. Therefore, it is difficult to evaluate the diversity of recently introduced cultivars.

CROP WILD RELATIVE ASSESSMENT

The CWR are taxa related to species of direct socio-economic importance, which includes the progenitors of crops. According to modern concept of wild relatives, under CWR we should understand all species related to any cultivated plants, as well as to wild species of ornamental, food, fodder and forage, medicinal plants, condiments, forestry species and plants used for industrial purposes, such as oils and fibers i.e. to all plants of economic importance (Laguna, 2004). Although, "classical" definition of CWR is restricted only to species related to cultivated crops, including such important field crops as wheat, barley, rye, oats, sorghum, common and Italian millet, grain and forage legumes (such as Phaseolus, Vicia, Vigna, Lens, Lathyrus, Cicer) and some vegetables and industrial crops.

The flora of the Caucasus harbours a remarkable concentration of economically important plants, particularly CWRs such as cereals, legumes, fruits, vegetables, herbs and technical plants like flax. The list of CWRs in Georgia was published in Plant Genetic Resources (PGR) Forum - CWR Catalogue of Europe and the Mediterranean (Maxted et al., 2008). This catalogue listed 1784 species of vascular plants, representing 43% of the 4130 vascular plant species found in Georgia. These are mainly wild species that also have considerable economic importance providing food, fuel, timber, forage, hay and habitats for animal life. A large number of taxa used in folk and scientific medicine are also included among economically valuable plants. However, this list is not detailed enough to assess the economic value of CWRs representing the same species or direct ancestors of crop plants. There is no information on the status of endangered and endemic species. Thus, we developed a general description of vegetation types and separated CWR endemic species and species genetically closely related to crops.

Flora and Vegetation

Georgia is located between 41°02' and 43°34' latitude north and between 40° and 46°43' longitude east. It borders the Russian Federation to the north, Turkey and Armenia to the south, Azerbaijan to the east, and has approximately 310 km of coastline along the Black Sea to the west. Georgian territory (69.700 km2) covers two separate mountain systems: the Greater Caucasus Range which trends north-west to east-southeast between the Black Sea and Caspian Sea; and the Lesser Caucasus Mountains, which run parallel to the greater range at a distance to the south that averages about 100 kilometres. Two thirds of the country is mountainous with an average height of 1.200 m.a.s.l., with the highest peaks of Mount Shkhara (5.184 m.a.s.l.) in the western Greater Caucasus and Mount Didi Abuli (3.301 m.a.s.l.) in the Lesser Caucasus. Colchis, Kartli and Alazeni valleys and Iori plateau represent intermontane lowlands located between these two mountain systems. Geologically, the Caucasus consists of Meso- to Cenozoic deposits. Ancient Precambrian and Paleozoic formations are rare (Neidze et al., 2008). The Likhi Range divides the country into eastern and western halves that differ in climate and landscapes. Western Georgia has a humid subtropical climate with annual precipitation ranging from 1000–4000 mm. Temperatures fluctuate between the winter averages of 2.8° to 6.7° C and the summer averages of 22.7° to 23.8° C. Eastern Georgia has a more continental climate, due to the barrier of the Likhi Range, which bars the warm Black Sea winds from this area. The temperatures vary from the January averages of 0-2.2° C to the July mean of 27.8° C. Annual precipitation is considerably less in eastern Georgia and ranges from 400– 1600 mm. Southern Georgia has a continental climate. The local winters are cold. The frosts are - 25° C and in July temperatures rise to 40° C. Annual rainfalls are usually less than 600- 1000 mm (Neidze et al., 2008). Soil types vary in Georgia. The most widespread types in the lowlands of western Georgia are bog, podzolic, red, yellow and hilly piedmonts, which are mainly acidic. Mountainforest and mountain-meadow soils occur in higher elevations. Chestnut and chernozem soils are widespread in the lowlands of eastern Georgia and are characterized by neutral pH. Brown humid-sulphates, saline soils of steppes and semi-deserts, as well as intermediate forest-steppe and mountain-meadow soils occur in semi-desert areas. Alluvial soils are found along the rivers throughout Georgia. Brown soils are typical for the Georgian forest zone in the range of 800-2000 m a.s.l. (Sabashvili, 1970). Western Georgia's landscape ranges from see-level swamps and lowland temperate rainforests to eternal subnival zone and glaciers, while the eastern part of the country contains temperate forests and semi-arid plains in lower elevations and alpine and subnival zones. Main rivers are R. Mtkvari, R. Rioni, R. Enguri and R. Alazani. There are 70 natural lakes and 11 artificial reservoirs. The lower section of the Rioni River is located in the Colchis valley and was naturally occupied by marshes and lagoons, but in 1960s this area was the site of a large reclamation-drainage project and it was converted to agricultural land. The majority of the forests that covered the Colchis plain are now virtually gone; the exceptions are those included in national parks and reserves. The Mtkvari River basin, which includes the major parts of southern and eastern Georgia, is drier. It is covered with semi-arid vegetation and temperate forests. Forests, in total, amount to 40% of Georgia's territory while the alpine zone accounts for roughly 10% of the land. Much of the natural habitat in the lowland areas of western Georgia has disappeared over the last 100 years because of agricultural development and urbanization. The vegetation of Georgia belongs to three floristic provinces – Euxine, Caucasian and Armeno-Iranian (Takhtajian, 1986). The Euxine and Caucasian provinces belong to the Circumboreal Region, the Boreal Subkingdom and the Holarctic Kingdom and the ArmenoIranian Province belongs to the Irano-Turanian Region, the Tethyan (Mediterranean) Subkingdom and the Holarctic Kingdom. There are following vegetation zones: 1. Colline zone (0-400 m a.s.l.), which includes coastal and halophytic habitats in western Georgia and dry open woodlands and semi-deserts in eastern Georgia; 2. Lower montane zone (400-800 m a.s.l.) is used as arable land. The natural vegetation in western Georgia is represented by small remnant areas of Colchic broad-leaved mixed forest. Oak-hornbeam forests and dry scrublands occur in eastern Georgia; 3. Middle montane zone (800-1500 m a.s.l.) is primarily used for agriculture. Broad-leaved mixed forests, mountain xerophytes scrublands, and mountain steppes are represented; 4. Upper montane zone (1200-2050 m a.s.l.) is covered by beech and broadleaf-coniferous mixed forests; 5. Subalpine zone (1900-2400[2500] m a.s.l.) is a treeline ecotone, with tall herbaceous vegetation, shrublands and polydominant subalpine grass and herb meadows used as pastures or arable land; 6. Alpine zone (2500–2900 m a.s.l.) has alpine meadows and snowbed communities. Vegetation is mostly used for grazing and is of considerably lower quality than the subalpine vegetation, both by biomass volume and typological diversity; 7. Subnival zone (2900-3300 m a.s.l.) is patchy highest limits of vegetation. 8. Nival zone (3300-5184 m a.s.l.) covered by glaciers. 9. Azonal vegetation type is represented by fragments of wetlands rich in boreal type flora, halophytic desert vegetation and rocky areas (Nakhutsrishvili, 1999). Flora of Georgia is represented by 4,130 species of vascular plants. Among them are 79 pteridophytes, 17 gymnosperms and 4034 angiosperms (Nakhutsrishvili, 1999). The 10 leading families are Asteraceae (538 species), Poaceae (332 species), Fabaceae (317 species), Rosaceae (238), Brassicaceae (183), Scrophulariaceae (179), Apiaceae (177), Lamiaceae (149), Caryophyllaceae (135) and Liliaceae (129). High endemism is characteristic of the Caucasus and represents one of the world's hot spots of biodiversity. Out of all, 1304 (32.3%) species are endemics of the Caucasus ecoregion and 261 (6.6%) are endemics of Georgia (Schatz et al., 2009). There are 17 endemic genera in the flora of the Caucasus. Most of them are represented by one species: Agasyllis latifolia (M. Bieb.) Boiss., Alboviodoxa elegans (Albov) Woronow, Charesia akinfievii (Schmalh.) E. Busch, Cladochaeta candissima (M. Bieb.) DC., Gadellia lactiflora (M. Bieb.) Schulkina, Mandenovia komarovii (Manden.) Alava, Paederotella pontica (Rupr. ex Boiss.) Kem.-Nath., Petrocoma hoefftiana (Fisch.) Rupr., Pseudobetckea caucasica (Boiss.) Lincz., Pseudovesicaria digitata (C. A. Mey.) Rupr., Sredinskya grandis (Trautv.) Fed., Symphyoloma graveolens C. A. Mey., Trigonocaryum involucratum (Steven) Kusn., Woronowia speciosa (Albov) Juz. Two genera contain two species of each: Chymsydia agasylloides (Albov) Albov, C. colchica (Albov) Woronow, Grossheimia macrocephala (Muss.- Puschk. ex Willd.) Sosn. & Takht., G. polyphylla (Ledeb.) Holub. One endemic genus is represented by 5 species: Kemulariella caucasica (Willd.) Tamamsch., K. rosea (Steven ex M. Bieb.) Tamamsch., K. abchasica (Kem.-Nath.) Tamamsch., K. tugana (Albov) Tamamsch., and K. colchica (Albov) Tamamsch. For conservation action to be effective, it is important to understand not just the needs of individual species, but also the context in which conservation efforts will need to take place. Therefore, it is important to evaluate the conservation values of the species that contribute most to human health and to develop conservation measures to avoid their extinction.

Diversity of Crop Wild Relatives in Georgia

Flora of the Caucasus region is rich as there are high concentrations of economically important and edible plants, particularly wild crop relatives such as grapevine, wheat, barley, rye, oats, seed and forage legumes, fruits and vegetables. The Caucasus is considered to be the centre of evolution for many unique life forms and is a natural museum for rich genetic resources (Vavilov, 1992). We identified the number of species of the genera that are traditional crops in Georgia (Tab.4). A total of 20 plant families, 76 genera and 479 species were identified as wild relatives of ancient crops in Georgia. Most of these plant species are closely related genetically to landraces and might be their progenitor species.

Table 4: List of wild relatives of ancient crops in Georgia

Families	Number of genera	Number of species	Genera with number of species
Apiaceae	8	17	Anethum (1), Apium (2), Carum (5), Coriandrum (1), Daucus (1), Foeniculum (1), Pastinaca (5), Petroselinum (1)
Asparagaceae	1	3	Asparagus (3)
Asteraceae	3	16	Cichorium (1), Lactuca (7), Scorzonera (8)
Betulaceae	1	6	Corylus (6)
Brassicaceae	5	20	Brassica (4), Lepidium (8), Raphanus (2), Rorippa (4), Sinapis (2),
Cannabaceae	2	3	Cannabis (2), Humulus (1)
Chenopodiaceae	2	3	Beta (2), Spinacia (1)
Cornaceae	1	1	Cornus (1)
Fabaceae	10	154	Cicer (1), Lathyrus (20), Lens (3), Lotus (5), Medicago (21), Onobrychis (19), Pisum (1), Trifolium (40), Trigonella (10), Vicia (34)
Grossulariaceae	2	4	Grossularia (1), Ribes (3)
Juglandaceae	1	1	Juglans (1)
Lamiaceae	4	19	Mentha (4), Origanum (1), Satureja (3), Thymus (11)
Liliaceae	2	39	Allium (36)
Linaceae	1	12	Linum (12)
Moraceae	2	3	Ficus (1), Morus (2)
Poaceae	16	64	Aegilops (7), Agropyron (2), Avena (8), Brachypodium (3), Cynosorus (2), Elymus (4), Elytrigia (9), Echinochloa (3), Hordeum (5), Hordelymus (1), Panicum (5), Psathyrostachis (1), Secale (5), Setaria (6), Sorghum (1), Taeniatherum (2)
Punicaceae	1	1	Punica (1)
Rosaceae	12	110	Amygdalus (1), Cerasus (4), Crataegus (8), Cydonia (1), Fragaria (3), Malus (1), Mespilus (1), Prunus (2), Pyrus (11), Rosa (30), Rubus (36), Sorbus (12)
Staphyleaceae	1	2	Staphylea (2)
Vitaceae	1	1	Vitis (1)
Total: 20	76	479	

CWR are commonly defined in terms of wild species related to agricultural and horticultural crops. As such a broad definition of a CWR would be any wild taxon belonging to the same genus as a crop. A working definition of a CWR was provided by N. Maxted and colleagues (Maxted et al., 2006): "A crop wild relative is a wild plant taxon that has an indirect use derived from its relatively close genetic relationship to a crop; this relationship is defined in terms of the CWR belonging to gene pools 1 or 2, or taxon groups 1 to 4 of the crop". According to gene pool concept three gene pools are distinguished as follows: (1) Primary Gene Pool (GP-1) within which GP-1A are the cultivated forms and GP-1B are the wild or weedy forms of the crop; (2) Secondary Gene Pool (GP-2) which includes the coenospecies (less closely related species) from which gene transfer to the crop is possible but difficult using conventional breeding techniques; (3) Tertiary Gene Pool (GP-3) which includes the species from which gene transfer to the crop is impossible, or if possible requires sophisticated techniques, such as embryo rescue, somatic fusion or genetic engineering. The taxon group concept is used to establish the degree of CWR relatedness of a taxon. Application of the taxon group concept assumes that taxonomic distance is positively related to genetic distance. CWR rank of taxon groups is defined as follows: (1) Taxon Group 1A – crop; (2) Taxon Group 1B – same species as crop; (3) Taxon Group 2 – same series or section as crop; (4) Taxon Group 3 – same subgenus as crop; (5) Taxon Group 4 – same genus; (6) Taxon Group 5 – same tribe but different genus to crop (Maxted et al., 2006). Diversity and Genetic Erosion of Ancient Crops and Wild Relatives of Agricultural Cultivars for Food: Implications for Nature Conservation in Georgia (Caucasus)

Thus, the combined use of the gene pool and taxon group concept proposed above provide the most pragmatic means available to determine whether a species is a CWR and how closely related a CWR is to its crop. We have determined 66 species of CWR belonging to 43 genera and 17 families, which can be assigned as Primary (GP-1) and Secondary Gene Pool (GP-2) and Taxon Group 1 and 2 (Tab.5). Seventeen (25.75%) are wild species but used as crops by collecting in the natural habitats and they were identified as GP1A. The same CWR species as crop were 19 (28.8%). Different species were 30 (45.45%), but representing direct progenitors whose genome is involved in the evolution of cultivars. Almost the same numbers were obtained during taxon group classification: TG1A - 17 species (25.75%), TG1B - 16 (24.25%), TG2 - 30 (45.45%), TG 5 - 3 (4.55%). The last 3 species belonging to the Taxon Group 5 are Aegilops, a wild relative of wheat. Goatgrass (Aegilops tauschii) is considered to be a donor of D genome of bread wheat genomic constitution = AABBDD (Petersen et al., 2006). The distribution area of this species is wide, however, D genomes of all forms of T. aestivum were found to be most closely related to accessions collected in Georgia, Armenia, Nakhitshevan and Azerbaijan (Dvorák et al., 1998). Thus, the germplasm of the populations of goatgrass in the South Caucasus needs conservation and should be preserved both in situ in protected areas and ex situ in seed collections. Barley is one of the oldest crops to be domesticated from its wild progenitor Hordeum spontaneum

(Badr et al., 2000, Kilian et al., 2006). We have found H. spontaneum in Georgia in three different places. This species was not included in the list of "Flora of Georgia' and it is a new species for Georgia.

Table 5: Gene pool and taxon group of wild relatives to Georgian ancient crops GP-Gene Pool; TG-Taxon Group

Family	Crop	Taxon	GP	TG
Apiaceae	Daucus carota	Daucus carota L.	GP1B	TG1B
Apiaceae	Coriandrum sativum	Coriandrum sativum L.	GP1B	TG1B
Asparagaceae	Asparagus officinalis	Asparagus caspius Schult. & Schult. fil.	GP1B	TG1B
Asparagaceae	Asparagus officinalis	Asparagus officinalis L.	GP1B	TG1B
Asparagaceae	Asparagus officinalis	Asparagus verticillatus L.	GP1B	TG1B
Betulaceae	Corylus avellana	Corylus avellana L.	GP1B	TG1B
Betulaceae	Corylus avellana	Corylus iberica Wittm. ex Kem.-Nath.	GP2	TG2
Betulaceae	Corylus avellana	Corylus colchica Albov	GP2	TG2
Brassicaceae	Brassica oleracea	Brassica juncea (L.) Czern.	GP2	TG2
Brassicaceae	Brassica oleracea	Brassica napus L.	GP2	TG2
Brassicaceae	Brassica oleracea	Sinapis arvensis L.	GP2	TG2
Cannabaceae	Cannabis sativa	Cannabis sativa L.	GP1A	TG1A
Cannabaceae	Humulus lupulus	Humulus lupulus L.	GP1A	TG1A
Chenopodiaceae	Beta vulgaris	Beta maritima L.	GP2	TG2
Fabaceae	Pisum sativum	Pisum elatius M. Bieb.	GP1B	TG1B
Fabaceae	Cicer arietinum	Cicer caucasica Bornm.	GP2	TG2
Fabaceae	Lathyrus sativus	Lathyrus tuberosus L.	GP2	TG2
Fabaceae	Lens culinaris	Lens nigricans (M. Bieb.) Webb & Berth.	GP2	TG2
Fabaceae	Lens culinaris	Lens ervoides (Brign.) Grande	GP2	TG2
Fabaceae	Lens culinaris	Lens culinaris Medik. subsp. orientalis (Boiss.) Ponert	GP1B	TG1B
Fabaceae	Vicia faba	Vicia johannis Tamamsh.	GP2	TG2
Fabaceae	Vicia faba	Vicia narbonensis L.	GP2	TG2
Fabaceae	Vicia sativa	Vicia sativa L.	GP1A	TG1A
Grossulariaceae	Ribes rubrum	Ribes alpinum L.	GP2	TG2
Grossulariaceae	Ribes rubrum	Ribes caucasicum M. Bieb.	GP2	TG2
Juglandaceae	Juglans regia	Juglans regia L.	GP1A	TG1A
Lamiaceae	Satureja hortensis	Satureja laxiflora K. Koch	GP2	TG2
Lamiaceae	Satureja hortensis	Satureja spicigera (K. Koch) Boiss.	GP2	TG2
Linaceae	Linum usitatissimum	Linum bienne Mill.	GP1B	TG1B
Linaceae	Linum usitatissimum	Linum usitatissimum L.	GP1A	TG1A
Moraceae	Morus alba	Morus alba L.	GP1A	TG1A
Moraceae	Morus nigra	Morus nigra L.	GP1A	TG1A
Moraceae	Ficus carica	Ficus carica L.	GP1A	TG1A
Poaceae	Triticum aestivum	Aegilops cylindrica Host	GP1B	TG5
Poaceae	Triticum aestivum	Aegilops triuncialis L.	GP2	TG5
Poaceae	Triticum aestivum	Aegilops tauschii Coss.	GP1B	TG5
Poaceae	Hordeum hexastichon	Hordeum bulbosum L.	GP1B	TG2
Poaceae	Hordeum distichon	Hordeum spontaneum K. Koch	GP1B	TG1B
Poaceae	Avena sativa	Avena barbata Pott ex Link	GP2	TG2
Poaceae	Avena sativa	Avena sterilis L.	GP2	TG2
Poaceae	Secale cereale	Secale strictum subsp. anatolicum (Boiss.) K. Hammer	GP2	TG2
Poaceae	Secale cereale	Secale strictum subsp. kuprijanovii (Grossh.) K. Hammer	GP2	TG2
Poaceae	Secale cereale	Secacle cereale L. subsp. segetale Zhuk.	GP1A	TG1A
Poaceae	Panicum miliaceum	Panicum capillare L.	GP2	TG2
Poaceae	Panicum miliaceum	Panicum sumatrense Roth	GP2	TG2

Family	Crop	Taxon	GP	TG
Poaceae	Panicum miliaceum	Panicum dichotomiflorum Michx.	GP2	TG2
Poaceae	Setaria italica	Setaria viridis (L.) P. Beauv.	GP2	TG2
Poaceae	Setaria italica	Setaria verticillata (L.) P. Beauv.	GP2	TG2
Poaceae	Setaria italica	Setaria glauca (L.) P. Beauv.	GP2	TG2
Poaceae	Setaria italica	Setaria intermedia Roem. & Schult.	GP2	TG2
Punicaceae	Punica granatum	Punica granatum L.	GP1A	TG1A
Rosaceae	Pyrus communis	Pyrus caucasica Fed.	GP1B	TG1B
Rosaceae	Pyrus communis	Pyrus balansae Decne.	GP1B	TG1B
Rosaceae	Malus domestica	Malus orientalis Uglitzk.	GP2	TG2
Rosaceae	Cydonia oblonga	Cydonia oblonga Mill.	GP1B	TG1B
Rosaceae	Prunus domestica	Prunus domestica L. subsp. insititia (L.) C. K. Schneid.	GP1A	TG1A
Rosaceae	Prunus domestica	Prunus spinosa L.	GP1B	TG1B
Rosaceae	Prunus cerasifera	Prunus cerasifera Ehrh. var. divaricata (Ledeb.)L.H.Bailey	GP1A	TG1A
Rosaceae	Cerasus avium	Cerasus avium (L.) Moench	GP1B	TG1B
Rosaceae	Cornus mas	Cornus mas L.	GP1A	TG1A
Rosaceae	Mespilus germanica	Mespilus germanica L.	GP1A	TG1A
Rosaceae	Rubus idaeus	Rubus idaeus L.	GP1A	TG1A
Rosaceae	Amygdalus communis	Amygdalus georgica Desf.	GP2	TG2
Staphyleaceae	Staphylea pinnata	Staphylea pinnata L.	GP1A	TG1A
Staphyleaceae	Staphylea colchica	Staphylea colchica Steven	GP1A	TG1A
Vitaceae	Vitis vinifera	Vitis vinifera subsp. sylvestris (C.C.Gmel.) Hegi	GP1B	TG1B

It is assumed that H. spontaneum might have evolved from H. bulbosum by fixation of the genes controlling self-compatibility and annual habit (Cass et al., 2005). This last species is widespread in Georgian regions. Most fruit trees in Georgia are wild in forests and have cultivars domesticated from these wild ancestors. An economically important Georgian fruit crop is grape, which has a wild relative species Vitis vinifera subsp. sylvestris. We have found 9 populations of wild grapevine and conducted studies to reveal genetic and morphological relations between wild grapevine and landraces in Georgia (Ekhvaia et al., 2010). Many fruits are domesticated in the Caucasus from wild ancestors representing Primary Gene Pool (GP-1B) to be the wild or weedy forms of the crops (Tab.5). The fruit crops (GP1A) and ancestor species (GP-1B) are the following: Pome fruits - pear (Pyrus communis, P. caucasica), apple (Malus domestica, M. orietalis), quince (Cydonia oblonga); stone fruits - plum (Prunus domestica, P. domestica var. insititia, P. spinosa), Myrobalan (Prunus vachushti), sourplum (Prunus cerasifera var. divaricata), cherries (Cerasus avium, C. vulgaris), Cornel cherry (Cornus mas), medlar (Mespilus germanica), mulberry (Morus alba, M. nigra), pomegranate (Punica granatum); berries - red raspberry (Rubus idaeus), currant (Ribes rubrum, R. nigra, R. alpinum, R. biebersteinii), fig (Ficus carica), bladdernut (Staphylea pinnata, used flowers for marinade), and nuts - such as hazelnut (Corylus avellana), almond (Amygdalus communis), and walnut (Juglans regia), etc. We evaluated CWRs endemic for the Caucasus (Tab.6). The endemic species of the same genus as crop were calculated. The number of endemic species from the total 479 CWRs of agricultural cultivars for food is 114 (23.8%).

DOMESTICATION EVENTS

Domestication of Grapevine in Georgia

The grapevine was among the first fruits to be cultivated in Georgia (Javakhishvili, 1987). There are many arguments to confirm the fact that domestication events of grape took place in Georgia. One of the indicators is archaeological evidence. The 1,5-Myr-old petrified specimen of wild grapevine leaf was found in Georgia in the Meskheti province (Fig.4A). A confirmation for long lasting cultivation of grapevine in Georgia is archaeological remains of berries and seeds of domesticated grapes dated ~6.000 BC (in the vicinity of village Shulaveri [Fig.1], southeast Georgia; Ramishvili 1988). Other archaeological evidence of prehistoric winemaking was found in proximity of the Caucasus in northern Iran at the Hajji Firuz Tepe site in the northern Zagros Mountains dated to about 5.400–5.000 BC (McGovern, 2003) and in the Levant where archaeological findings are dated from ca. 4.000-3.200 BC (Zohary & Hopf 2000). Georgian traditions based on winemaking and grape culture to a high degree might be considered to be a second indicator of Caucasian origin of the grapevine. However, the primary scientific argument should be premised on N. I. Vavilov's (1992) idea that the centres of origin of cultivars should be characterized by high genetic and morphological variability of both wild and cultivated taxa. Five hundred is a very high number of known autochthonous grapevine varieties found in such a small territory (Javakhishvili, 1987; Ketskhoveli et al., 1960). These cultivars showed high morphological variability of leaf form, colour and shape of berries and shape and structure of pips. They are adapted to wide array of climatic conditions (Ketskhoveli et al., 1960).

Each province of Georgia possesses unique grapevine cultivars adapted to local climate. The landraces of western Georgia grow in humid subtropical climate and other cultivars are adapted to moderate climates in eastern Georgia. Several local cultivars are growing in high mountain regions in Meskheti and Svaneti up to 1500 m a.s.l. Besides the cultivars, there is high morphological and genetic diversity of wild grapevine populations in the Caucasus. All five haplotypes detected by using cpDNA microsatellite markers have been found in the Caucasian ecoregion suggesting that this area is possibly the centre of origin of both wild and cultivated grapevines (Grassi et al., 2006). However, only one provenance from Georgia has been analyzed in this study despite the number of populations of wild grapevine found in Georgia today that display morphological diversity (Ramishvili, 1988). We carried out a detailed comparative morphometric study of nine populations of wild grapevine, V. vinifera subsp. sylvestris, growing in the four river basins of the Ajaristskali, Mtkvari, Alazani and Iori located in western, southern and eastern Georgia.

The results reveal high morphological diversity of wild grapevine growing in Georgia. Morphological characters such as shape of leaf blade, number of lobes, pubescence type, coloration of internodes, leaves and berry skin, leaf vein lengths and angles between them and form of petiole sinus show high variability both within and among populations. The variability was related to the skin colour of berries. Some wild grapes had white berries, most had blue-black coloration. White-fruited phenotype is considered to be determined by the variation present in the gene VvmybA1, a transcriptional regulator of anthocyanin biosynthesis (This et al., 2006).

Table 6: One hundred fourteen endemic species of the Caucasus ecoregion related to the ancient crops in Georgia (Schatz et al., 2009)

Family	Taxon	Family	Taxon
Apiaceae	Carum alpinum (M. Bieb.) Benth. & Hook. ex B. D. Jacks.	Lamiaceae	Thymus nummularius M. Bieb.
Apiaceae	Carum grossheimii Schischk.	Lamiaceae	Thymus tiflisiensis Klokov & Des. - Shost.
Apiaceae	Carum porphyrocoleon (Freyn & Sint.) Woronow	Liliaceae	Allium albovianum Vved.
Apiaceae	Pastinaca armena Fisch. & C. A. Mey.	Liliaceae	Allium candolleanum Albov
Apiaceae	Pastinaca aurantiaca (Albov) Kolak.	Liliaceae	Allium gramineum K. Koch
Apiaceae	Pastinaca pimpinellifolia M. Bieb.	Liliaceae	Allium gunibicum Miscz. ex Grossh.
Asparagaceae	Asparagus caspius Schult. & Schult. fil.	Liliaceae	Allium kunthianum Vved.
Asteraceae	Scorzonera charadzeae Papava	Liliaceae	Allium leucanthum K. Koch
Asteraceae	Scorzonera czerepanovii R. Kam.	Liliaceae	Allium otschiauriae Tscholokaschvili
Asteraceae	Scorzonera dzhawakhetica Sosn. ex Grossh.	Liliaceae	Allium ponticum Miscz. ex Grossh.
Asteraceae	Scorzonera ketzkhowelii Sosn. ex Grossh.	Liliaceae	Allium szovitsii Regel
Asteraceae	Scorzonera kozlowskyi Sosn. ex Grossh.	Linaceae	Linum hypericifolium Salisb.
Asteraceae	Scorzonera seidlitzii Boiss.	Poaceae	Elymus buschianus (Roshev.) Tzvelev
Betulaceae	Corylus abchasica (Kem.-Nath.) Kem.-Nath.	Poaceae	Elymus troctolepis (Nevski) Tzvelev
Betulaceae	Corylus colchica Albov	Poaceae	Elytrigia gracillima (Nevski) Nevski
Betulaceae	Corylus egrissiensis Kem.-Nath.	Poaceae	Elytrigia sinuata (Nevski) Nevski
Betulaceae	Corylus imeretica Kem.-Nath.		
Betulaceae	Corylus kachetica Kem.-Nath.	Poaceae	Secale strictum subsp. anatolicum (Boiss.) K. Hammer
Betulaceae	Corylus x fominii Kem.-Nath.	Poaceae	Secale strictum subsp. kuprijanovii (Grossh.) K. Hammer
Fabaceae	Cicer caucasicum Bornm.	Rosaceae	Amygdalus georgica Desf.
Fabaceae	Lathyrus colchicus Lipsky	Rosaceae	Crataegus caucasica K. Koch

Family	Taxon	Family	Taxon
Fabaceae	*Lathyrus cyaneus* (Steven) K. Koch	Rosaceae	*Crataegus eriantha* Pojark.
Fabaceae	*Lotus caucasicus* Kuprian. ex Juz.	Rosaceae	*Pyrus demetrii* Kuthatheladze
Fabaceae	*Medicago dzawakhetica* Bordz.	Rosaceae	*Pyrus eldarica* Grossh.
Fabaceae	*Medicago hemicycla* subsp. *medidaghestanica* Sinskaya	Rosaceae	*Pyrus fedorovii* Kuthatheladze
Fabaceae	*Onobrychis angustifolia* Chinth.	Rosaceae	*Pyrus georgica* Kuthatheladze
Fabaceae	*Onobrychis biebersteinii* Sirj.	Rosaceae	*Pyrus ketzkhovelii* Kuthatheladze
Fabaceae	*Onobrychis cyri* Grossh.	Rosaceae	*Pyrus oxyprion* Woronow
Fabaceae	*Onobrychis grossheimii* Kolak. ex Fed.	Rosaceae	*Pyrus sachokiana* Kuthatheladze
Fabaceae	*Onobrychis iberica* Grossh.	Rosaceae	*Pyrus salicifolia* Pall.
Fabaceae	*Onobrychis kachetica* Boiss. ex Huet.	Rosaceae	*Pyrus takhtadzhianii* Fed.
Fabaceae	*Onobrychis kemulariae* Chinth.	Rosaceae	*Rosa buschiana* Chrshan.
Fabaceae	*Onobrychis komarovii* Grossh.	Rosaceae	*Rosa didoensis* Boiss.
Fabaceae	*Onobrychis meschetica* Grossh.	Rosaceae	*Rosa doluchanovii* Manden.
Fabaceae	*Onobrychis oxytropoides* Bunge	Rosaceae	*Rosa ermanica* Manden.
Fabaceae	*Onobrychis petraea* (M. Bieb. ex Willd.) Fisch.	Rosaceae	*Rosa galushkoi* Demurova
Fabaceae	*Onobrychis sosnowskyi* Grossh.	Rosaceae	*Rosa hirtissima* Lonacz.
Fabaceae	*Onobrychis transcaucasica* Grossh.	Rosaceae	*Rosa irysthonica* Manden.
Fabaceae	*Trifolium fontanum* Bobrov	Rosaceae	*Rosa kozlowskii* Chrashan.
Fabaceae	*Trifolium ruprechtii* Tamamsch. & Fed.	Rosaceae	*Rosa marschalliana* Sosn.
Fabaceae	*Trifolium sintenisii* Freyn	Rosaceae	*Rosa oplisthes* Boiss.
Fabaceae	*Vicia abbreviata* Fisch. ex Spreng.	Rosaceae	*Rosa ossethica* Manden.
Fabaceae	*Vicia alpestris* Steven	Rosaceae	*Rosa oxyodon* Boiss.
Fabaceae	*Vicia antiqua* Grossh.	Rosaceae	*Rosa prilipkoana* Sosn.
Fabaceae	*Vicia caucasica* Ekutim.	Rosaceae	*Rosa pulverulenta* M. Bieb.
Fabaceae	*Vicia ciliatula* Lipsky	Rosaceae	*Rosa teberdensis* Chrshan.
Fabaceae	*Vicia grossheimii* Ekutim.	Rosaceae	*Rosa transcaucasica* Manden.
Fabaceae	*Vicia iberica* Grossh.	Rosaceae	*Rosa tuschetica* Boiss.
Fabaceae	*Vicia sosnowskyi* Ekutim.	Rosaceae	*Sorbus buschiana* Zinserl.
Grossulariaceae	*Ribes biebersteinii* Berland. ex DC	Rosaceae	*Sorbus caucasica* Zinserl.
Lamiaceae	*Satureja bzybica* Woronow	Rosaceae	*Sorbus caucasigena* Kom.
Lamiaceae	*Thymus caucasicus* Willd. ex Ronniger	Rosaceae	*Sorbus colchica* Zinserl.
Lamiaceae	*Thymus collinus* M. Bieb.	Rosaceae	*Sorbus migarica* Zinserl.
Lamiaceae	*Thymus coriifolius* Ronniger	Rosaceae	*Sorbus subfusca* (Ledeb.) Boiss.
Lamiaceae	*Thymus grossheimii* Ronniger	Rosaceae	*Sorbus fedorovii* Zaikonn.
Lamiaceae	*Thymus karjaginii* Grossh.	Rosaceae	*Sorbus velutina* (Albow) C.K. Schneid.
Lamiaceae	*Thymus ladjanuricus* Kem.-Nath.	Staphyleaceae	*Staphylea colchica* Steven

The wild ancestor, however, should be considered to be black colour grapevine, which is most common in Georgian wild populations, eventhough the mutation leading to the white-fruited wild grapevine has been found in other Georgian wild populations (Ramishvili, 1988). On a phenotypical basis of our investigation (Ekhvaia & Akhalkatsi, 2010) it can be confirmed that the infraspecific evolution of Vitis vinifera subsp. sylvestris has produced three population groups south of the Great Caucasus Range. Overall, there are three phenetically distinct morphometric groups of western, southern and eastern Georgian wild grapevines. These three groups can be readily distinguished by the length of main leaf veins and lengths of nectaries in male flowers. This conclusion differs from the classical classification of Georgian wild and cultivated grapevine (Ramishvili, 1988) that considers two centres of origin of grapes in the South Caucasus region: (1) an Alazani origin with whole eastern and southern Georgia and adjacent territories of Azerbaijan and Armenia, and (2) a Colchic centre of origin which includes the entire western Georgian region with the Black Sea coastal zone. Our data clearly show a separated group in the southern Georgian population located in the territory of historical Tao-Klarjeti, a region of Georgia with many aboriginal grapevine cultivars. Therefore, it is of high importance to study aboriginal grape varieties in the place of its supposed domestication and to determine genetic relations among native grapevine cultivars and local wild populations.

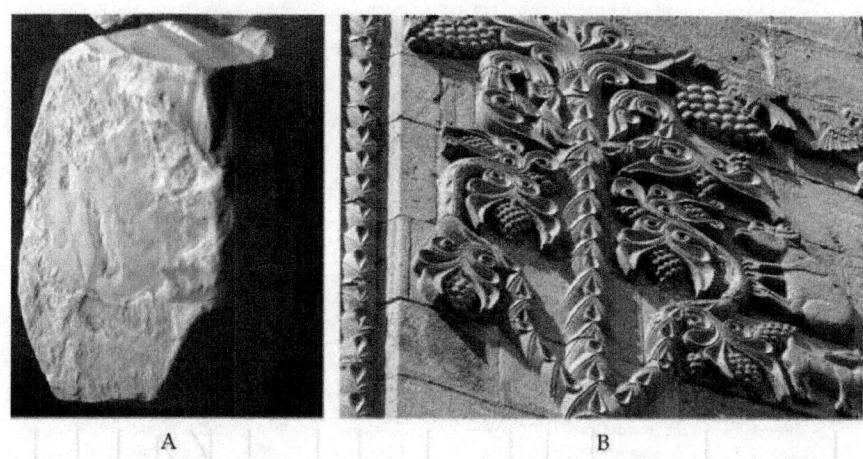

Figure. 4: A- The 1,5-Myr-old petrified specimen of wild grapevine leaf. National Museum of Akhaltsikhe, Georgia; B-Stone carving on the medieval church Ananuri, Mtskheta-Mtianeti region, Georgia. Photos by Maia Akhalkatsi.

Molecular study based on nuclear microsatellite (SSR) markers revealed close genetic relationships between wild grape and local cultivars in Georgia (Ekhvaia et al., 2010, 2011). Twenty-four Georgian autochthonous and 45 accessions of wild V. vinifera subsp. sylvestris were analyzed at 17 microsatellite loci (VrZAG21, VrZAG47, VrZAG62, VrZAG64, VrZAG79, VrZAG83, VVMD7, VVMD24, VVMD25, VVMD27, VVMD28, VVMD32, VVMD34, VVS2, VVS4, scu04vv and scu14vv). Six accessions of the American rootstocksFercal ('Cabernet -Sauvignon' x Vitis berlandieri), Telecki 5C (V. berlandieri x V. riparia), Malegue 44-53 (V. riparia X [V. cordifolia X V. rupestris]), Couderc 3309 (Riparia tomenteuse x Rupestris Martin), cultivar V. labrusca 'Odessa' and V. riparia naturalized in Georgia were used as outgroup. Thirty-seven accessions of wild grapevine were collected from different regions of Georgia and 8 wild accessions were sampled in Turkey's Artvin province adjacent to Georgia. All individuals within the studied populations of wild grapevine were identified as dioecious plants with male or female flowers. Genotype analysis at the most studied loci showed that Georgian cultivated and wild grapevine was characterized by high level of genetic variability. Genetic structure also was analyzed using F statistics. The low level of genetic differentiation (Fst=0.03) between Georgian cultivated and wild grapevines demonstrates that in situ domestication of wild germplasm took place within local populations. This means that autochthonous Georgian cultivars should be originated from local wild grapevine (Ekhvaia et al., 2010). The dendrogram generated using Dice coefficient identified eight major clusters within the 75 different genotypes defined at 0.22 similarity level (Fig.5). Clusters A consist of 13 Georgian local cultivars and 8 wild grapevine accessions from different regions of Georgia. One example confirming the genetic linkage between cultivated and local wild grapevine is placement of the famous Georgian red cultivar 'Tavkveri' in cluster A, where it is closely linked to the wild accession WT46Tbilisi5 (GS value 0.96; Fig. 5) due to identical alleles at 33 out of 34 alleles. Thus, the hypothesis that this cultivar could have been selected from local wild grapevine can be considered, especially because like wild grapevine 'Tavkveri' is characterized by presence of functionally female flowers. The fact that the 5 ancient Georgian cultivars 'Chvitiluri', 'Kachichi', 'Shonuri', 'Saperavi' and 'Uchakhardani' fall within cluster B (Fig.5), which mainly contains wild accessions allowed us to suppose that these cultivars were derived from the earliest local domestication events. Cluster G shows the genetic similarity of most ancient Georgian cultivars 'Meskhuri Shavi' and 'Krikina' adapted to high mountain climate conditions in Meskheti. In conclusion, it should be mentioned that the Georgian cultivated and wild grapevines represent a unique and interesting genetic resources that is characterized by a high similarity level between wild and cultivated grapevines. The admixture found among local Georgian cultivars and wild grapevine indicates the possibility that these cultivars are derived from ancestral domestication of local wild types. Thus, the obtained data are supporting that Georgia is one of the oldest centres of domestication of grapevine and harbour of valuable genetic resources for grape breeding. It should be mentioned that wild grapevine populations occurring nowadays on the territory of Georgia are threatened by different impacts in their natural habitats and need to be protected. The confirmation of threatened status of the Georgian wild grapevine might be detected low level of heterozygous individuals found for the most of the studied loci, which reflects the isolated status and the reduced number of individuals in the wild populations. Therefore, it is necessary to conserve wild forms and aboriginal cultivars of grape for the maintenance of genetic variability and to avoid genetic erosion of valuable genetic resources for grape breeding in Georgia.

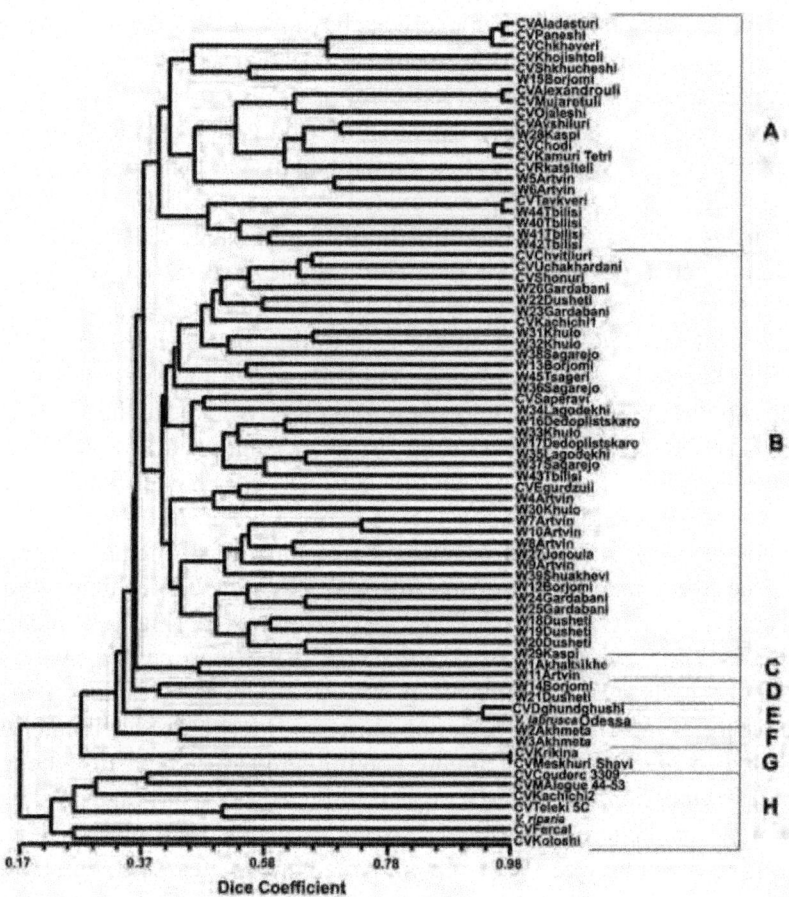

Figure. 5: Dendrogram of 75 accessions: Twenty-four Georgian local cultivars (Vitis vinifera subsp. vinifera), 45 Georgian wild grapevine (V. vinifera subsp. sylvestris), 4 rootstock cultivars (Couderc 3309, Fercal, MAlegua 44-53 and Teleki 5C), introduced cultivar V. labrusca 'Odessa' and naturalized V. riparia constructed by unweighted arithmetic average clustering (UPGMA) method based on Dice's coefficient of shared SSR polymorphisms.

Domestication of Pear in Georgia

Many pear cultivars occur in Georgia from pre-historic period indicating the early domestication event of this cultivated fruit tree (Javakhishvili, 1987). In total, 11 species of wild pear occur in Georgia (Kuthatheladze, 1980). They are distributed in different regions of Georgia, what is caused by the variable geographical relief and habitat diversity of the country. Pyrus caucasica Fed., the endemic species of the Caucasus is most widespread among the wild pears of Georgia and is considered as main progenitor species of local pear cultivars (Khomizurashvili, 1973). Moreover, P. caucasica and P. pyraster (L.) Burgsd. are regarded as the main wild progenitors, from which the cultivated European pear (P. communis L.) has probably evolved (Zohary & Hopf, 2000; Volk et al., 2006; Yamamoto & Chevreau, 2009). According to the literature data (Khomizurashvili, 1973) introduced cultivars of pear from Europe and Russia appeared in Georgia at the end of 19th century and before there were existed only the local cultivars. However, we assume that the process of cultivar introduction might have started much earlier, as Georgia had cultural contacts to Asian and European countries since antique period. It is also remarkable that Greece is considered to be a first provider of selective cultivars of pear to the ancient world (Jackson, 2003) and earliest relations between Georgia and Greece should be considered as possible way of introduction of European pear cultivars in the Caucasus. The local Georgian names of the cultivated pear Mskhali and wild Caucasian pear Panta exists in all Georgian dialects; they do not have analogues in any other languages (Javakhishvili, 1987). The Georgian names of cultivated and wild pears are linked with geographic objects such as mountains (Skhaltbis Range in Kartli, Mt. Mskhal-Gori in Kakheti's Kavkasioni), rivers (R. Skhaltba), or villages (Pantiani, Skhalta, Skhlobani, etc.; Javakhishvili, 1987). The name of wild Caucasian pear Panta is used among cultivars called 'Panta Mskhali', i.e. cultivar with name of wild pear. Moreover, the classification of Georgian pear cultivars (Khomizurashvili, 1973) contains a group of landraces with the same name. This classification system divides Georgian cultivars into four groups: 'Gulabi', 'Panta Mskhali', 'Kalos Mskhali', and 'Khechechuri'. The name of each group represents the name of a cultivar, which is considered to be a typical representative of a group. In the 'Gulabi' group, there are included both local and introduced cultivars to have most high economic values, big juicy fruits with sweet taste. The 'Panta Mskhali' group contains local varieties with small fruits becoming black after maturation. This is a character feature of wild Caucasian pear. The 'Kalos Mskhali' group includes local cultivars having bigger fruits than the second group. The 'Khechechuri' group matures in late autumn with juicy fruits containing a big amount of stone cells. According to N. Khomizurashvili (1973), the last three groups are originated by direct domestication of wild pear in Georgia. Although, some signs of selective breedings are remarkable as well. Relationships between wild P. caucasica and local cultivars are mirrored by a high morphological variability of leaf and fruit forms. This idea was for the first time confirmed by statistical methods of taxonomic identification and relationships among taxa in our study (Asanidze et al., 2011). We decided to conduct comparative morphometric study of cultivars recently occurred in Georgia and reveal their relationship to local wild pear species. The results have to determine local cultivars originated by direct domestication events in ancient time and discriminate from cultivars, which will have relationships with other wild species - P. pyraster or P. pyrifolia, considered as wild ancestors of European and Far East pear cultivars respectively. We carried out the investigation to determine morphological characteristics of leaves, young shoots and fruits differentiating local and introduced pear cultivars of Georgia to reveal the relationships between cultivars and wild ancestor species of pear by statistical methods used in plant morphology (Asanidze et al., 2011). A total of 214 wild and cultivated pear trees have been sampled in natural habitats, living collections and peasant grounds in different regions of Georgia. Wild pear species were determined according to Sh. Kuthatheladze (1980). The pear accessions evaluated in this study consisted of Caucasian endemic P. caucasica Fed. (=P. communis subsp. caucasica (Fed.) Browicz; N=100), P. balansae Decne. (= P. communis L.; N=8) and P. pyraster (L.) Burgsd. (=Pyrus communis L. subsp. pyraster (L.) Ehrh.; N=3), which has been obtained from Germany, Hessen, in surrounding of v. Erda. Eightyone individuals of 26 Georgian local and 22 individuals of 9 introduced cultivars (total 103 individuals of 35 cultivars) have been collected. Some of them are sampled in the collection of the Institute of Horticulture, Viticulture and Oenology, village Skra, Gori district, Georgia; local cultivars were sampled in the collection of aboriginal cultivars of Biological Farming Association Elkana in village Tsnisi, Akhaltsikhe district, Georgia. Many local cultivars are collected in peasant house gardens in different regions of Georgia. The individuals were evaluated by 27 morphological traits, which included one landmark analysis data, 12 leaf and shoot descriptors and 14 fruit descriptors. Morphological characters have been taken as recommended by International Union for the Protection of New Varieties of Plants (UPOV, 2000) for P. communis and J. Voltas and colleagues (Voltas et al., 2007), which delimited differences between wild and cultivated taxa of the genus Pyrus based on morphometric analysis. In total 21 morphological traits have been analysed by multivariate analysis. The Principal Components Analysis and Hierarchical Cluster Analysis (HCA) methods revealed the distance or similarity measure to be used in clustering with the Ward's method as amalgamation rules. According to HCA's results, pear cultivars, analysed in this study, are clustered into two groups (Fig.6). The first group A contains local cultivars related to P. caucasica and P. balansae by 21 morphological traits of leaves and fruits. Especially close Euclidean similarity distance is revealed between wild Caucasian pear, Panta in Georgian and a cultivar named 'Panta mskhali', which confirms etymological and taxonomic similarity within these taxa. P. balansae shows very close similarity distance with 'Tsvrili mskhali' and 'Korda'. Very closely related group of local cultivars to wild Georgian pears contains: 'Bebani', 'Samariobo', 'Tavrejuli',

'Kvichicha', 'Khinos mskhali' and 'Akiro'. The other group: 'Shavmskhala', 'Nenes mskhali', 'Borbala', 'Majara', 'Shakara' and 'Kartuli mskhali', is more distanced from wild pears, but located in the same cluster. We assume that these local cultivars must have been originated by direct domestication of wild Caucasian and Balanse's pears in Georgia. The second cluster B (Fig.6) contains introduced cultivars of pear originated in European countries and some old Georgian cultivars. The group B reveals relationship with wild pear - P. pyraster, which is distributed in Europe and does not reach Georgian territory.

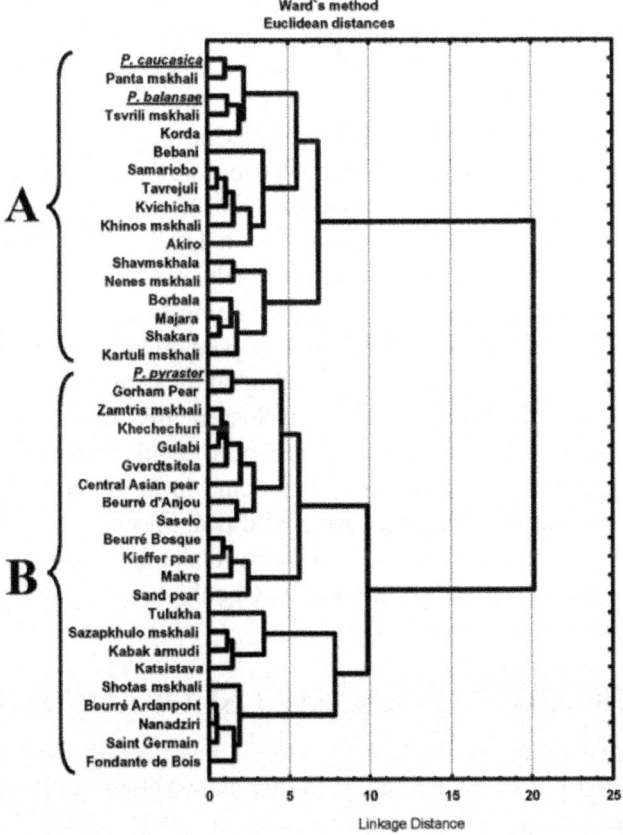

Figure. 6: HCA dendrogram of Euclidean distance with the Ward's method show-ing the relationships between the 35 cultivars and 3 wild species of pear based on 20 morphological traits of leaf, shoot and fruit and 20 landmark harmonics of mature leaf; The taxa in the dendrogram are clustered into two main groups - A and B. (N=214).

The area of distribution is up to the middle of Turkey. The most cultivars from intermediate group B are more widespread in Georgia than the local cultivars originated by direct domestication of wild pears from group A. Two local pears 'Gulabi' and 'Khechechuri' are the most widespread among all local pears of Georgia and there are two or more varieties of them in each localities of the country. Moreover, Georgian name of cultivar 'Gulabi', which is also used to be a name of local pear group in classification of N. Khomizurashvili (1973), means pear in Persian. We suggest that local cultivars from cluster B (Fig.6) might have appeared in Georgia very long time ago and were improved by local population using breeding procedures.

Leaf margin shape is the main morphological trait that differentiates Caucasian pear (P. caucasica) from European pear (P. communis). Leaf margins are entire in P. caucasica and serrate in P. communis. This theory was proved by the statistical analysis of the collected samples for the present study. Nowadays, P. pyraster is considered as the wild pear of Europe and cultivars are named as P. communis (Yamamoto & Chevreau, 2009). 'Communis' or the 'Common pear' group of cultivars has become the name of the cultivated pears of Europe, however, the structure and the diversity of the wild and cultivated pears of this group is not studied in details and needs further genetic and molecular investigations. Thus, the results of this study have shown that some local cultivars of Georgia are direct domesticated from the native wild pear species - P. caucasica and P. balansae. The other local cultivars might be obtained due to selective works by breeding of local landraces with introduced cultivars from different countries in historically different periods. The molecular study of these taxa will clear in more details origin of these cultivars. The results confirm the hypothesis that some local cultivars of Georgia are directly domesticated from the native wild pear species - P. caucasica and P. balansae. The other local cultivars might be obtained due to selective works by breeding of local landraces with introduced cultivars from different countries in historically different periods.

TRADITIONAL SUSTAINABLE USE OF ANCIENT CROPS

Since ancient times, agriculture in Georgia has been divided in two zones: lowlands (0-1300 m a.s.l.) and high-mountains (1300-2200 m a.s.l.). This classification was based on production of wine (Javakhishvili, 1987). Winemaking was always the major branch of agriculture in Georgia. Wine was exported from Georgia since ancient times. The vineyards were cut off to reduce income for exporting the wine in neighbour countries during the occupation of the country by the Muslim nations. The exchange of agricultural products took place between lowland and high-mountain regions and not only within the Georgians, but also with North Caucasians. This tradition remained till the end of 20th century when Dagestanian people from Didoeti region visited Kakheti lowland in eastern Georgia in late autumn. They have exchanged agricultural products from high-mountains to lowland crops. In 20th century, they brought from Dagestan cows, cheese and potato and have exchanged them on Kakhetian wine, schnapps from grapevine called chacha, bread wheat flour, fruits and vegetables.

A   B   C

Figure. 7: A - Satsnekheli - stone construction for the pressing of grapes in Nekresi monastery (IV century AD), Kakheti; B - Clay vessels for wine storage Kvevri, Ikalto monastery (XII century AD), Kakheti; C- Red wine Saperavi and boghlortso (wheat bread into red wine) in clay cups on Georgian table. Photos by Maia Akhalkatsi.

The information about the traditional use of wine is remained in folklore and ethnographic studies collecting this knowledge. The problem in reducing of written information was again the wars, destruction of settlements and burning of manuscripts during historical times. Therefore, the information on traditional agriculture is based on both literature data and interviews of local people obtained during our field trips in different provinces of Georgia. The ancient stone construction satsnekheli for the pressing of grape might be found near many historical monuments (Fig.7A) even in ancient caves of Chachkari near Vardzia monastry complex. Each family in lowland regions have a room for winemaking in houses called marani in which the clay vessels are buried in soil named kvevri (Fig.7B) where the wine is made and stored by traditional Georgian technology. One of kvevri in each marani was called zedashe and the wine in it might be used only in religious rituals since it belonged to the God. Zedashe was filled by schnapps called Araki derived from different fruits e.g. wild pear - P. caucasica, in high mountain regions and used for rituals (Fig. 8A-C). Wine and bread are ritual accessories of Christian religion and first cross entered in Georgia by St. Nino from Cappadocia in 4th century AD was made from vines. However, the folklore knowledge let us know that grapevine was a ritual plant in ancient religion and represented a tree of the God of sun. The remnants of grapevine images are often demonstrated as stone carvings on Christian churches (Fig.4B) and ancient golden and silver cups and Jewellery. Wine was used not only as alcohol drink but in religious rituals and as traditional food. Boghlortso (Fig.7C) was a healthy food prepared by wheat bread placed within a red wine and used by the whole family members including children and chronic invalid and consumptive people. Bread is served with all Caucasian meals. It is the same ritual food like wine in Georgia. Two landraces of bread wheat - Triticum aestivum var. erythrospermum and T. aestivum var. lutescens are used for religious rituals in Svaneti (Girgvliani, 2010). The flour of these cultivars is preserved separately from other reserves of bread wheat flour and used on religious holydays. Milled faba bean and kenaf seeds are added to the bread flour for baking ritual bread. There are barley cultivars: H. vulgare var. pallidum in Svaneti and H. vulgare var. nutans in Meskheti, used for traditional bread preparation added to the T. carthlicum 'Dika' flour.

Figure. 8: A- Murtaz Chankseliani collected wild pear (Pyrus caucasica) in forest, R. Kheledula gorge, Lower Svaneti; B - Wild Caucasian pear for preparing of alcohol schnapps Araki; C - Distillation equipment for preparing of pear schnapps. Photos by Maia Akhalkatsi.

The healthy quality of food in Georgia is connected with usage of fruit and herb sauces for roasted and fried meet. Sour plum sauce (P. cerasifera var. divaricata) is always added to spitroasted chicken and pork. Many herbs and spices are added to meet meals. This should be understand to be a modern direction in diet works when an alkaline food is recommended to neutralize acid food, which is everyday meal including meet, sugar and bread. Modern civilization eats considerably more acid-forming foods than alkalizing foods. According to well-known naturopath P. Airola (1984), acidosis, or over-acidity in the body tissues, is one of the basic causes of diseases, especially the arthritic and rheumatic diseases. There is in the internet now a lot of information on alkaline food. The lists of alkaline products show that fruits and vegetables have highest pH most of which are traditional cultivars in Georgia. There are alkaline vegetables such as - alfalfa, barley, beet greens, beets, cabbage, carrot, celery, cucumber, eggplant, garlic, green beans, green peas, herbs, lentils, lettuce, onions, radishes, spices, spinach, watercress, wild greens, etc. which are the traditional cultivars in Georgia. Alkaline traditional fruits are - apple, apricot, berries, cherries, sour cherries, figs, grapes, grapefruit, nectarine, peach, pear, strawberries, watermelon, seeds and nuts, etc. To maintain health, the diet should consist of 60% alkaline forming foods and 40% acid forming foods. To restore health, the diet should consist of 80% alkaline forming foods and 20% acid forming foods (Airola, 1984). Therefore, we should think about sustainability of traditional use of crops and wild plant species in the past time. The fact that the nature remained undisturbed in the country centuries long and population was characterized by much higher healthy features as current situation exists in Georgia should be explained as occurrence of healthy food and nature tolerant use of the plant resources.

THREATS

Threats and Conservation of Ancient Crops

There are several reasons for the genetic erosion of ancient cultivars and the wide distribution of new varieties of introduced crops. First of all, new cultivars have higher yields and are therefore preferred both as a source of food for local people and as a cash crop that determines local income. The second reason why local peasants began to prefer cultivating genetically modified (GM) plants may be explained by introduction of new diseases into Georgian agricultural fields in recent years, causing harm primarily to ancient cereals and vegetables. However, the introduction of new parasites has revealed that endemic forms of Georgian crop plants contain valuable selective disease-resistant material for genetic engineering. The tetraploid and hexaploid endemic wheat species T. timopheevii and T. zhukovskyi, for example, are characterized by a high level of resistance to a new race (TTKS, commonly known as Ug99) and many other races of Puccinia graminis f. sp. tritici due to the wheatstem rust resistance gene Sr36 (Tsilo et al., 2008). T. carthlicum is characterized by immunity to diseases, a short growing period, and resistance to cold. Intensive Genetic erosion of ancient crops started in Georgia since 1950s which was also a period of intense selection work in breeding stations in the whole of the Soviet Union, e.g. the highly productive awnless wheat cultivar Bezustaja I developed in Russia has been sown in all wheat fields in Georgia since the 1970s, and this variety eventually replaced Georgian endemic wheat species. Recently new breeder's varieties of wheat and other cereals are introduced from different countries. The process of genetic erosion of ancient crop varieties has not been a great concern for the mountain areas of Georgia, which until the 1990s acted as a depository of ancient crops. One important consideration that explains why ancient cultivars were conserved longer in mountainous regions than in the lowlands is that the local population preserved their traditional ways of life and socioeconomic structures. The traditional agricultural system is characterized by dependence on local genetic resources and locally developed technologies. Even today, peasants in mountain villages use an ox-drawn sledge made of wood for loading and transporting cereals and a threshing sledge on threshing floors to thresh wheat, oats, rye, and barley. Traditional agricultural equipment makes it possible to cultivate areas even on steep slopes and at high elevations, where modern tractors cannot be used. Moreover, some old landraces of wheat and barley are used to prepare bread and beer for religious rituals. Substitution of these landraces by others would go against centuries-old traditions (Akhalkatsi et al., 2010). Despite these conditions that support the maintenance of ancient landraces, many endemic and native representatives of crop plants are currently in danger of extinction and face severe problems of genetic erosion in all mountainous regions of Georgia. While agrobiodiversity is declining rapidly in many areas of the world due to anthropogenic pressure (Körner et al., 2007), including population growth, in Georgia the main reason for genetic erosion of ancient crop varieties is demographic decline in mountain regions due to harsh economic conditions and lack of modern infrastructure (Nakhutsrishvili et al., 2009). The shift from ancient cultivars to modern high-yielding crops such as maize and potato, which took place in the lowland areas much earlier, began in mountain villages only in the last 20 years. Greater income from marketing allows families to stay in mountain villages. Moreover, the economic security of the traditional farming systems in these mountain regions appears to be in jeopardy when traditional agriculture is replaced by cattle breeding, which causes abandonment of cultivated fields and their transformation into pastures. Several research centres maintain ex situ germplasm collections of Georgia, such as gene banks and living collections. According to the National Biodiversity Action Plan of Georgia (Jorjadze, 2005), international nature conservation institutions and Georgian scientific and nongovernmental organizations have taken care to preserve the genetic resources of local cultivars. Several gene banks and living collections occur in Georgia. There is one biggest genebank located at the Georgian Institute of Farming established in 2004 through support of International Centre for Agricultural Research in the Dry Areas (ICARDA). They owned a total 3057 accessions of local and introduced cultivars and CWRs in 2010. The other 5 gene banks are located in different research institutes unified with Agrarian University in 2011. Total number of germplasm accessions is 6286 in Georgian gene banks. However, the material kept in ex situ collections are not sufficient and need more contribution. Many seed banks worldwide contain about 7000 accessions of germplasm of Georgian cultivars and crop wild relatives. A recently initiated project, "Mountain Biodiversity in the Caucasus and its Functional Significance," supported by the Swiss National Science Foundation Program SCOPES, will build an electronic biodiversity archive for Georgia, and include data on mountain plant biodiversity in Georgia. Because it will be built in compliance with Global Biodiversity Information Facility (GBIF) standards, it will contribute to the Global Mountain Biodiversity Assessment (GMBA) mountain portal with GBIF (www.mountainbiodiversity.org). A research agenda concerned with the use of georeferenced mountain biodiversity data for science and management was developed at a GMBA workshop in Kazbegi, Central Caucasus, in July 2006 (Körner et al., 2007). Such a database of the plant species of the Caucasus will become a prominent entry in the GBIF database and highlight the current status of plant genetic resources in Georgia. It should be emphasized that establishment and maintenance of ex situ collections and databases is just a first step in the conservation process of ancient crop varieties. The next step should be return of conserved seed material to the fields of local farmers. From 2004 to 2009, the Global Environmental Facility/United Nations Development Fund (GEF/UNDP) project "Recovery, Conservation and Sustainable Use of Georgia's Agro-Biodiversity" was carried out with the aim of conservation and sustainable use of threatened local plant genetic resources in the oldest historical mountainous region of Georgia, SamtskheJavakheti. This project enabled establishment of sources of primary seed and planting material for threatened crops and fruit varieties, and assisted farmers in accessing markets for organic products from such crops as lentil, grass pea, chickpea, faba bean, common millet, Italian millet, etc. Another project was the return of the Georgian wheat variety T. aestivum var. ferrugineum 'Akhaltsikhis Tsiteli Dolis Puri' in Meskheti province, where it was sown on 10 ha and produced bread that was introduced in shops featuring organic products in Tbilisi as of 2008. Afterward, this project was supported by the Georgian church, which expressed an interest in cultivating ancient crops on monastery grounds. However, these attempts have been realized only on a small scale and not in larger areas of the country. In our opinion, the major activity of the corresponding governmental institutions should be directed on supporting local farmers in reintroducing ancient crops on the market and maintain maximum diversity of the target taxon's gene pool. The importance of agricultural achievements not should be oriented only on high yield of crops but the traditional foods to which people have adapted a long time determines their healthy lifestyle. Thus, conservation and reintroduction of ancient cultures to modern agriculture can insure longevity of people.

Threats and Conservation Needs of Crop Wild Relatives

The natural populations of many species of CWRs are increasingly at risk. The primary causes of diversity loss of wild plant species are habitat loss, degradation and fragmentation. Many cereal CWRs, including relatives of wheat and millet species, occur in arid or semi-arid lands and are severely affected by over-grazing and desertification. The forest species are affected by habitat disturbances because of illegal forest cutting occurring in 1990s in Georgia. Climate change is having significant impacts of species distributions and survival in a concrete habitat. One of the most important threats to the diversity of CWRs are genetic erosion and pollution. The threat of genetic pollution or introgression, either from genetically modified organisms (GMOs) or from conventionally bred crops, to wild species has become an increasing risk to the in situ genetic conservation of crop wild relatives. Another problem is that many species of important CWR occur in centres of plant diversity and crop diversity located mainly in developing countries, which often lack resources to invest in the necessary conservation activities. South Caucasus and Georgia, in particular, is the centre of origin and diversity of many of the world's important crop plants. There are several international projects realized by the ICARDA, the International Plant Genetic Resources Institute (IPGRI), US Department of Agriculture (USDA), United Nations Environmental Program (UNEP), etc. contributed in undertaking efforts in monitoring and conservation of plant. Although, additional resources are urgently needed in such areas of high diversity to identify priority species for conservation, determine the necessary conservation activities, monitor the status of key species, improve the use of these valuable resources. Habitat disturbances are main threats leading to the extinction of rare and endangered plant species. Deforestation took place during last decades in Georgia and caused habitat degradation. The fact detected with the population of wild grapevine has revealed the threat to the riparian forests, which is situated along rivers in very close proximity of settlements and local people uses the resources of this forest in a highest degree. We have detected that some trees were cut representing the support of clambering wild grapevine and the individuals were lying on the earth, which will cause its drying up and death. More great scale cuttings in dark coniferous forests lead to arising of forest openings with high irradiation leading to drying up the underground cover of mosses and lichens, which drastically changes habitat and determines disappearing of natural species adapted to this habitat. Overgrazing of meadows and pastures was a problem in Soviet period, when several million head of sheep were grazing summer pastures of mountainous regions of Georgia. However, now the number of cattle is reduced and does not threaten much the rare species in their natural habitats. In spite of this fact, grazing affect survival of rare species such as Hordeum spontaneum, which was found on road side and during the next visit it was grazed completely. Such disturbances as habitat degradation due to road and pipeline construction works threatens the populations but has temporary effect. These types of disturbances are especially threatening the rare and endangered species of high conservation value. The best way of in situ preservation of genetic diversity of valuable plants is creation of nature reserves on the territories, where natural populations of CWR occur. The first nature reserve of Georgia was established in Lagodekhi in 1912. In present, the protected areas occupy 7% of the country's territory, which is equal to 495.892 ha. There are 16 nature reserves, 9 national parks, 12 managed resource protected areas, 14 natural monuments and 2 protected landscapes in Georgia. The problem remains for the species, which are growing in rural habitats and on arable lands mixed with field crops have different assessment to threats. These species are depending in their existence to the monitoring of arable lands, which crop will be sown, how will be transformed filed crop to pasture or hay meadow, or what kind of herbicides and mineral fertilizers will be used in the field. The maintenance of wild populations growing as weeds in cultivated fields depends on sustainable management of agriculture in the region. The governmental institutions should control the processes which might bring to the genetic erosion of CWRs having high value of conservation. In this case the legislation bases should be effective to control local farmers not affect CWRs with ecologically unsuitable for this species actions in the field e.g. use of fertilizers or introduction of new crops leading to changing in technology of field cultivation methodology and leading to disturbances of wild weed species of high conservation value. Ex situ conservation of the germplasm of CWRs is very valuable material for improvement of crop quality and their resistance against fungal and microbial disease. It will be of interest to collect their seed material and distribute to genbanks, which will contribute to provide necessary germplasm to research centres dealing with the genetic engineering. The Tbilisi Botanical Garden and Institute of Botany has two collections of seeds. One is collection of rare endemic plant seeds, which is collected in the framework of the Millennium project managed by Kew Royal Botanical Garden, UK. The second is collection of aboriginal crop varieties collecting in different regions of Georgia. These program works together with IPK, Gatersleben Germany, where the analogy of the collected material is kept at the gene bank. The living collections of CWRs are very few. Botanical Gardens in Tbilisi and Batumi have some small collections of CWRs collected in the frameworks of international collaborative projects. However, maintenance of the collections after finishing the projects is impossible and they are cancelled in several years. The plant genetic resources documentation in Georgia is mostly computerized. There are several databases, which include all information and passport data for accessions of field crops, but so far they have no free access. Most threats to biodiversity are the results of human actions, which are expressed in the overuse of natural resources for fuel, fodder, manure, grazing and collecting of ornamental and medicinal plants. This process leads to the loss of genetic diversity including crop wild relatives.

The in situ protection measures are not easy to implement and, thus, the accent should be directed on ex situ conservation.

CONCLUSIONS

Very old archaeological findings, cultural heritages and so far existing high morphological and genetic diversity of ancient crops and their wild relatives show that Georgia has very old agricultural traditions that have preserved to our times. The fact that large-scale genetic erosion of the ancient landraces in Georgia has reached extreme levels from 1950s and almost all the local varieties of cereals (wheat, barley and millet), legumes (peas, lentils, common vetch and faba bean), and grapes are now disappear from the farms requires special analyses and development of conservation measures. Only the gene banks and living collections hold germplasm of landraces extinct in the farms. An assessment of the effectiveness of current conservation strategies to protect the diversity of ancient crops in Georgia reveals a gap in the reintroduction of conserved germplasm to the fields of local farmers. In our opinion, the corresponding governmental institutions responsible for conservation of biodiversity should refocus the strategy to require complementary in situ and ex situ conservation actions to maintain maximum diversity of the target taxon's gene pool by supporting local farmers in reintroducing ancient crops on the market and thereby filling this gap. Moreover, at present, neither field crop genebank nor live collections of the permanent crops have sufficient land and equipment in Georgia, as well as funding to carry out ex situ conservation at the modern level. Storage of the in situ collections should be improved through upgrading the present storage of the field crop genebank facilities. There is a need to improve public awareness of importance of ex situ conservation. Popularity of the data obtained by scientists should be distributed among the local population so that they themselves have contributed to the preservation of national heritages. The results of scientific investigations that some crops represent local cultivars and even domesticated landraces in this area means that this is connected with lifestyle of the local population. The fact that longevity of life in the Caucasus was very high and centenarians lived to 120 years and more should be understand that a healthy diet of mankind is not only amount of calorie but the combination of food of high quality. The modern alkaline diet almost completely coincides with the traditional Georgian cuisine. Therefore, we must appreciate the importance of conservation of local varieties to ensure the health of local people.

The data obtained in our investigations (Ekhvaia &Akhalkatsi, 2010; Ekhvaia et al., 2010, 2011; Asanidze et al., 2011) indicate importance of CWR species in Georgia as many of them represent direct ancestors of local cultivars. The fact that wild grape shows high genetic relation to local varieties of grape indicates that winemaking represents an ancient culture in Georgia, which is expressed even in religious rituals of the nation. Wild grapevine and pear representing the wild ancestors of local varieties are under threat because of wood cutting. Many other CWRs are in the same position. The legislation of species conservation is applied to rare and endemic species and in situ conservation is maintained only at protected areas. However there is no legislation that can protect CWRs growing in rural and urban areas and representing weed species. No actions of conservation are undertaken to protect these species. Many CWRs (wild wheat, rye, coriander, etc.) are grown in cultivated lands of local farmers. Many years, wild wheat species were mixed with local varieties of wheat and barley but now they are disappeared. The events which are protecting them are traditional cultivation technology of the landraces to which the local weeds are adapted by their life strategy and propagation character. The threats here will be change of traditional crops to the new varieties, which will need different cultivation events. This might lead to disappearance of the CWRs from the cultivated beds. At present, CWRs ex situ collections are almost absent in Georgia. The problem of genetic erosion of landraces and their wild relatives needs active contributions by national policies and comprehensive measures are urgently needed to avoid the complete loss of ancient crop genetic resources in Georgia. International nature conservation institutions and Georgian scientific and nongovernmental organizations should show more activity to the restoration of ancient crops, which defined the healthy life of Georgians.

ACKNOWLEDGEMENT

Research on agrobiodiversity in Georgia is supported by grant IZ73Z0_128057, in the framework of the Swiss National Science Foundation (SNSF) program SCOPES (Scientific Cooperation between Eastern Europe and Switzerland). The expeditions were funded by the Global Environmental Facility/United Nations Development Fund (GEF/UNDP) Project GEO/01/G41/A/1G/72. We are grateful to K. Pistrick, R. Fritsch, F. Blattner, and M. Gurushidze from IPK Gatersleben for supporting our investigations.

REFERENCES

1.   Adler, D. S. & Bar-Oz, G. (2009). Seasonal patterns of prey acquisition during the Middle and Upper Palaeolithic of the southern Caucasus. In: The Evolution of Hominid Diets: Integrating approaches to the study of Palaeolithic subsistence, Hublin, J.-J. & Richards, M. (Eds), pp. 127-140, Springer, ISBN: 9781402096983, Leipzig.

2. Airola, P. (1984). How to Get Well: Handbook of Natural Healing. Health Plus Pub, (March 1984), ISBN:0932090036, Sherwood, OR.

3. Akhalkatsi, M., Ekhvaia, J., Mosulishvili, M., Nakhutsrishvili, G., Abdaladze, O. & Batsatsashvili, K. (2010). Reasons and processes leading to the erosion of cropgenetic diversity in mountainous regions of Georgia. Mountain Research and Development, Vol.30, No.3, (August 2010), pp.304-310. ISSN: 0276-4741.

4. Asanidze, Z., Akhalkatsi, M. & Gvritishvili, M. (2011). Comparative morphometric study and relationships between the Caucasian species of wild pear (Pyrus spp.) and local cultivars in Georgia. Flora, Vol.206, No.11, (November 2011), article in press.doi:10.1016/j.flora.2011.04.010.

5. Badr, A., Müller, K., Schäfer-Pregl, R., El Rabey, H., Effgen, S., Ibrahim, H. H., Pozzi, C., Rohde, W. & Salamini, F. (2000). On the origin and domestication history of barley (Hordeum vulgare). Molecular Biology and Evolution Vo.17, No.4, (April 2000), pp.499–510. ISSN: 0737-4038.

6. Bakhtadze, K.E. (1947). Biologia, selektsia i semenovodstvo chainogo rastenija (Biology, Selection and Seed Productivity of Tea Plants). Nauka, Moscow. (in Russian).

7. Batonishvili, V. (1991). Aghtsera samefosa sakartvelosi (Geographic Description of Georgian Kingdom). (Ed. 7), Ganatleba, Tbilisi. (In Georgian).

8. Bedoshvili, D. (2008). National Report on the State of Plant Genetic Resources for Food and Agriculture in Georgia. Available from:

9. http://www.pgrfa.org/gpa/geo/Georgian report on State of PGR Sep 29, 2008.pdf; Accessed on 21 March 2010.

10. Bockelman, H. E., Erickson, C. A. & Goates, B. J. (2002). National Small Grains Collection: wheat germplasm evaluations. Annual Wheat Newsletter, Vol.48, (10 August, 2002), pp.273-286.

11. Casas, A. M., Yahiaoui, S., Ciudad, F. & Igartua, E. (2005). Distribution of MWG699 polymorphism in Spanish European barleys. Genome, Vol.48, No.1, (February 2005), pp.41–45. ISSN: 0831-2796.

12. Davies, P. (2003). An Historical Perspective from the Green Revolution to the Gene Revolution. Nutrition Reviews, Vol.61, No.6, pp.124–134. ISSN: 0029-6643, doi:10.1301/nr.2003.jun.S124-S134.

13. Dekaprelevich, L. (1947). Saqartvelos martsvlovani kulturebis dziritadi jishebi (Main cultivars of cereals in Georgia). Proceedings of State Selective Station, Georgia, Vol.2, No.1, pp.5-47. (In Georgian).

14. Dekaprelevich, L. & Menabde, V. (1929). K izucheniu polevykh kultur

zapadnoi Gruzii. I. Racha. (Study of cereal cultivars in Georgia. I. Racha). Scientific Papers of the Applied Sections of the Tbilisi Botanical Garden. Vol.6, No.2, pp. 219-252. (In Russian).

15. Dvorák, J., Luo, M.-C. & Yang, Z.-L. (1998). Genetic Evidence on the Origin of Triticum aestivum L. In: The Origins of Agriculture and Crop Domestication. Damania, A.B., Valkoun, J., Willcox, G. & Qualset C. O. (Eds). Proceedings of the Harlan Symposium, pp. 254-267, ISBN: 92-9127-084-9. Aleppo, Syria, 10-14 May 1997.

16. Ekhvaia, J. & Akhalkatsi, M. (2010). Morphological variation and relationships of Georgian populations of Vitis vinifera L. subsp. sylvestris (C. C. Gmel.) Hegi. Flora, Vol.205, No.9, (September 2010), pp.608-617. ISSN: 0367-2530, doi:10.1016/j.flora.2009.08. 002.

17. Ekhvaia, J., Blattner, F. R. & Akhalkatsi, M. (2010). Genetic diversity and relationships between wild grapevine (Vitis vinifera subsp. sylvestris) populations and aboriginal cultivars in Georgia. Proceedings of 33rd World Congress of Vine and Wine, 8th General Assambley of the OIV, Tbilisi, Georgia, June 2010, Available from: <http://www.oiv2010.ge/ index. php?page=5&lang=0>

18. Ekhvaia, J., Blattner, F.R., Gurushidze, M. & Akhalkatsi, M. (2011). Relationships between wild grapevine (Vitis vinifera subsp. sylvestris) populations and native grape cultivars in Georgia. Proceedings of BioSystematics Berlin 2011. ISBN: 978-3-921800-68-3, pp.111, Berlin, Germany, 21–27 February 2011.

19. Eliava, G. M. 1992. Dzvelkolkhuri unikaluri vazis jishta katalogi (Catalog of ancient grape varieties of Colchis). (Ed.2), Kutaisi Press N6, Georgia. (In Georgian).

20. Eritzjan, A. A. (1956). K izucheniu izmenchivosti sostava pchenits v Gruzii (Study of variability of wheat field composition). Proceedings of Tbilisi Botanical Institute, Vol.18, No.1. pp.251-277. (In Russian).

21. Finlayson, C. (2005). Biogeography and evolution of the genus Homo. Trends in Ecology and Evolution. Vol. 20, No.8, (August 2005), pp.457-463. ISSN: 0169-5347.

22. Fox, W. Sh. (2004). Living longer with phytomedicines from the Republic of Georgia. Woodland Publishing, ISBN-10: 158054388X, Health & Fitness.

23. Gabunia, L. & Vekua, A. A. (1995). Plio-Pleistocene hominid from Dmanisi, East Georgia, Caucasus. Nature, Vol.373, No.6514, (9 February 1995), pp.509-512, ISSN: 0028-0836, doi:10.1038/373509a0.

24. Gabunia, L., Vekua, A., Swisher, C. C., Ferring, R., Justus, A., Nioradze,

M., Ponce de Leon, M., Tappen, M., Tvalchrelidze, M. & Zollikofer Ch. (2000). Earliest Pleistocene hominid cranial remains from Dmanisi, Republic of Georgia: taxonomy, geological setting, and age. Science, Vol.288, No. 5578, (July 5, 2002), pp. 85-89. ISSN: 0036-8075.

25. Garson, L.K. (1991). The Centenarian Question: Old-Age Mortality in the Soviet Union, 1897 to 1970. Population Studies, Vol.45, No.2, pp. 265-278. ISSN: 0032-4728.doi:10.1080/0032472031000145436.

26. Girgvliani, T. (2010). Zemo Svanetis khorblis aborigenuli formebis istoria. (The history of aboriginal forms of wheat varieties of the Upper Svaneti). Artanuji, ISBN:978-9941-421-20-4,Tbilisi. (In Georgian).

27. Grant, K., Russel, Ch. & Everill, P. (2009). Anglo-Georgian Expedition to Nokalakevi. Interim report on excavations in July 2009. Available from: http://www.nokalakevi.org/Downloads/Reports/English/Site/2009.pdf.

28. Grassi, F., Labra, M., Imazio, S., Ocete Rubio, R., Failla, O., Scienza, A. & Sala, F. (2006).

29. Phylogeographical structure and conservation genetics of wild grapevine.

30. Conservation Genetics, Vol.7, No.6, (December 2006), pp.837–845. ISSN: 1566-0621.

31. Heun, M., Schäfer-Pregl, R., Klawan, D., Castagna, R., Accerbi, M., Borghi, B. & Salamini, F.

32. (1997). Site of einkorn wheat domestication identified by DNA fingerprinting.

33. Science, Vol.278, No.5341, (14 November 1997), pp.1312-1314. ISSN:0036-8075.

34. Jackson, J. (2003). The Biology of Apples and Pears, Series: The Biology of Horticultural Crops.

35. Cambridge University Press, ISBN: 0521380189, Cambridge.

36. Javakhishvili, I. (1987). Sakartvelos ekonomiuri istoria (Economic History of Georgia). (Ed. 2), Vol.5., Metsniereba, SB:2927, Tbilisi. (In Georgian).

37. Jorjadze, M. 2005. Agrobiodiversity, biotechnology and biosafety. In: National Biodiversity Strategy and Action Plan—Georgia. NACRES and Fauna and Flora International (FFI). Polygraph, ISBN: 99940-716-6-1. Tbilisi.

38. Ketskhoveli, N. (1928). Masalebi kulturul mtsenareta zonalobis shesastavlad kavkasionze. (Materials on zonal distribution of cultivated plants in the Greater Caucasus). Agricultural National Committee Press, Tbilisi. (In Georgian).

39. Ketskhoveli, N. (1957). Kulturul mtsenareta zonebi sakartveloshi (Zones of cultivated plants in Georgia). Georgian Academy of Sciences Press, Tbilisi. (In Georgian).

40. Ketskhoveli , N., Ramishvili, M. & Tabidze, D. 1960. Sakartvelos ampelograpia. (Amphelography of Georgia). Georgian Academy of Sciences Press, Tbilisi. (In Georgian).

41. Khomizurashvili, N. (1973). Sakartvelos Mekhileoba (Horticulture of Georgia). Metsniereba, Tbilisi (in Georgian).

42. Kilian, B., Özkan, H., Kohl, J., von Haeseler, A., Barale, F., Deusch, O., Brandolini, A., Yucel, C., Martin, W. & Salamini, F. (2006). Haplotype structure at seven barley genes: relevance to gene pool bottlenecks, phylogeny of ear type and site of barley

43. domestication. Molecular Genetics and Genomics, Vol. 276, No.3, (September, 2006), pp. 230-241, ISSN: 1617-4615. doi 10.1007/s00438-006-0136-6.

44. Kobakhidze, A. (1965). Sakartvelos samartsvle-parkosani mtsenareebi: Phaseolus L. (cereals of Georgia: Phaseolus L.). In: Botanika (Botany), Ketskhoveli, N. (Ed), pp.87-89, Metsniereba, Tbilisi. (In Georgian).

45. Kobakhidze, A. (1974). Sakartvelos samartsvle-parkosan mtsenareta botanikur-sistematikuri shestsavlisatvis. (Botanical-systematic study of cereals in Georgia). In: Botanika (Botany), Ketskhoveli, N. (Editor), pp. 58--190, Metsniereba, Tbilisi. (In Georgian).

46. Körner, Ch., Donoghue, M., Fabbro, T., Häuse, Ch., Nogués-Bravo, D., Arroyo, M.T.K., Soberon, J., Speers, L., Spehn, E.M., Hang Sun, Tribsch, A., Tykarski, P. & Zbinden N. (2007). Creative use of Mountain Biodiversity Databases: The Kazbegi Research Agenda of GMBA-DIVERSITAS. Mountain Research and Development, Vol.27, No.3, (August 2007), pp.276–281. ISSN: 0276-4741.

47. Kuthatheladze, Sh. (1980). Pyrus L. In: Sakartvelos Flora (Flora of Georgia), Ketskhoveli, N. (Ed.), Vol. 6. Metsniereba, ISBN:5-520-00325-7, Tbilisi. (in Georgian).

48. Kvavadze, E., Bar-Yosef, O., Belfer-Cohen, A., Boaretto, E., Jakeli, N., Matskevich, Z. & Meshveliani, T. (2009). 30,000-Year-Old Wild Flax Fibers. Science, Vol.325, No.5946, (11 September 2009), pp.1359, ISSN: 0036-8075.

49. Laguna, E. (2004). The plant micro-reserve initiative in the Valencian Community (Spain) and its use to conserve populations of crop wild relatives. Crop Wild Relative, Issue 2, (July 2004), pp.10-13. ISSN 1742-3627.

50. Maghradze, D., Failla, O., Bacilieri, R., Imazio, S., Vashakidze, L., Chipashvili, R., Mdinaradze, I., Chkhartishvili, N., This, P. & Scienza, A. (2010). Georgian Vitis germplasm: usage, conservation and investigation. Bulletin de l'OIV, Vol.83, No.956-958, (October-December, 2010), pp. 485-496. ISSN: 0029-7127.

51. Maisaia, I., Shanshiashvili, T. & Rusishvili, N. (2005). Agriculture of Colchis. Metsniereba, ISBN:99940-785-3-4, Tbilisi. (In Georgian).

52. Maxted, N., Ford-Lloyd, B. V., Jury, S., Kell, Sh. & Scholten, M. (2006). Towards a definition of a crop wild relative. Biodiversity and Conservation. Vol.15, No.8, (July 2006), pp.2673-2685. ISSN: 0960-3115

53. Maxted, N., Dulloo, E., Ford-Lloyd, B. V., Iriondo, J. M & Jarvis, A. (2008). Gap analysis: a tool for complementary genetic conservation assessment. Diversity and Distributions, Vol.14, No.6, (November 2008), pp.1018–1030. ISSN:1472-4642. doi: 10.1111/j.1472-4642.2008.00512.x.

54. McGovern, P.E. (2003). Ancient wine: the search for the origins of viniculture. Princeton University Press, ISBN: 0691070806, Princeton, New Jersey.

55. Melikishvili, G. (Ed). (1970). Sakartvelos istoriis narkvevebi (Historical essays of Georgia). Sabchota Sakartvelo, Tbilisi. (In Georgian).

56. Menabde, V. (1938). Sakartvelos kerebi (Barleys of Georgia). Georgian Academy of SciencesPress, Tbilisi. (In Georgian).

57. Menabde, V. (1948). Pshenitsi Gruzii (Wheats of Georgia). Georgian Academy of Sciences Press, Tbilisi. (In Russian).

58. Menabde, V. (1961). Kartuli khorblebi da mati roli khorblis saerto evolutsiashi (Georgian wheat cultivars and their role in wheat evolution). Proceedings of Tbilisi Botanical Institute, Vol.21, No.1. pp.229-259. (In Georgian).

59. Nakhutsrishvili, G. (1999). The vegetation of Georgia (Caucasus). Braun-Blanquetia, Vol.15, pp.1-74. ISSN:0393-5434.

60. Nakhutsrishvili, G., Akhalkatsi, M. & Abdaladze, O. (2009). Main threats to the mountain biodiversity in Georgia (the Caucasus). Mountain Forum Bulletin, Vol.9, No.2, (July 2009), pp.18–19. ISSN 1815-2139.

61. Neidze, V., Bokeria, M., Kharadze, K., Kurtubadze, M. & Garsenishvili, L. (2008). Sakartvelos geographia (Geography of Georgia). LogosPress, ISBN: 978-99940-64-48-9, Tbilisi. (In Georgian).

62. Ovchinnikov, I. V., Gotherstrom, A., Romanova, G. P., Kharitonov, V. M., Liden, K. & Goodwin, W . (2000). Molecular analysis of Neanderthal

DNA from the northern Caucasus. Nature, Vol.404, No.6777, (30 March 2000), pp.490–493, ISSN: 0028-0836.doi:10.1038/35006625.

63. Petersen, G., Seberg, O., Yde, M. & Berthelsen, K. (2006). Phylogenetic relationships of Triticum and Aegilops and evidence for the origin of the A, B, and D genomes of common wheat (Triticum aestivum). Molecular Phylogenetics and Evolution, Vol.39 , No.1, (April 2006), pp.70–82. ISSN: 1055-7903.

64. Pistrick, K., Akhalkatsi M., Girgvliani, T. & Shanshiashvili, T. (2009). Collecting plant genetic resources in Upper Svaneti (Georgia, Caucasus Mountains). Journal of Agriculture and Rural Development in the Tropics and Subtropics, Supplement 92, pp.127-135.ISSN:1613-8422.

65. Ramishvili, R., (1988). Dikorastushchii vinograd Zakavkazia (Wild grape of the South Caucasus). Ganatleba, ISBN: 5-505-00690-6, Tbilisi. (In Russian).

66. Sabashvili, M. (1970). Niadagmtsodneoba (Soil Sciences). Tbilisi University Press, Tbilisi. (In Georgian).

67. Schatz, G., Shulkina, T., Nakhutsrishvili, G., Batsatsashvili, K., Tamanyan, K., Ali-zade, V., Kikodze, D., Geltman, D. & Ekim, T. (2009). Development of Plant Red List Assessments for the Caucasus Biodiversity Hotspot. In: Status and Protection of Globally Threatened Species in the Caucasus. Zazanashvili, N. & Mallon, D. (Editors).

68. Contour Ltd. pp. 188-192. Tbilisi. ISBN 978-9941-0-2203-6. Available from: <www.assets.panda.org/downloads/ cepf_caucasus_web_1.pdf >Takhtajan, A. (1986). Floristic Regions of the World. University of California Press, First English Language Edition, (September 1986), ISBN: 0520040279, Berkeley, CA.

69. This, P., Lacombe, T. & Thomas, M.R. (2006). Historical origins and genetic diversity of wine grapes. Trends in Genetics Vol.22, No.9, (September 2006), pp.511–519. ISSN: 0168-9525.

70. Tsilo, T. J., Yue Jin & Anderson, J. A. (2008). Diagnostic Microsatellite Markers for the Detection of Stem Rust Resistance Gene Sr36 in Diverse Genetic Backgrounds of Wheat. Crop Science, Vol.48, No.1, (January–February 2008) pp.253–261. ISSN: 0011-183X.

71. Tushabramishvili, N. (2011). The main problems of Palaeolithic of Georgia and the complex studies of the Western Georgian cave sites. In: Khornabuji, Pitskhelauri, K. (Ed), pp. 39-41, Ilia State University Press, ISBN: 1512-2999, Tbilisi.

72. UPOV (2000). Guidelines for the conduct of tests for distinctness, uniformity and stability. Pear (Pyrus communis L.). Geneva. Date

of access: 2000-04-05. Available from: <http://www.upov.int/en/publications/tg-rom/tg015/tg_15_3.pdf>

73.   Vavilov, N. I. (1992). Origin and geography of cultivated plants. Cambridge University Press, ISBN: 0-521-40427-4, Cambridge.

74.   Volk, G. M., Richards, C. M., Henk, A. D., Reilley, A. A., Bassil, N. V. & Postman, J. D. (2006). Diversity of wild Pyrus communis based on microsatellite analyses. Journal of the American Society for Horticultural Science, Vol.131, No.3, (May 2006), pp.408–417.ISSN: 0003-1062.

75.   Voltas, J., Pemán, J. & Fusté, F. (2007). Phenotypic diversity and delimitation between wild and cultivated forms of the genus Pyrus in North-eastern Spain based on morphometric analyses. Genetic Resources and Crop Evolution, Vol.54, No.7,(November 2007), pp.1473–1487. ISSN:0925-9864.

76.   Yamamoto, T. & Chevreau, E. (2009). Pear genomics. In: Genetics and Genomics of Rosaceae., Folta, K.M. & Gardiner, S.E. (Eds.), Springer, ISBN:0387774904, New York.

77.   Zhizhizlashvili, K. & Berishvili, T. (1980). Zemo Svanetis kulturul mtsenaretashestsavlisatvis (Study of cultivated plants in Upper Svaneti). Bulletin of Georgian Academy of Sciences, Vol.100, No.2, pp.417-419. ISSN:0132-1447. (In Georgian).

78.   Zohary, D., & Hopf, M. (2000). Domestication of plants in the Old World. (3rd edition.) Oxford University Press, ISBN 0-19-850356-3, New York.

CITATION

CHAPTER 1

Luciano Lourenço Nass, Mário Sérgio Sigrist, Cláudia Silva da Costa Ribeiro and Francisco José Becker Reifschneider, Genetic resources: the basis for sustainable and competitive plant breeding, http://dx.doi.org/10.1590/S1984-70332012000500009

CHAPTER 2

Rukhsar Ahmad Dar,Mushtaq Ahmad,Sanjay Kumar,Monica Reshi, Agriculture germplasm resources: A tool of conserving diversity, DOI: 10.5897/SRE2015.6206.

CHAPTER 3

Kerri L SteenwerthEmail author, Amanda K Hodson, Arnold J Bloom, Michael R Carter, Andrea Cattaneo, Colin J Chartres, Jerry L Hatfield, Kevin Henry, Jan W Hopmans, William R Horwath, Bryan M Jenkins, Ermias Kebreab, Rik Leemans, Leslie Lipper, Mark N Lubell, Siwa Msangi, Ravi Prabhu, Matthew P Reynolds, Samuel Sandoval Solis, William M Sischo, Michael Springborn, Pablo Tittonell, Stephen M Wheeler, Sonja J Vermeulen, Eva K Wollenberg, Lovell S Jarvis and Louise E Jackson, Climate-smart agriculture global research agenda: scientific basis for action, DOI: 10.1186/2048-7010-3-11.

CHAPTER 4

Mohammed Kasso and Mundanthra Balakrishnan, "Ex Situ Conservation of

Biodiversity with Particular Emphasis to Ethiopia," ISRN Biodiversity, vol. 2013, Article ID 985037, 11 pages, 2013. doi:10.1155/2013/985037.

CHAPTER 5

Pouta E, Tienhaara A and Ahtiainen H (2014) Citizens' preferences for the conservation of agricultural genetic resources. Front. Genet. 5:440. doi: 10.3389/fgene.2014.00440

CHAPTER 6

Jikun Huang and Qinfang Wang, Agricultural Biotechnology Development and Policy in China, http://www.agbioforum.org/v5n4/v5n4a01-huang.htm.

CHAPTER 7

K. Nirmal Babu, G. Yamuna, K. Praveen, D. Minoo, P.N. Ravindran and K.V. Peter (2012). Cryopreservation of Spices Genetic Resources, Current Frontiers in Cryobiology, Prof. Igor Katkov (Ed.), ISBN: 978-953-51-0191-8, InTech, DOI: 10.5772/35401.

CHAPTER 8

Maia Akhalkatsi, Jana Ekhvaia and Zezva Asanidze (2012). Diversity and Genetic Erosion of Ancient Crops and Wild Relatives of Agricultural Cultivars for Food: Implications for Nature Conservation in Georgia (Caucasus), Perspectives on Nature Conservation - Patterns, Pressures and Prospects, Prof. John Tiefenbacher (Ed.), ISBN: 978-953-51-0033-1, InTech, DOI: 10.5772/30286.

INDEX

A

Abnormal plants 219
Accessions conserved 23, 30, 40, 41
Agricultural Production Systems Simulator (APSIM) 77
Akaike information criterion (AIC) 162
alternative specific constants (ASC) 164
Alternative-specific constants (ASC) 162
Archaeological data 234
Artificial habitat 138

B

Bayesian information criterion (BIC) 162
Biological Diversity 131, 134, 135

C

Canopy temperature (CT) 56
Captive breeding 137, 143, 150, 153
choice experiment (CE) 156
Climate-smart agriculture (CSA) 48
Conservation requirements 202
Consultative Group on International Agricultural Research (CGIAR) 15
Crop wild relatives (CWR) 236

E

Environmental enrichment 137

F

field gene banks (FGB) 32

G

Genetically Modified Organisms (GMO) 175
Genetic resources 1, 3, 4, 5, 6, 7, 8, 10, 11, 12, 13, 15, 16, 17, 18, 19
Genetic Resources of the Netherlands (CGN) 37
Genetic variability 3, 8, 9, 10, 16
Georgia occurred 238

I

Instituto Agronômico de Campinas (IAC) 8
International Food Policy Research Institute's (IFPRI) 73
International Model for Policy Analysis of Agricultural Commodities and Trade (IMPACT) 73
Intimately linked 1

K

Key Science Engineering Program
(KSEP) 182

L

liquid nitrogen (LN) 34
Low carbon fuel standard (LCFS) 73

M

Ministry of Agriculture (MOA) 176, 182
Ministry of Science and Technology
[MOST] 178
Murashige and Skoog (MS) 220

N

National Active Germplasm Sites
(NAGS) 39, 40
National Bureau of Plant Genetic Re-
sources (NBPGR) 210, 219, 222
nongovernmental organizations (NGOs)
139
Normalized Difference Vegetation Index
(NDVI) 68

P

Plant genetic resources 201
Plant Genetic Resources for Food and
Agriculture (PGRFA) 41
plant genetic resources (PGR) 29
Plant genetic resources (PGR) 4
Plant Genetic Resources (PGR) 249

Plant regeneration 202, 215, 216
Pollen storage 35
polymerase chain reaction (PCR) 221
Primary somatic embryos (PEs) 205

S

Socio-economic 249
Special Foundation of Transgenic Plants
Research and Commercialization
(SFTPRC) 181
State Development and Planning Com-
mission (SDPC) 178
State Science and Technology Commis-
sion [SSTC] 178
Successfully cryopreserved 204, 212,
219, 220
Svalbard Global Seed Vault (SGSV) 24,
36
System-Wide Genetic Resources Pro-
gramme (SGRP) 15

T

Trinity College Dublin (TCD) 35

V

vapour pressure deficit (VPD) 77

W

willingness to pay (WTP) 155, 156

Z

Zoological gardens 134, 135